Elements of Physical Hydrology

Elements of Physical Hydrology

SECOND EDITION

George M. Hornberger

Patricia L. Wiberg

Jeffrey P. Raffensperger

Paolo D'Odorico

Johns Hopkins University Press   Baltimore

© 2014 Johns Hopkins University Press
All rights reserved. Published 2014
Printed in Canada
9 8 7 6 5 4 3 2

Johns Hopkins University Press
2715 North Charles Street
Baltimore, Maryland 21218-4363
www.press.jhu.edu

Library of Congress Cataloging-in-Publication Data
Elements of physical hydrology / George M. Hornberger . . . [et al.]. — 2nd ed.
     p. cm.
   Includes bibliographical references and index.
ISBN 978-1-4214-1373-0 (hardcover : alk. paper) — ISBN 978-1-4214-1396-9
(electronic) — ISBN 1-4214-1373-6 (hardcover : alk. paper) — ISBN 1-4214-
1396-5 (electronic)   1. Hydrology.   I. Hornberger, George M.
   GB661.2.E44 2014
   551.48—dc23   2013036530

A catalog record for this book is available from the British Library.

*Special discounts are available for bulk purchases of this book. For more
information, please contact Special Sales at 410-516-6936 or specialsales
@press.jhu.edu.*

Johns Hopkins University Press uses environmentally friendly book materials,
including recycled text paper that is composed of at least 30 percent post-
consumer waste, whenever possible.

Contents

Preface

The world population is expected to grow to 9 billion by 2050. Food production will have to grow even faster than the population because, as countries develop and achieve a higher standard of living, diets tend to include more animal protein and thus there is a concomitant increased consumption of grain by animals. Agriculture consumes the greatest amount of fresh water of all sectors, so we anticipate increased stress on water resources as a result of increased food production. The growth in population will also result in increases in the use of energy, which entails the use of water as well. Furthermore, these changes will occur amidst large-scale changes in the use of land resources (e.g., increased urbanization) and changes in climate caused by the burning of fossil fuels. It is clear that dealing with water resources issues will be a key activity for regions, nations, and the global village as these changes unfold. Arguably, university students need to acquire knowledge about the hydrological cycle just to be informed citizens of the world in the coming decades.

This book is based on the premise that students of environmental science must learn the quantitative physical basis of hydrology if they are to appreciate the scientific approach to understanding observed phenomena and the basis for achieving solutions to water resources problems. The text is not a catalog of observations nor is it a compendium of applied techniques of engineering hydrology. Rather, it presents a basic coverage of physical principles and how these allow one to grasp the essential elements of hydrological processes. One tenet that we hold is that an understanding of fundamental fluid mechanics is essential to the study of hydrology. Our aim is to provide an integrated coverage of flows of water on and beneath the Earth's surface with the underpinning of a knowledge of basic fluid mechanics.

This book originated from lectures in a course on Physical Hydrology at the University of Virginia that began more than 25 years ago. Following publication of the first edition of our book in 1998, many colleagues elsewhere told us that they found it useful, but that there were a few topics that were underrepresented. Thus, in addition to updating material throughout the book, in this second edition we include a good bit of new material, including a new chapter on Ecohydrology and a completely new final chapter on water-climate-energy-food interactions. The changes notwithstanding, the book is still designed to accompany an undergraduate course in physical hydrological science. It is aimed at upper-level undergraduates majoring in environmental or Earth sciences. The coverage presupposes a modest background in calculus and physics.

We use several conventions in an attempt to make the book "user friendly." Each chapter has introductory and concluding remarks that seek to place the material presented in the chapter in the context of some contemporary environmental issue. Terms that appear in boldface at their first occurrence are found in the Glossary. Terms that appear in italics deserve emphasis. Supporting material is contained in three appendixes: a review of units and dimensions (Appendix 1); a tabulation of certain properties of water (Appendix 2); and a review of some elementary statistical concepts (Appendix 3).

Over the years we have benefited from ideas and data shared by many colleagues and students. We are grateful to all who have contributed to our education, but will refrain from attempting to produce a comprehensive list. We do want to acknowledge specifically colleagues who either have taught the course at UVA that stimulated the first edition of the book in the past or who are currently in the teaching rotation for the course: John Albertson, Keith Beven, Keith Eshleman, John Fisher, Peter German, Aaron Mills, Matt Reidenbach, and Todd Scanlon. In addition, we especially thank Margot Bjoring for the preparation of the figures for this edition of the book.

Elements of Physical Hydrology

1 The Science of Hydrology

1.1 Introduction

The role of water is central to most natural processes. Water transports sediment and solutes to lakes and oceans, thereby shaping the landscape. The global energy balance is influenced strongly by the high capacity of water for storing thermal energy and the large amount of heat required to change water from liquid to vapor and vice versa. The abundance of water in the atmosphere and oceans makes it an important regulator of climate. Water vapor is the most important of the greenhouse gases. Life depends on water.

Hydrology, literally "water science," encompasses the study of the occurrence and movement of water on and beneath the surface of the Earth, the properties of water, and its relationship with the living and material components of the environment. Ultimately, many hydrological questions involve the transport of dissolved nutrients, energy, sediment, or contaminants. The starting point for investigations of transport must be the *physical* processes of water movement.

Hydrological science has an important place in discussions of natural resources. Water resources, especially freshwater resources, are the subject of intense scrutiny and speculation. In arid regions, the fair allocation and wise use of freshwater resources are significant challenges facing governments and people, affecting relations between nations, states, cities, and individual users. As a resource, water appears unlimited. However, the twentieth century saw a tremendous growth in the use of water, as well as an increase in the threat of its contamination, and the trends have continued in the new millennium.

Hydrological science has aspects related to "curiosity-driven" questions and to "problem-driven" questions. The first aspect relates to questions about how the Earth works, and specifically about the role of water in natural processes. The second relates to using scientific knowledge to provide a sound basis for the proper use and protection of water resources.

1.2 Hydrology and Water Resources

Hydrological science has both curiosity-driven and problem-driven origins that stretch back to antiquity (Biswas, 1972). Many of the great ancient philosophers of Greece and Rome speculated on hydrological phenomena. Vitruvius, writing during the second half of the first century BC, often is credited with first recognizing that groundwater is derived primarily from infiltration of rain and snowmelt, rather than upwelling of subterranean water from great depths. By applying hydrological and hydraulic principles, the ancients constructed great hydraulic works such as the ancient Arabian wells, the Persian kanats, the Egyptian and Mesopotamian irrigation projects, the Roman aqueducts, the ancient water civilizations of Sri Lanka, and the Chinese irrigation systems, canals, and flood control works. During the Renaissance, Leonardo da Vinci wrote what is likely the earliest complete statement of the hydrological cycle, including the notions of condensation and evaporation and the dissolution of rock minerals (after Eagleson, 1970):

> Whence we may conclude that the water goes from the rivers to the sea and from the sea to the rivers, thus constantly circulating and returning, and that all the sea and rivers have passed through the mouth of the Nile an infinite number of times. The conclusion is that the saltiness of the sea must proceed from the many springs of water which, as they penetrate the earth, find mines of salt, and these they dissolve in part and carry with them to the ocean and other seas, whence the clouds, the begetters of rivers, never carry it up.

During the eighteenth and nineteenth centuries, hydraulic experiments flourished and, until about 1950, pragmatic considerations dominated hydrology. Primarily due to the development and availability of digital computing, theoretical approaches in hydrology have increasingly allowed hydrological theories to be subjected to rigorous mathematical analysis.

Freshwater resources are needed to meet the needs of humans, livestock, commercial enterprises, agriculture, mining, industry, and thermoelectric and hydroelectric power. In today's world, the necessity of solving water-supply problems has become obvious, with many regions exhibiting signs of looming water shortages (Figure 1.1).

Most of the human consumption of freshwater resources is associated with food production. (We use *consumption* to refer to water that is used in ways that return it to the atmosphere rather than to a stream, river, or groundwater. For example, thermoelectric power plants may withdraw large quantities of water from a river for cooling, but then return the bulk of that water to the stream, albeit at a higher temperature.) Relative to the consumption of water for food production, drinking, household, and industrial uses of water are overall smaller (Figure 1.2). We need much more water to produce the food we eat than the amount of water we use for drinking or other activities. The global water crisis is more likely causing hunger than thirst. A report from the office of the U.S. Director of National Intelligence, Global Water Security (http://www.dni.gov/index.php/about/organization/national-intelligence-council-nic-publications) concludes:

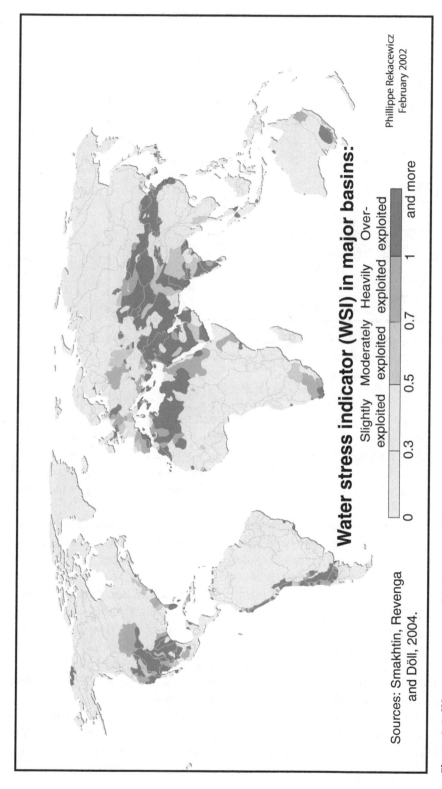

Figure 1.1 Water resources are highly stressed in many parts of the world. The water stress indicator is the fraction of available water appropriated for use by humans. Fractions greater than one indicate the use of fossil groundwater.
http://www.grida.no/graphicslib/detail/water-scarcity-index_14f3

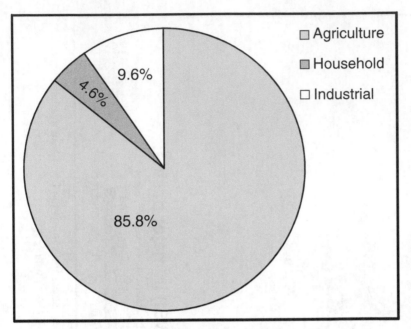

Figure 1.2 Relative contributions of agriculture, household, and industrial consumption to the human appropriation of freshwater resources.
Data from Hoekstra and Chapagain 2008.

"Between now and 2040, fresh water availability will not keep up with demand absent more effective management of water resources. Water problems will hinder the ability of key countries to produce food and generate energy, posing a risk to global food markets and hobbling economic growth." The need for improved water management, based on scientific knowledge, is of critical importance in the coming decades (Jury and Vaux, 2007).

Although the rate of increase in water use in some areas has slackened over the past several decades, the total consumption of water globally has continued to increase (Figure 1.3). Population growth, depletion or deterioration of freshwater resources, and changing demands (most notably the tendency to adopt more water-demanding diets in emergent countries,–i.e., an increase in meat consumption) will tend to further stress water resources in the future for many countries. Falkenmark et al. (2007) indicate their view of the severity of the issue. "We are on the verge of a new and more serious era of water scarcity, and it is clear that we will face increasingly complex challenges. Water supply to different sectors will become more challenging as supplies of blue water (e.g. water in rivers and aquifers) become overstretched, while a scarcity of green water (e.g. water in the soil) will limit food and biomass production."

World population is expected to increase in the next 50 years, stressing water resources. Issues related to the *quality* of water supplies have occupied an increasingly important niche in hydrology. It is estimated that 80% of all diseases and over one-third

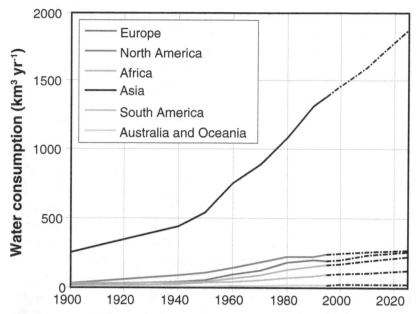

Figure 1.3 Water consumption grew during the twentieth century and is projected to continue increasing in the future.

Data from Shiklomanov 1999.

of all deaths in developing countries result from the consumption of contaminated water. Provision of basic sanitation and water treatment in much of the world is still lacking. In 1980, the United Nations launched the International Drinking Water Supply and Sanitation Decade, with the goal of clean water and sanitation services to those without them. Despite enormous effort, expense, and progress, at the close of the decade 1.8 billion people still had no access to sanitation services, and nearly 1.3 billion people still lacked access to clean water. Population growth wiped out the progress achieved through this effort (Gleick, 1993).

In the developed world, the availability of safe drinking water and adequate sewage treatment is taken for granted, but other water quality issues abound. These include *eutrophication* of surface waters (deleterious effects due to excess nutrient supply), contamination of groundwater with a variety of organic compounds and metals, and the acidification of surface waters from acid rain. The ability of streams, rivers, lakes, and estuaries to dilute contaminants to safe levels and to purify the water through natural processes (their **assimilative capacity**) depends on the quantity of water flowing in them.

Understanding surface-water flow is a requisite for water quality studies. Similarly, knowledge of subsurface flow is necessary for understanding the movement of pollutants underground. To predict or evaluate the effects of liquid-waste disposal in deep injection wells or the use of solid waste as fill for reclamation of strip mines, one must appreciate the mechanics of water flow in rock and soil. The massive effort now under way in the United States to clean up sites where groundwater has been contaminated represents another challenge for hydrological science.

Major floods are the most dramatic and visible of hydrological phenomena. The number of people affected by floods and the number of lives lost have increased in recent decades (Figure 1.4). With pressure for increased use of floodplains, the prediction and control of floods remain among the most important applications of hydrology.

Addressing hydrological challenges, such as those related to floods or groundwater contamination, requires a firm understanding of the basic principles of the physics and chemistry of water. Hydrological science uses the fundamentals of the basic sciences and mathematics to develop explanations (models) of observed phenomena. One of the basic problems in hydrology is a description of water motion. One goal of physical hydrology is to identify the paths of water movement on and beneath the surface of the Earth. Using physical theory and associated mathematical models, hydrologists seek to describe quantitatively the motion of water in the natural environment. As Jury and Vaux (2005) put it, "There is little question that science must play a critical role in forming a successful solution to the world's emerging water problems." They go on to conclude that "a new science of sustainability will be needed if the prospects for managing and solving the world's emerging water problems are to be bright." Knowledge of physical hydrology will be a critical component of this new science.

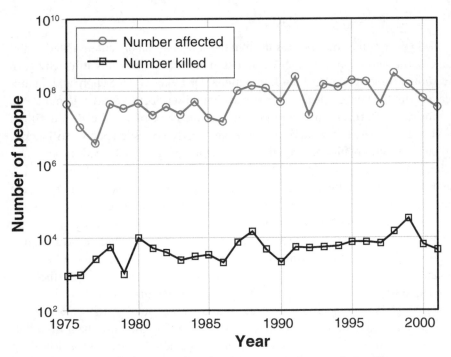

Figure 1.4 Losses during river flooding have trended upward over the past several decades.

Redrawn from Jonkman 2005.

1.3 The Hydrological Cycle

The motion of water can be described at many different scales. The fundamental concept of hydrology is the **hydrological cycle**—the global-scale, endless recirculatory process linking water in the atmosphere, on the continents, and in the oceans. We can think of this recirculatory process in terms of reservoirs or compartments that store water (the oceans, atmosphere, etc.) and the movement of water between them. Within the various compartments of the hydrological cycle, water can be stored in any one of three separate phases or states: gas (vapor), liquid, or solid. For example, water in the atmosphere can exist as vapor (the concentration of water vapor is expressed as humidity), in liquid (cloud droplets, rain drops), or in solid phase (ice crystals, snowflakes). Similarly, all three phases of water can be found on and below the land surface. Movement of water from one compartment to another can occur in any of the three phases. For example, the movement of water between the oceans and atmosphere occurs in vapor phase (evaporation from the ocean surface), liquid phase (rain onto the ocean surface), and solid phase (snowfall onto the ocean surface).

Solar energy drives the hydrological cycle; gravity and other forces also play important roles. The dynamic processes of water vapor formation and transport of vapor and liquid in the atmosphere are driven largely by solar energy. Precipitation and the flow of water on and beneath the Earth's surface are driven primarily by gravity. Within partially dry soil, gravitational and other forces are responsible for the movement of water.

The hydrological cycle can be considered to "start" anywhere, but let us consider atmospheric water first (Figure 1.5). As hydrology is concerned mainly with water at or near the Earth's surface, from our point of view the dominant process involving atmospheric water is the precipitation of water on the land surface. The portion of the precipitation that reaches the land surface as solid precipitation (mostly snowfall) can be retained temporarily on vegetation or ground surfaces, or accumulate in seasonal *snowpacks* or in permanent snowpacks known as glaciers. Considering liquid precipitation (rain), a portion also can be retained temporarily on vegetation surfaces or in surface depressions, a portion enters into the soil (*infiltration*) and a portion flows over the land surface first into small rivulets and ultimately into larger streams and rivers. This last process is called *surface runoff,* which can be augmented by runoff during periods of snowmelt (snowmelt runoff). The portion of rainfall that infiltrates into the soil can also follow one of several paths. Some of the water *evaporates* from the soil and some is returned to the atmosphere by plants (*transpiration*). We often refer to the total evaporation and transpiration from vegetated land surfaces as *evapotranspiration.* The remaining water continues to move downward through the soil and recharges the saturated portion of the subsurface, becoming *groundwater.* At lower elevations, groundwater discharges into streams and rivers or directly to the ocean (*groundwater runoff*). Water evaporates from the surface of the oceans and thereby replenishes the water in the atmosphere. Thus we have returned to the particular compartment that we considered first, atmospheric water.

Looking at the relative magnitude of these water fluxes (Figure 1.6), we observe that over the continents evapotranspiration is on average smaller than precipitation. The fraction of precipitation exceeding evapotranspiration contributes to surface and

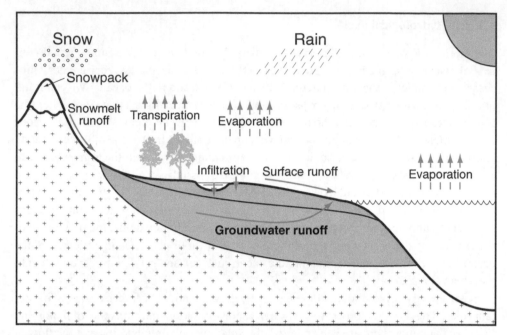

Figure 1.5 Mechanisms of water movement within the hydrological cycle. Water movement from the atmosphere to the oceans and continents occurs as precipitation, including rain, snow, sleet, and other forms. On the continents, water may be stored temporarily but eventually returns to the oceans through surface and groundwater runoff or to the atmosphere through evapotranspiration.

groundwater runoff. Conversely, over the oceans evaporation exceeds precipitation. Therefore, precipitation alone is not sufficient to remove from the atmosphere above the oceans all the water vapor coming from ocean evaporation. In fact, the atmospheric circulation transports part of this water vapor on land masses, thereby allowing continental precipitation to exceed the evapotranspiration rate.

It is instructive to analyze the water cycle at different scales. We have seen that at the *global scale* water reaches Earth's surface as precipitation. This water is transported over land and in the oceans until it is returned to the atmosphere as water vapor in the processes of evaporation and transpiration. Similarly, water vapor is transported by the atmospheric circulation until it contributes to precipitation (Figure 1.7a). At a *local scale*, it is unlikely that the same water molecules that evapotranspire from an area of, say, a few hectares will contribute to the precipitation on the same area. Most of the local precipitation is contributed by water brought into the area by atmospheric transport; the excess of rainfall with respect to evapotranspiration (locally) generates runoff (Figure 1.7b). At a *regional scale* (e.g., a few thousand square kilometers) precipitation can be contributed both by moisture coming from regional evapotranspiration and by water vapor brought into the region by atmospheric transport (Figure 1.7c). Thus at the regional scale the water cycle may have an important internal component, whereby part of regional evapotranspiration contributes to a substantial part of the regional precipitation. Known

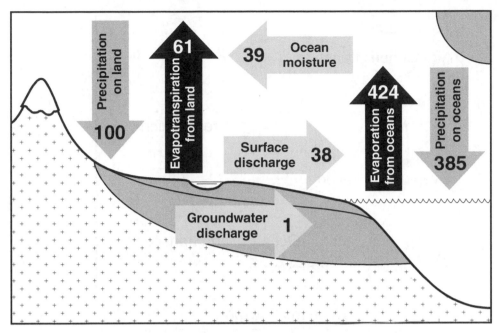

Figure 1.6 Flows within the hydrological cycle. Units are relative to the annual precipitation on the land surface ($100 = 119{,}000$ km^3 yr^{-1}). Black arrows depict flows to the atmosphere, gray arrows depict flows to land or oceans, and blue arrows indicate lateral flows.
Data from Maidment 1993.

as **precipitation recycling**, this phenomenon is usually quantitatively expressed in terms of the recycling ratio, which is the fraction of the total regional precipitation that is contributed by water evapotranspiring from the same region. The recycling ratio is overall greater during the growing season, when evapotranspiration is more intense, and it increases with the region size. Studies on precipitation recycling provide important insights into the impact of terrestrial vegetation on the water cycle. This research has shown how the removal of forest vegetation (e.g., deforestation) may affect regional precipitation by changing (decreasing) transpiration and precipitation recycling.

1.4 The Water Budget

The hydrological cycle can be described quantitatively by applying the principle of **conservation of mass**, which often is referred to as a water balance or **water budget** when used in this way. A simple statement of conservation of mass for any particular compartment (usually referred to as a *control volume*) is that the time rate of change of mass stored within the compartment is equal to the difference between the inflow rate and the outflow rate. For example, if we are adding two grams of water to a bucket every minute and one gram of water is leaking out each minute, then the mass stored within the bucket is increasing at the rate of one gram per minute. Symbolically, we can write this as:

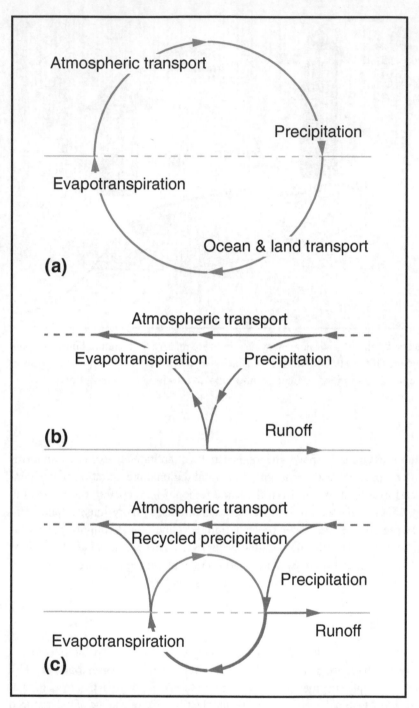

Figure 1.7 The water cycle at the global (*a*), local (*b*), and regional (*c*) scales. Modified after Eltahir and Bras 1996.

$$\frac{dM}{dt} = I' - O', \tag{1.1}$$

where M = mass within the control volume [M]; t = time [T]; I' = mass inflow rate [M T^{-1}]; and O' = mass outflow rate [M T^{-1}]. The expressions in square brackets are the mass-length-time dimensions associated with the defined quantity; for example, the dimensions of I' are mass per time or M T^{-1}. (See Appendix 1 for a discussion of units, dimensions, significant figures, and unit conversions.)

In many instances, the **density** of water can be taken as approximately constant and the conservation law expressed in terms of volume. The terms involving mass in Equation 1.1 can be expressed in terms of density times volume and density can then be canceled from both sides of the equation. Thus, Equation 1.1 can be rewritten:

$$\frac{dV}{dt} = I - O, \tag{1.2}$$

where V = volume of water within the control volume [L^3]; I = volume inflow rate [L^3 T^{-1}]; and O = volume outflow rate [L^3 T^{-1}].

1.4.1 The global water budget

We can construct a global water budget by applying the principle of mass conservation (Equation 1.2), using the continents as our control volume. The quantity V is then the volume of water stored on or within the continental land masses. Inflow is precipitation and outflow consists of evapotranspiration (evaporation and transpiration combined) and runoff (both surface water and groundwater). Note that in addition to ignoring density variations we must express the rate of evaporation or transpiration-outflows of water vapor from the continents to the atmosphere-in "liquid water equivalent," or LWE, units. Otherwise, density is varying (water vapor is much less dense than liquid water), and mass, rather than volume, is the conserved quantity.

If we consider only *average annual conditions* for this water budget, the dV/dt term in Equation 1.2 becomes negligible. That is, over a period of years the average amount of water stored as ice (ice caps and glaciers), as surface water (rivers and lakes), and as subsurface (groundwater) does not change significantly. Over much longer time periods such as centuries or millennia this may not be true if there is a dramatic shift in climatic conditions. If there is no change in storage over time, we say that the system is at *steady state*. For any given control volume at steady state, a completely general water budget equation can be written (using bars over the terms to indicate that they are annual average quantities):

$$\frac{d\overline{V}}{dt} = 0 = \overline{p} + \overline{r}_{si} + \overline{r}_{gi} - \overline{r}_{so} - \overline{r}_{go} - \overline{et}, \tag{1.3}$$

where $\overline{V}$ = average volume of water stored, and assumed to be constant; $\overline{p}$ = average precipitation rate; $\overline{r}_{si}$ = average surface water inflow rate; $\overline{r}_{gi}$ = average groundwater inflow rate; $\overline{r}_{so}$ = average surface water outflow rate; $\overline{r}_{go}$ = average groundwater outflow rate; and

$\overline{et}$ = average evapotranspiration rate. All terms in the equation have dimensions of volume per time [$L^3 T^{-1}$]. For the continents, we will simplify Equation 1.3 by neglecting the inflows and outflows of groundwater, because they tend to be very small. We also will neglect surface water inflows, because surface water flows from the continents to the oceans, and will refer to surface water outflows as runoff, r_s. With these simplifications, Equation 1.3 becomes:

$$\frac{d\overline{V}}{dt} = \overline{p} - \overline{r}_s - \overline{et} = 0 \tag{1.4}$$

or

$$\overline{p} = \overline{r}_s + \overline{et}, \tag{1.5}$$

where $\overline{p}$ = average precipitation rate [$L^3 T^{-1}$]; $\overline{r}_s$ = average surface runoff rate [$L^3 T^{-1}$]; and $\overline{et}$ = average evapotranspiration rate [$L^3 T^{-1}$].

To quantify the global hydrological cycle, we can examine the relative sizes of the various storage compartments and the magnitudes of the various flows to and from these compartments (Figure 1.6). Nearly 97% of all water on the Earth is stored in the oceans, while only about 0.001% is stored in the atmosphere (Table 1.1). Considering only freshwater (defined as having a concentration of total dissolved solids less than 0.5 parts per thousand and considered potable), which accounts for about 2.5% of the total storage, 69.6% is contained in the polar icecaps and glaciers while 30.1% is contained in groundwater. The freshwater contained in lakes, streams, rivers, and marshes represents only 0.26% of all freshwater and 0.008% (80 drops in a million) of all water on Earth. Another useful concept for thinking about the size of the various reservoirs in relation to the flows of water into and out of them is the residence time. The **residence time**, T_r [T], is a measure of how long, *on average*, a molecule of water spends in that reservoir before moving on to another reservoir of the hydrological cycle. The residence time is easily calculated for systems at steady state, when the inflow and outflow rates are identical:

$$T_r = \frac{V}{I}. \tag{1.6}$$

The residence time has dimensions of time, because volume divided by volume per time is time. The residence time provides an indication of the time scales for flushing a solute out of that particular reservoir. Water in the oceans has a residence time approaching 3000 years, while water in the atmosphere has a residence time of only 0.02 years or about 8 days; the residence time of water in rivers is 0.05 years or about 17 days (Table 1.1).

The water budget for all the land areas of the world is: $\overline{p}$ = 800 mm, $\overline{r}_s$ = 310 mm, and $\overline{et}$ = 490 mm (Figure 1.6). (Note that we now are referring to the volumes divided by the areas being considered. It is sometimes more convenient to use depth rather than total volume, because the volumes can be quite large; also, we are probably more familiar with the statement, "20 mm of precipitation was recorded at Smith Airport," than "Smith Air-

Table 1.1. Sizes and residence times for major reservoirs in the hydrological cycle

	Volume (km³)	Percentage of total	Percentage of freshwater	Residence time (yr)
Water in land areas	47,971,710	3.5		
Lakes				
Fresh	91,000	0.007	0.26	**(All surface water:)**
Saline	85,400	0.006		4.0
Rivers	2,120	0.0002	0.006	
Marshes	11,470	0.0008	0.03	
Soil moisture	16,500	0.0012	0.05	
Groundwater				**(All subsurface water:)**
Fresh	10,530,000	0.76	30.1	20,000
Saline	12,870,000	0.93		
Biological water	1,120	0.0001	0.003	
Icecaps and glaciers	24,364,100	1.76	69.6	
Atmosphere	12,900	0.001	0.04	0.02
Oceans	1,338,000,000	96.5		2,650
Total	1,385,984,610	100	100	

Source: Maidment (1993).

port received 20,000 m³ of water".) On average, 39% of precipitation to the continents runs off and 61% is returned to the atmosphere through evapotranspiration. In other words, the **runoff ratio** ($\bar{r}_s/\bar{p}$) is equal to 0.39. The balance is, of course, affected by many topographic and climatic factors and the budgets for individual continents can be quite different from the average (Table 1.2). The budget for North America is $\bar{p} = 670$ mm, $\bar{r}_s = 290$ mm, and $\overline{et} = 380$ mm. Thus, in North America 43% of precipitation runs off and 57% evapotranspires on average.

1.4.2 The catchment water budget

A global view of the hydrological cycle is not appropriate for discussion of details of hydrological processes. Just as the water budget for each continent can differ from the global average, we expect the local water budget to vary from place to place within a continent, and most problems require the use of control volumes much smaller than continental scale. Although any volume can be defined as the control volume for the application of Equations 1.1 and 1.2, hydrologists typically choose the volume to be a catchment (other terms used in the United States are watershed and drainage basin). The catchment is a fundamental hydrological unit. A **catchment** is an area of land in which water flowing across the land surface drains into a particular stream or river and ulti- mately flows through a single point or outlet on that stream or river; thus, the catchment is defined relative to a specific location along a water course and the associated land area

Table 1.2. Average annual water budget for the continents (excluding Antarctica)

Continent	Area (km²)	$\bar{p}$ (mm)	$\bar{r}_s$ (mm)	$\bar{et}$ (mm)	Runoff ratio, $\bar{r}_s/\bar{p}$
Africa	30.3×10^6	690	140	550	0.20
Asia	45×10^6	720	290	430	0.40
Australia	8.7×10^6	740	230	510	0.31
Europe	9.8×10^6	730	320	410	0.44
North America	20.7×10^6	670	290	380	0.43
South America	17.8×10^6	1650	590	1060	0.36

Source: L'vovich (1979).

Note: Values for average annual precipitation, runoff, and evapotranspiration are reported as depths of water over each land area. The total volume may be calculated by converting the depths to km and multiplying by the land areas. Several estimates of these quantities exist, all of which are uncertain to some degree; Gleick (1993) presents a summary.

can be considered to "catch" the water that flows past that point (Figure 1.8). Clearly, then, any number of catchments can be defined for a particular river (corresponding to any location along the river).

We can define a catchment by specifying when a point is part of the catchment and when it is not. A point is within a catchment if surface water hypothetically flowing from that point ultimately appears at the river station defining the catchment outlet; the point is not within the catchment if surface water flows from it into another river or into the same river below the given river station. The boundary separating regions which do and do not contribute water to that river station is called a **divide**. A catchment is then defined as all points that potentially can contribute surface water to a particular river station.

As an example, consider the Potomac River Basin near Washington, D.C. A point on the western slope of the Blue Ridge Mountains near Waynesboro, Virginia, drains into the South River, which is tributary to the Shenandoah River, which is tributary to the Potomac. The selected point is therefore within the Potomac River catchment. On the other hand, a point on the eastern slope of the Blue Ridge near Charlottesville, Virginia, ultimately drains into the James River and is, thus, not within the Potomac catchment. A portion of the crest of the Blue Ridge Mountains forms the divide between the Potomac and James River catchments. (Note that in the United Kingdom the term watershed is synonymous with divide rather than catchment. Also, in the United States the term catchment is sometimes used to refer to a small drainage basin, while the term river basin refers to a large drainage basin.)

Let us consider application of Equation 1.2 to a catchment. In this case, V is the volume of water stored on or beneath the surface of the catchment. One inflow is precipitation falling on the catchment. By definition, there is no surface water inflow into the catchment (this is the primary reason for using the catchment concept in hydrology, as discussed).

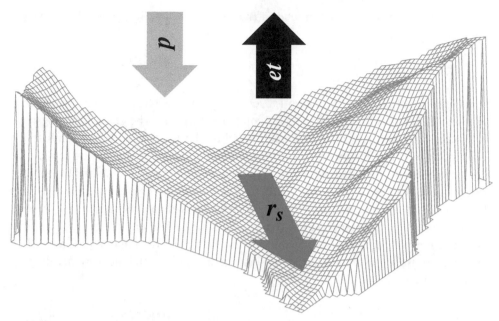

Figure 1.8 The catchment. The boundary of the catchment is referred to as a divide. If the catchment has been properly delineated, there should be no surface-water inflows or outflows across the divide, except at the outlet. In this case, the major inflow is precipitation (p), and the major outflows are evapotranspiration (et) and surface-water outflow through the catchment outlet (r_s). The topography of the land surface controls where divides are drawn. In the figure, two mountain peaks and their adjacent ridges constitute the divide.

There may be both inflow and outflow of groundwater across catchment boundaries. However, if the assumption is made that the groundwater divide coincides with the surface water divide, then groundwater inflows can be neglected in the formulation of the water budget equation. This assumption is often valid but should be evaluated for each application. Outflows consist of loss to the atmosphere (via evapotranspiration) and discharge at the river station chosen in defining the basin. The conservation equation for the catchment may be written:

$$\frac{dV}{dt} = p - r_s - r_g - et, \tag{1.7}$$

where V = volume of water stored in the catchment [L^3]; p = precipitation rate [$L^3\ T^{-1}$]; r_s = rate of surface runoff [$L^3\ T^{-1}$]; r_g = net rate of groundwater runoff [$L^3\ T^{-1}$]; et = rate of evapotranspiration [$L^3\ T^{-1}$].

We can illustrate the use of Equation 1.7 by considering long-term average conditions. For questions involving long-range regional planning, average quantities are often adequate. In this case, changes in volume of water stored in the catchment can be neglected.

If, in addition, we assume net groundwater runoff to be negligibly small, the budget equation becomes:

$$\bar{p} - \bar{r}_s - \overline{et} = 0. \tag{1.8}$$

As an example of the use of Equation 1.8, let us estimate the average annual evapotranspiration in the James River Basin above Scottsville, Virginia. The groundwater systems in this area are highly localized (i.e., controlled by the small-scale topography) so the assumption of negligible groundwater flow is likely to be valid. According to the U.S. Geological Survey, the James at Scottsville drains an area of 11,834 km^2 and the average **discharge** is 144.5 m^3 s^{-1}. The U.S. Weather Bureau climatological records report data that indicate that average precipitation on the catchment is about 1080 mm yr^{-1}. The terms in Equation 1.8 must be expressed in comparable units, so we will express the river discharge in mm yr^{-1} distributed over the catchment; that is, the volume of flow will be converted to the equivalent depth of water over the entire catchment by dividing by catchment area. Thus,

$$\bar{r}_s \frac{mm}{yr} = 144.5 \frac{m^3}{s} \times \frac{1}{1.1834 \times 10^{10} \ m^2} \times 3.154 \times 10^7 \frac{s}{yr} \times \frac{1000 \ mm}{m} \tag{1.9}$$

Performing the indicated calculation we find that $\bar{r}_s$ is 385 mm yr^{-1} over the basin area. Equation 1.8 indicates that $\overline{et} = \bar{p} - \bar{r}_s = 695$ mm yr^{-1} or that about 64% of the rain that falls in this humid region of the United States is returned to the atmosphere through evapotranspiration. It is small wonder that evaporation control is of major concern in arid regions of the world.

The use of a water budget for annual average conditions is straightforward and provides useful information about climatological mean values. One reason that these applications of the conservation laws are simple and straightforward is that in dealing with annual averages we have been able to neglect temporal changes in storage. We have not had to consider how soils store and release water or how channel storage changes with the passage of a flood wave. An understanding of hydrological phenomena at time scales of hours, days, or even months requires that changes in storage be described and that the relationship between the forces applied to water "particles" and the motion of these "particles" be studied. Thus, the *dynamics* of water flow must be studied.

1.5 Concluding Remarks

Hydrology may be broadly defined as "water science." Therefore, one's background, interests, or motives are probably in some way responsible for one's perception of hydrology. In truth, hydrology can be about a rather large variety of subjects. Those whose background is in Earth science will probably already be familiar with the important role of groundwater in a variety of geological processes, or how river flow influences and is influenced by the topography. Others may be interested in issues of water resources, protection, or water rights. Because water is essential to life, biologists and ecologists also need to understand temporal and spatial distributions of water near the land surface.

Increasingly, hydrologists are collaborating with other Earth scientists to solve problems. Such collaboration is often difficult, because each specialist will have his or her own unique language and methods, but ultimately it is beneficial. Today, scientific and practical problems exist that cross disciplinary boundaries. Examples include storage of radioactive waste (involving geoscientists, hydrologists, physicists, and engineers) and global climate change (involving atmospheric scientists, ecologists, environmental chemists, and hydrologists). Scientists are working across traditional disciplinary boundaries to mutual benefit. Throughout this book, we discuss examples of how hydrological problems involve other disciplines. Chapter 11 concludes with a discussion of key challenges facing hydrologists now and in the future, including the impacts of humans and climate on the water cycle and problems related to ensuring the quantity and quality of water needed for drinking, agriculture, and energy as human populations grow and climate changes.

This book is organized around the unifying concepts of the hydrological cycle and the catchment as a basic unit of study. The goal of physical hydrology is to explain phenomena of water flow in the natural environment by application of physical principles. Solutions to many hydrological problems require an understanding of the dynamics of water motion. Thus, much of the remainder of this book is devoted to the quantitative description of components of the hydrological cycle based on models that arise from fluid dynamics.

1.6 Key Points

- Hydrology is defined as the study of the occurrence and movement of water on and beneath the surface of the Earth, the properties of water, and water's relationship to the biotic and abiotic components of the environment. {Section 1.1}

- Hydrological science has both curiosity-driven and problem-driven aspects that stretch back to antiquity. The importance of water as a resource remains one of the central reasons for studying hydrological processes. {Section 1.2}

- The fundamental concept of hydrology is the hydrological cycle: the global-scale, endless recirculatory process linking water in the atmosphere, on the continents, and in the oceans. This cyclical process is usually thought of in terms of reservoirs (e.g., oceans, atmosphere) and the volumetric flows of water between them. {Section 1.3}

- Within the hydrological cycle, the dynamic processes of water vapor formation and transport of vapor and liquid in the atmosphere are driven by solar energy, while precipitation and many of the various flows of water at or beneath the Earth's surface are driven primarily by gravitational and capillary forces. {Section 1.3}

- For a given control volume, a water budget may be constructed based on the principle of conservation of mass: the time rate of change of mass stored within the control volume is equal to the difference between the mass inflow rate and the mass outflow rate. If density is constant, then $dV/dt = I - O$. {Section 1.4}

- A catchment is defined as an area of land in which water flowing across the land surface drains into a particular stream or river and ultimately flows through a single point or outlet on that stream or river. {Section 1.4.2}

- Catchments are delineated on the basis of land-surface topography. The boundary of a catchment is called a divide. {Section 1.4.2}

- On an average annual basis, the water budget for a catchment ($\bar{p} - \bar{r}_s - \bar{et} = 0$) indicates that precipitation is balanced by surface water runoff and evapotranspiration, $\bar{p} = \bar{r}_s + \bar{et}$. {Section 1.4.2}

1.7 Example Problems

Problem 1. Precipitation is typically measured as a volume [L^3] per unit area [L^2], which has dimensions of length [L]. In the United States, the average annual precipitation varies from minimum at Death Valley, California (1.6 inches), to a maximum on Mt. Waialeale on the island of Kauai in Hawaii (460 inches) What is the average annual precipitation (in millimeters, mm) at each of these locations?

Problem 2. In the United States, stream discharge is often measured in units of cubic feet per second ($ft^3 \ s^{-1}$, or "cfs"). In most other countries, discharge is measured in cubic meters per second ($m^3 \ sec^{-1}$). What is the equivalent flow (in $m^3 \ s^{-1}$) of 18.2 $ft^3 \ s^{-1}$? (You might want to review Appendix 1 on units, dimensions, and conversions.)

Problem 3. In an average year, 1.0 meter of precipitation falls on a catchment with an area of 1000 (or 10^3) km^2.

A. What is the volume of water received during an average year in cubic meters?

B. In gallons?

Problem 4. The polar ice caps (area $= 1.6 \times 10^7 \ km^2$) are estimated to contain a total equivalent volume of $2.4 \times 10^7 \ km^3$ of liquid water. The average annual precipitation over the ice caps is estimated to be 5 inches per year. Estimate the residence time of water in the polar ice caps, assuming their volume remains constant in time.

Problem 5. In an average year, a small (area $= 3.0 \ km^2$) agricultural catchment receives 950 mm of precipitation. The catchment is drained by a stream, and a continuous record of stream discharge is available. The total amount of surface-water runoff for the year, determined from the stream discharge record, is $1.1 \times 10^6 \ m^3$.

A. What is the volume of water (in m^3) evapotranspired for the year (assume no change in water stored in the catchment)?

B. What is the depth of water (in mm) evapotranspired for the year (again, assuming no change in water stored in the catchment)?

C. What is the runoff ratio ($\bar{r}_s/\bar{p}$) for the catchment?

1.8 Suggested Readings

Oki, T., and S. Kanae. 2006. Global Hydrological Cycles and World Water Resources. *Science* 313:1068–1072.

Univ. of Oregon. Global Water Balance Animations. http://geography.uoregon.edu /envchange/clim_animations/

Vörösmarty, C. J., P. Green, J. Salisbury, and R. B. Lammers. 2000. Global Water Resources: Vulnerability from Climate Change and Population Growth. *Science* 289:284–288.

2 Precipitation and Evapotranspiration

2.1 Introduction

Chapter 1 introduced the concept of the hydrological cycle and briefly described some of the important processes involved in the motion of water at or near the Earth's surface. This chapter explores two processes of great importance to the fields of hydrology and meteorology: precipitation and evapotranspiration. Precipitation is the primary input of water to a catchment. Evapotranspiration is often the primary output of water from a catchment. A detailed treatment of the precipitation process is most often a subject of meteorology and is beyond the scope of this book. Our discussion of precipitation will focus on methods for directly or indirectly quantifying precipitation inputs to catchments and on describing these inputs using statistical techniques. Evapotranspiration, on the other hand, depends both on properties of the land surface and the state of the near-surface air and is, therefore, well within the domain of hydrology. Our concern with these processes is in knowing the rates, timing, and spatial distribution of these water fluxes between the land and the atmosphere.

    A significant portion of precipitation that falls on vegetated catchments is intercepted by and temporarily stored on the surfaces of vegetation (for example, within dense forest canopies). The remainder falls through to the land surface, where it may infiltrate into the soil or run off. Water stored on the surfaces of plants and at the surface of bare soil, as well as at the surface of open water bodies, can be returned directly to the atmosphere by **evaporation**. Water that infiltrates into the soil may be taken up by roots, converted from liquid to vapor in micropores on leaf surfaces (stomata) of plants, and released to the atmosphere in a process referred to as **transpiration**. As it is very difficult to keep track of the fractions of water lost by the various pathways, we adopt the convention of referring to the combined processes of evaporation and transpiration as evapotranspiration. Evapotranspiration represents a dominant outflow of water from most catchments and accounts for approximately two-thirds of precipitation over most continental land masses. Understanding the physical processes and factors that govern the rate, timing, and distribution of this flux of water from the Earth's surface is, therefore, an important goal of hydrological science. In practice, direct measurements of evaporation from surfaces (such as leaves, soils, or ponds) at the scale of the catchment are unavailable, and we must rely on indirect measures of evapotranspiration or inferences from empirical techniques and the use of proxy variables. Later in this chapter we explore the process of evapotranspiration and discuss some techniques to calculate or infer rates of this flux from the catchment.

There are a number of reasons for studying the processes of precipitation and evapotranspiration. The most apparent reason derives from the catchment annual water budget. Equation 1.8 tells us that for a known or specified value of p, there are infinite combinations of et and r_s that will satisfy Equation 1.8. What processes ultimately govern the way in which these two annual outflows are proportioned in a particular catchment? What controls the relative proportioning on a seasonal or individual storm basis? How might changes in land use and climate influence the water budget for a particular catchment (or for an entire continent)? Will clear-cutting a forest in a catchment cause an increase or decrease in et or in r_s?

The historical development of hydrological concepts indicates the importance of understanding both precipitation (the ultimate supply of freshwater) and evapotranspiration (which can reduce the supply of freshwater and affect its quality by concentrating impurities). The earliest measurements of precipitation, attributed to Kautilya, an Indian chancellor of the exchequer during the fourth century BC (Biswas, 1972), were used as a basis for taxation because agricultural production was presumed to be proportional to rainfall amounts. Knowledge of precipitation and evapotranspiration is crucial to water resources and agricultural questions, especially in arid regions.

One historical example illustrates just how important it is to know the magnitude of hydrological fluxes. In the early part of last century, rapid growth in the western and southwestern United States led to efforts to "reclaim" the desert, mostly through management of the Colorado River. To apportion the flow of the Colorado among the states that would use the water (the Colorado River Compact of 1922), it was necessary to determine the amount of water available each year (i.e., $\bar{p} - \overline{et}$). This was done by averaging the annual discharge measured at a single point on the Colorado River over the available period of record (1896–1921), which turned out to be about 1,233 m^3 each year. Unfortunately, this period of time turned out to be a particularly *wet* era (or, the following years were particularly *dry*). From 1922 to 1976, the average annual discharge of the Colorado River at the gaging station was 1,020 m^3. When the budget was calculated for the Colorado River Compact of 1922 and the water apportioned among the states, there was not enough water to go around! It should be clear that the temporal and spatial patterns of precipitation and evapotranspiration within the Colorado River basin strongly influence water availability and, hence, its use and management.

2.2 Precipitation

Precipitation is defined as the deposition of liquid water droplets and ice particles that are formed in the atmosphere and grow to a sufficient size so that they are returned to the Earth's surface by gravitational settling. Precipitation is ordinarily classified according to the phase it is in when it reaches the surface of a collector or the ground: (a) solid (snow and ice crystals, including sleet and hail) or (b) liquid (rain and freezing rain). Some classifications also consider a class of mixed precipitation (liquid and solid). Precipitation is the dominant deposition mechanism by which atmospheric moisture is cycled from the atmosphere to both oceans and continents. Other mechanisms (i.e., direct deposition of dew and fog) may be important in some instances, especially in coastal mountains.

There are three primary steps in the generation of precipitable water in the atmosphere: creation of saturated conditions in the atmosphere, condensation of water vapor into liquid water, and growth of small droplets by collision and coalescence until they become large enough to precipitate. Saturated conditions occur when the air (which is a mixture of gases, including water vapor) has the maximum water vapor content it can hold without the emergence of condensation (e.g., dew or fog). Water vapor content can be expressed in terms of water vapor density (or absolute humidity), which is the volume of water vapor per volume of air. However, in micrometeorology and hydrology, the partial pressure of water vapor, or **vapor pressure**, e [M L^{-1} T^{-2}], which is proportional to the vapor density, is typically used as a measure of how much water vapor is in the air. The **saturation vapor pressure**, e_{sat} [M L^{-1} T^{-2}], is the value of e for saturated conditions. This value strongly depends on the air temperature; that is, warm air can hold more water vapor than can cool air, or, conversely, cool air cannot hold as much water vapor as warm air (Figure 2.1). Thus, cooling an air mass tends to produce saturated conditions, because e_{sat} is reduced. Cooling typically occurs when an air mass is lifted vertically. Uplift can happen in a number of ways, such as when (a) air masses rise over mountains or other topographic features (referred to as orographic uplift), (b) warm air masses rise above cooler air masses at fronts (frontal uplift), or (c) when heating of the Earth's surface (especially during the summer) makes air near the surface less dense so that it rises (convection) and cools, often producing thunderstorms. Condensation is simply the phase change whereby water vapor becomes liquid or solid water. It requires not only the creation of saturated conditions within the air, but also the presence of condensation nuclei, small particles (size of 0.001–10 µm), such as dust, pollutants, smoke from burning biomass, sea salts or previously formed water droplets or ice particles. Cloud condensation nuclei provide surfaces on which the condensation of water molecules can start. Condensation may produce such small particles that they remain stable in the atmosphere. The white clouds observed on a fair day, for example, are composed of water droplets that are too small to precipitate. It is the **coalescence** of small droplets into larger drops, through

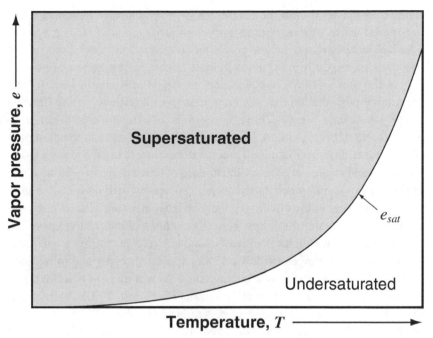

Figure 2.1 Temperature-dependence of the saturation vapor pressure, e_{sat}. Cooling of an undersaturated air mass (at a constant pressure) will produce saturated or supersaturated conditions, the first step in the precipitation process.

collision of small droplets with each other or with larger drops that gives rise to precipitable raindrops or ice crystals. To generate substantial amounts of precipitation, clouds also need to draw atmospheric moisture from the surrounding air.

Close to the land surface, locations suitable for condensation are provided by grass and tree canopies, which—with adequate cooling—allow for dew formation. Dew then either evaporates or drips down to the ground surface. Similarly, plant canopies may favor the deposition of fog droplets, a process known as **occult precipitation**. Canopy condensation and fog deposition may provide an important input of moisture in some landscapes, including areas of the west coast of the United States affected by frequent fog during the rainless summer season, cloud forests, and coastal deserts adjacent to cold ocean surfaces (e.g., the Namib and the Atacama deserts).

Most of the precipitation falling on the continental United States originates from the bordering oceans, even in the interior of the continent. Some studies estimate that up to 30 to 40% of the precipitation over large land areas is derived from local evapotranspiration, a process known as **precipitation recycling** (see Chapter 1). Although this indicates that local evapotranspiration does influence local precipitation, much of the precipitated water must be transported significant distances across the continents from the oceans. It is the large-scale motions of the atmosphere that are responsible for the broad patterns that are observed in annual precipitation.

Average annual precipitation onto the continents is geographically extremely vari-able, reflecting the influence of a number of important physiographic factors. For ex-ample, in the United States, average annual precipitation ranges from a minimum about 40 mm yr^{-1} at Death Valley, California (in the Mojave Desert), to a maximum of nearly 12,000 mm yr^{-1} at the summit of Mt. Waialeale on the island of Kauai in Hawaii. The lowest average annual precipitation that has been measured anywhere on the Earth is less than 1 mm yr^{-1} at Arica, Chile (the 59-year record includes a period of 14 consecu-tive years totally devoid of precipitation). In general, average annual precipitation onto the continents is a function of: (a) latitude (precipitation highest in latitudes of rising air—0° and 60° north and south—and lowest in latitudes of descending air—30° and 90° north and south); (b) elevation (precipitation usually increases with elevation, a phe-nomenon known as the **orographic effect**); (c) distance from moisture sources (precipi-tation is usually lower at greater distances from the ocean, a phenomenon known as **continentality**); (d) position within the continental land mass; (e) prevailing wind direc-tion; (f) relation to mountain ranges (windward sides typically cloudy and rainy, with leeward sides typically dry and sunny, a phenomenon known as **rain shadow**); and (g) relative temperatures of land and bordering oceans (Eagleson, 1970). Average an-nual precipitation onto the oceans is thought to be similarly variable.

Over much of the world precipitation is also extremely variable in time, with infrequent large storms (e.g., hurricanes and monsoons) delivering large portions of the total annual precipitation. As with the spatial variability of precipitation, the temporal variability has both random and persistent components. For example, although the arrival of individual hurricanes is not predictable far into the future, the season in which they arrive is consis-tent from year to year. Accurate forecasts of how many will appear, the amount of rainfall they will produce, and the path they will follow cannot be made for any given year.

2.2.1 Point measurements

Two problems arise in quantifying precipitation input to a given land area: how to mea-sure precipitation at one or more points in space and how to extrapolate these point mea-surements to determine the total amount of water delivered to a particular land area. A variety of instruments or gages are used to measure point precipitation (amount of precipitation deposited at a particular station representing a point in space). Point pre-cipitation usually is expressed in depth units (volume divided by collector cross-sectional area). Point measurement devices are generally of two types: non-recording (storage) gages and recording gages. The first category includes simple wedge- or funnel-shaped containers to collect precipitation over a period of time between observations. There are two common types of recording precipitation gages: weighing and tipping bucket gages. Weighing gages collect precipitation, typically through a funnel and record the weight of precipitation as a function of time. Tipping bucket gages consist of a container with a funnel at the top leading to a pair of small "buckets" attached to a fulcrum. When one bucket fills with water, it tips, emptying the water and moving the other bucket beneath the outlet of the funnel. The device records the *time* at which the bucket tips, and so the time over which a certain amount of precipitation fell (e.g., 0.25 mm for a bucket of a certain size) is recorded. In regions that receive snowfall, point gage data usually are expressed in liquid water equivalent (LWE), which is the equivalent amount of total precipitation if it had all fallen as rainfall. Most precipitation gages are equipped with windbreak devices or shields to minimize the measurement error caused by disruption of the airflow pattern around the gage.

Rainfall rates measured at a point are tremendously variable in time (Figure 2.2). The record of hourly precipitation over time (known as a **hyetograph**) for a station in Virginia illustrates that precipitation commonly is organized into discrete storm events of varying intensity and duration separated by interstorm periods of variable duration. Because we may be interested in longer periods of time (perhaps an annual water budget for a catchment), we can add hourly measurements together to derive daily or monthly hyetographs. Variability decreases as the period of reporting increases, as illustrated by the daily and monthly hyetographs for the Coweeta experimental watershed for the 1992 water year (Figure 2.3). (In the United States, the U.S. Geological Survey water year begins October 1 and runs through the end of the following September.) Both annual precipitation depths and temporal precipitation variability depend on local climatology. For example, in monsoon-prone regions most of the precipitation arrives in a certain season when large-scale

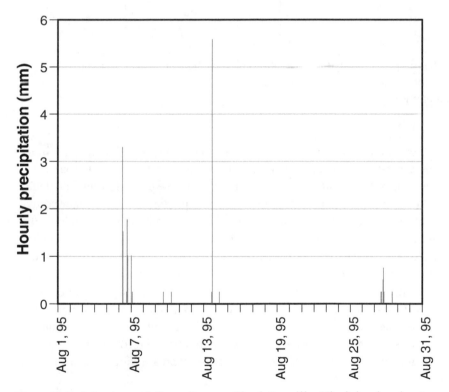

Figure 2.2 A hyetograph for a site near Charlottesville, Virginia, showing temporal variability in precipitation.
Data courtesy of Greg V. Jones.

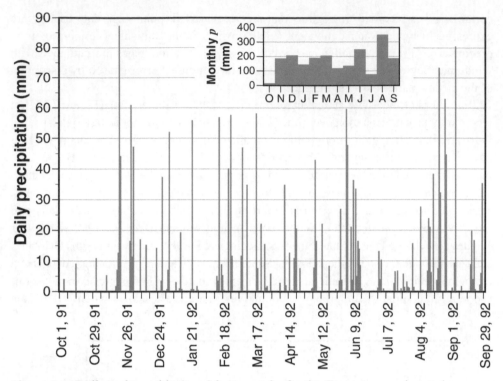

Figure 2.3 Daily and monthly (*inset*) hyetographs for the Coweeta experimental watershed, North Carolina, water year 1992.

Data courtesy of Wayne Swank.

atmospheric flows drive moist air from oceanic areas onto terrestrial areas. In rain forests and some coastal areas the precipitation is spread more evenly over the year.

The measurement of precipitation in the form of snow presents additional difficulty. Gages must be heated, for example, to reduce the precipitation to liquid water equivalent. In mountainous areas, in particular, it can be difficult to keep a standard gage running through the winter. Snow pillows, devices that record the pressure at the base of a snowpack, and snow surveys, in which teams periodically go into the field and take cores of snow along transects to estimate depth of snow, are among alternative measurement methods. The Soil Conservation Service in the United States maintains approximately 500 snowpack telemetry sites in remote mountainous areas of the western United States at which data from snow pillows are collected automatically and continuously. Nevertheless, point measurements of precipitation in the form of snow in mountainous areas are collected on a much sparser network than are rainfall data in lower-lying areas.

2.2.2 Spatial characteristics of precipitation and radar estimation

Thus far, we have discussed temporal variations in precipitation at a point or station in space. In hydrology we typically want to estimate the total volume of water delivered to a given area (e.g., a catchment) within some period of time. That is, we want to know the average precipitation depth over the area, i.e.,

$$P = \frac{1}{A} \iint\limits_A p(x,y)\,dx\,dy \qquad\qquad (2.1)$$

where P is the average depth of precipitation over the catchment, A is the catchment area, and $p(x,y)$ is the spatial distribution of cumulative precipitation for the time period of interest. Once P is determined, the total volume of water delivered to the area is simply PA. In many situations, hydrologists are faced with using gage measurements to estimate P. Given measurements made at discrete points within an area or catchment, how can the average precipitation for an area be estimated?

In the simplest and most common case there is only one rain gage in (or even near) the catchment of interest. Hence, by necessity the spatial variability of precipitation within the catchment is ignored and the single point measure is assumed to represent the mean value, such that the rate is simply multiplied by the area to estimate the total precipitation over the catchment. Given the great spatial variability in actual precipitation rates, this approach is likely to yield large errors. Unfortunately, with limited data there may be no other choice.

Over very large catchments or heavily instrumented research sites, several precipitation gages may be available (Figure 2.4a). In this case we might use a *weighted average* of these gage measurements to estimate P. The precipitation values at all gaging stations (p_i) are used, with each value weighted in some reasonable way. The simplest example is an arithmetic average that weights each gage equally. In some instances it may be more appropriate to weight one gage more heavily than others. For example, one gage might be located within the catchment that is the focus of a water resources investigation and all other gages might be some distance from the catchment. In this case we might prefer to

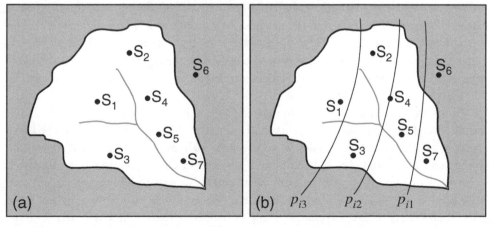

Figure 2.4 Depiction of gaging stations within a hypothetical catchment (*a*) and the construction of isohyets (*b*) to estimate the spatial distribution of precipitation.

assign weights on the basis of how important we think the gaging station is or how representative of precipitation over the catchment area that station is.

We might approximate the true distribution of precipitation by drawing "contour lines" of equal precipitation depths, called **isohyets** (Figure 2.4b). The **isohyetal method** uses the areas bounded by each pair of isohyets to determine P:

$$P = \frac{1}{A}\sum_{i=1}^{I} a_i \hat{p}_i, \tag{2.2}$$

where $\hat{p}_i = (p_{i-} + p_{i+})/2$, the average p for each subregion; p_{i-} = value of the lower magnitude isohyet bounding subregion i; p_{i+} = value of the higher magnitude isohyet bounding subregion i; a_i = area of subregion i. Again, the weights are fractional areas but this time the "contour" values of precipitation are used in place of the gage values. Alternatively, spatial interpolation algorithms may be used to determine the spatial distribution of precipitation based on point measurements from rain gages. These algorithms are included in most geographic information system (GIS) software packages.

Interpolation and weighting schemes that use rain gages are generally inadequate for defining average precipitation to a catchment on a storm-event basis. Even with fairly dense rain gage networks, the spatial distribution and intermittency of rainfall rarely is captured in the point-measurement data. For example, in a study involving a dense gage network and a NEXRAD (NEXt generation RADar), numerous storms for which radar estimates indicated hourly precipitation exceeding 50 mm were missed *completely* by the gage network (Smith, Seo, et al., 1996). As radar data become available on a routine basis, they may supplant the weighting schemes outlined above, or at least provide a rational basis for constructing isohyetal lines.

Weather radar has become an increasingly important tool for estimating the spatial distribution of rainfall, $p(x,y)$. Radar signals reflect from raindrops in the atmosphere and the characteristics of the reflected signal can be related to rainfall rates. Radar is far from an absolutely accurate measurement method, but it provides detailed information on the time and space distribution of rain and can be particularly valuable for heavy rainfall. The spatial distribution of total precipitation in a storm may depend in part on the path of the storm (Figure 2.5). We also see from these data the potential for significant spatial variability in precipitation over a catchment. Imagine the different estimates of total precipitation from a single gage located in the southern end of the catchment as compared with one in the northern end.

Measurements from satellites also may be used to infer precipitation rates and snow-accumulation patterns. Snow-cover maps are produced from data collected with NOAA's (National Oceanic and Atmospheric Administration) Advanced Very High Resolution Radiometer (AVHRR). These satellite data are used in conjunction with ground-based point measurements to determine spatial distributions of water in the snowpack (e.g., see DeWalle and Rango, 2008). Whereas the satellite observations estimate amounts of snow present on the ground, techniques are being developed to extract snowfall rates from the NEXRAD data. NEXRAD sites are being installed presently around the United States,

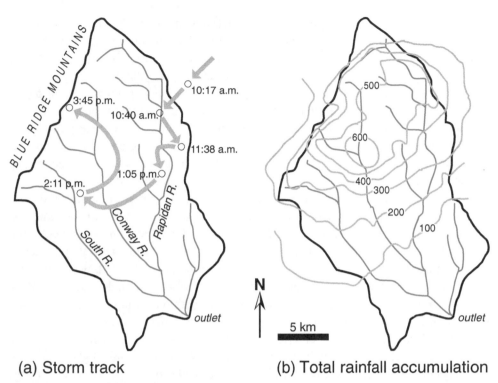

(a) Storm track (b) Total rainfall accumulation

Figure 2.5 Radar observations of the June 27, 1995, storm in the Rapidan River basin, central Virginia. NEXRAD tracked the center of the storm (*a*). Total rainfall accumulations (mm) were calculated based on the time series of radar images (*b*). This intense storm produced an average of 344 mm (13.5 inches) over the catchment and caused extensive flooding throughout the basin.

Smith et al. (1996).

providing coverage of most regions. With the data from these instruments being made available over the Internet, it is reasonable to anticipate that they will soon be applied widely in hydrology to estimate rainfall and snowfall rates distributed in space and time over catchments.

2.2.3 Temporal characteristics of precipitation

Precipitation at any particular point on the Earth's surface is subject to extreme temporal variation. One minute it may be raining heavily, while a few minutes later it may not be raining at all. Our ability to forecast this temporal variation even a few hours in advance is limited and our ability to forecast several days in advance is almost zero. The extreme uncertainty associated with precipitation forecasts suggests substantial randomness in the occurrence of precipitation at a point in space and implies the necessity of a probabilistic approach for characterizing the temporal variations in precipitation.

As noted previously, if we examine a typical short-term hyetograph (e.g., Figure 2.2), we see that precipitation is organized into discrete (storm) events of varying amounts (storm depths). Average **precipitation intensity** is the rate of precipitation over a specified time period, the precipitation depth divided by the time over which that depth is recorded. For example, the data in Figure 2.2 are reported for each hour (i.e., the hourly precipitation intensity). The hourly precipitation intensity for the period shown varies from zero to about 22 mm hr^{-1}. Average precipitation intensities depend on the time period over which the computation is done. That is, the variation in hourly precipitation intensity typically will be much greater than the variation in 6-hour intensity (average over a longer time), but less than that of 15-minute intensity (average over a shorter time). Average precipitation intensity for a storm is the total depth of precipitation for the storm divided by the storm length. In general, the longer the storm-event duration, the less the (average) storm intensity. However, the greater the storm duration, the greater the storm depth.

Hydrologists commonly employ a statistical technique known as **frequency analysis** to describe systematically the temporal characteristics of precipitation at a particular station. To illustrate the frequency analysis technique, we will consider annual precipitation amounts for several United States cities over a 64-year period (Table 2.1). We assume that the quantities in Table 2.1 (annual precipitation) are samples of a random variable. If we look at how often (the frequency with which) values within a certain range are encountered, we find that a plot of magnitude versus frequency displays a characteristic shape, a "bell-shaped" curve. In the case of the data from Seattle (Figure 2.6), we find that the most frequent values of annual precipitation are between 950 and 1000 mm. Higher and lower values occur with lower frequency. We also could normalize these data by dividing by the total number of observations (64) to determine the relative frequency. For example, annual precipitation values between 950 and 1000 mm occur with a relative frequency of 10/64, or 0.16. Alternatively, we might say that values within that range occurred 16% of the time. Note that, because annual precipitation is a continuous random variable, we need to be concerned with the probability of a range of values, rather than with one specific value. Also, note that these data represent yearly precipitation amounts. As we will discuss shortly, decreasing the averaging period (say, monthly or daily precipitation amounts) often causes the shape of the curve to become more skewed and look less like a bell.

Because annual precipitation amounts appear to follow a normal (or "Gaussian") distribution, the mean and the standard deviation (Figure 2.6, Table 2.1) are the only parameters that are required to describe the magnitude-frequency relationship. Once we have determined the parameters of the normal distribution (the mean and the standard deviation), we can determine the probability associated with a particular range of annual precipitation values (Appendix 3). As an example, we will calculate the probability that annual precipitation in any given year will exceed 1.0 m (this is referred to as the **exceedance probability**). The first step is to calculate the z-value, which is a way of normalizing the data to a distribution with zero mean and unit standard deviation. This is accomplished by finding the difference between the sample mean and the target value (1.0 m in this example) and dividing this value by the sample standard deviation:

$$z = \frac{x - \overline{x}}{s_x},$$ (2.3)

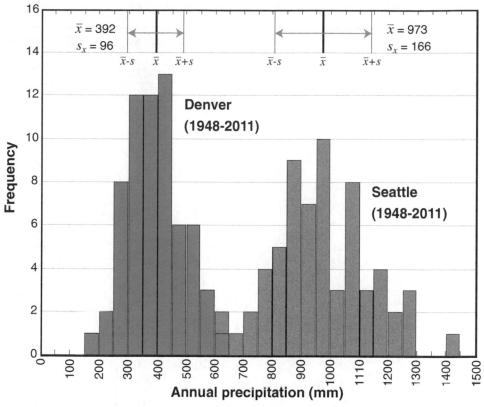

Figure 2.6 Absolute frequency of values of annual precipitation at Seattle (*gray bars*) and Denver (*blue bars*), with the calculated mean ($\bar{x}$) and standard deviation (s_x).

where x = the value of the variable (p) that we are interested in (e.g., 1.1 m); $\bar{x}$ = the mean of p; s_x = the standard deviation of p. For the Seattle data, the calculated z-value corresponding to x = 1.1 m is 0.77. For this z-value, the cumulative probability (Table A3.2 in Appendix 3) is 0.7794. This value is the probability that the annual precipitation will be *less than* 1.1 m, so we need to subtract 0.7794 from 1.0 (the total probability) to arrive at a value of 0.2206. Therefore, there is a 22% chance that annual precipitation will exceed 1.1 m in Seattle. Another way of looking at the same result is to say that precipitation in excess of 1.1 m occurs roughly every 4–5 years or with a **return period** (the inverse of the exceedance probability) of approximately 4–5 years.

Magnitude-frequency relationships for many hydrological variables are useful to hydrologists and water resources planners. For example, in the design of a reservoir for irrigation or hydropower it may be important to know how often part of the reservoir's storage capacity would remain unused. To that end, hydrologists use the methods described in this subsection to calculate the probability that the annual rainfall in a region is smaller than the minimum value required to reach full storage conditions in the course of the water year. Although annual precipitation often is described well by a normal distribution (Figure 2.6), not all hydrological parameters are (see Haan, 2002). In fact,

Table 2.1. Total annual precipitation (mm), 1948–2011, for selected cities in the United States

Year	Seattle	Santa Barbara	Denver	Charlottesville
1948	1164	230	321	1771
1949	825	326	428	1116
1950	1401	288	355	1090
1951	1024	350	495	1106
1952	604	693	342	1307
1953	1256	153	362	971
1954	1050	416	192	1066
1955	1185	468	408	1087
1956	937	274	349	1078
1957	881	420	549	1084
1958	1084	557	479	1057
1959	1182	209	420	1205
1960	998	408	381	1006
1961	1081	143	483	1412
1962	909	459	216	1248
1963	984	493	311	772
1964	1051	310	258	1076
1965	854	516	556	906
1966	971	306	276	1009
1967	904	457	593	1134
1968	1275	304	308	1012
1969	857	629	548	1402
1970	951	470	350	1140
1971	1098	265	279	1429
1972	1229	219	430	1678
1973	891	502	584	1249
1974	962	482	357	1106
1975	1130	315	395	1520
1976	678	405	342	1267
1977	834	304	264	809
1978	864	892	298	1265
1979	820	457	518	1558
1980	905	459	348	883
1981	900	459	320	926
1982	999	527	367	1317

Table 2.1. (*continued*)

Year	Seattle	Santa Barbara	Denver	Charlottesville
1983	1040	1035	514	1409
1984	939	239	419	1249
1985	639	306	415	1173
1986	975	494	308	844
1987	761	373	510	1237
1988	838	318	381	672
1989	881	110	393	1400
1990	1137	139	424	1294
1991	900	565	516	1119
1992	833	456	399	1143
1993	732	574	377	1204
1994	885	321	280	1349
1995	1082	721	408	1198
1996	1287	708	318	1373
1997	1100	470	606	1152
1998	1120	1024	453	1302
1999	1070	300	479	1154
2000	729	642	353	1095
2001	955	597	402	902
2002	798	234	216	1031
2003	1062	558	408	1895
2004	790	257	436	1133
2005	901	793	370	1151
2006	1231	467	321	1176
2007	991	186	443	880
2008	782	384	275	1159
2009	977	256	426	1328
2010	1195	636	299	1013
2011	925	718	480	1346
mean	973	438	392	1179
standard dev.	166	201	96	226

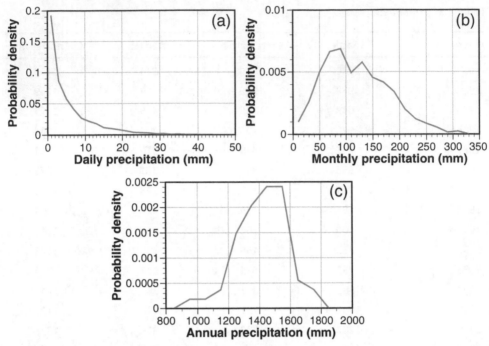

Figure 2.7 Probability densities for daily (*a*), monthly (*b*), and annual (*c*) precipitation at Valentia, Ireland.

Data courtesy of Gerard Kiely.

except for annual totals precipitation intensities (or amounts) typically do not follow a normal distribution. We have restricted our application of the normal distribution to annual precipitation amounts. If the precipitation averaging time is decreased to a much shorter period, the distribution begins to diverge from normal. This is shown in Figure 2.7 for a 50-year record of precipitation on the southern coast of Ireland. In this figure, probability density is analogous to a relative frequency in that it is a relative measure of how likely certain magnitudes of precipitation are. Notice the extreme skewness in the daily precipitation distribution, the moderate skewness in the monthly amounts, and the relative absence of skewness in the annual amounts. Hence, it would be incorrect to use the normal distribution to describe a variable such as daily precipitation at this location. It should be stressed that the probabilistic approach requires a significant amount of data (a long time series) to produce meaningful analyses.

2.2.4 Analysis of rainfall extremes

A number of hydrological applications require an analysis of extreme events and their frequency of occurrence. Urban planners, land developers, and insurance companies need to know the flooding frequency of areas subjected to different land uses. Similarly, in the design of bridges, canals, drainage systems or dams engineers need to refer to suitable hydrologic design conditions. It is important to know with what probability these conditions may be exceeded. Different exceedance probabilities correspond to different risks of damage to people and property so design decisions are made based on how much risk is tolerable in a given situation.

To explain how the exceedance probability of extreme rainfall events can be calculated, we need to revisit some of the concepts presented in the previous section. If X is the rainfall depth measured with a given resolution Δt (e.g., $\Delta t = 15$ min, 1 h, or 1 day) and X^* is a reference value of X (e.g., the design conditions) the time interval, τ, between two consecutive events exceeding X^* is known as the recurrence interval of events greater than X^* (see Figure 2.8). The average duration of the recurrence interval is defined as the return period, T_{return}, of X^* and is typically measured in years. The return period of an event, X^*, tells us that X^* is exceeded on average every T_{return} years. Based on this definition, the return period increases with increasing values of X^*. In other words events exceeding X^* become more rare as X^* increases. It should be stressed that the definition of return period is given as the mean recurrence interval, τ, which means that the rainfall event with return period of, say, 100 years (sometimes called "the 100-year rainfall") is exceeded *on average* (but not exactly) once every 100 years. We can have some 100-year intervals with more than one event greater than the hundred-year rainfall and others with none. It can be shown that if events with $X > X^*$ are statistically independent, $1/T_{return}$ is equal to the exceedance probability of X^* (i.e., Prob$[X > X^*] = 1/T_{return}$; Appendix 3).

The study of extreme rainfall events faces two challenges: First, extreme rainfall events are by definition rare and the rainfall records that are typically available to determine their

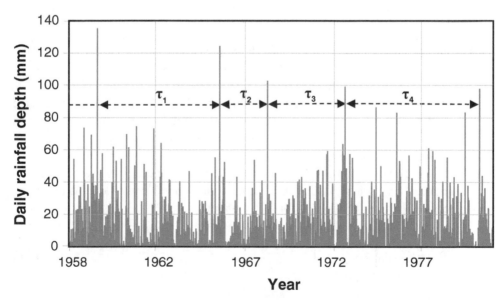

Figure 2.8 Record of daily precipitation depth, X, for the city of Turin. The return period of the extreme value $X^* = 88$ mm d^{-1} (*dashed horizontal line*) is the mean of the recurrence intervals τ_1, τ_2, and so on.

return period are often very short. For example, we might need to study events with the return period of 100 or 200 years, while having access to records much shorter than these return periods. Second, as noted in the previous section, event-scale precipitation does not exhibit the "bell-shaped" Gaussian distribution that we have considered in the case of annual precipitation. The properties of the distribution of X are in general unknown and depend on the time scale Δt at which X is measured (see differences between daily and monthly precipitation in Figure 2.7). Statistical theories, however, have shown that the extreme values of any random variable tend to follow only three possible classes of distributions. In particular, extreme precipitation is typically distributed according to the Gumbel (or "Type I") distribution.

To use extreme value distributions to analyze a precipitation record we need to extract from the "parent population," X, of precipitation the extreme values, X'. This can be accomplished by considering only the maximum annual value for each year of record. We then assume that X' has a Gumbel distribution and determine the return period of X^* (or its exceedance probability) as

$$Prob[X' \geq X^*] = \frac{1}{T_{return}} = 1 - \exp\left[-\exp\left(-\frac{X^* - \beta}{\alpha}\right)\right] \tag{2.4}$$

Equation 2.4 shows that the Gumbel distribution has only two parameters, α and β, that need to be estimated using the mean ($\overline{X'}$) and the standard deviation ($\sigma_{X'}$) of X' as

$$\alpha = \frac{\sqrt{6}}{\pi}\sigma_{X'} \quad \text{and} \quad \beta = \overline{X'} - 0.5772\alpha. \tag{2.5}$$

We can use the precipitation record from the city of Turin to examine this method. For every year we consider the maximum rainfall depth with a duration of 1, 3, 6, 12, and 24 hours (Table 2.2). To determine the return period of a rainfall depth, e.g., $X^* = 70$ mm with the duration of 3 hours, we first need to calculate the mean, $\overline{X'}$, and the standard deviation, $\sigma_{X'}$, of the annual maxima of that duration (Table 2.2, third column). Using equations (2.5) we determine the corresponding values of α and β for the 3h duration (see Table 2.2) and use Equation 2.4 to calculate the return period of X^*. We find that 70 mm of rainfall in 3 hours is exceeded on average once every 40 years ($T_{return} = 40$ yr). Alternatively, we might need to calculate the rainfall depth corresponding to a given return period. Solving Equation 2.4 for X^* we obtain

$$X^* = \beta - \alpha \ln\left[-\ln\left(1 - \frac{1}{T_{return}}\right)\right] \tag{2.6}$$

where $\ln()$ denotes the natural logarithm. Using Equation 2.6 we find that the rainfall depth with a 3 hour duration and a return period of 100 years is $X^* = 79.5$ mm. As expected, the rainfall depth associated with the same return period ($T_{return} = 100$ yr) increases with increasing rainfall duration (see Table 2.2).

Table 2.2. Maximum annual precipitation depths, X', for different durations, Δt, obtained from the rainfall record of the city of Turin (1927–2009)

Year	$\Delta t = 1\,h$	$\Delta t = 3\,h$	$\Delta t = 6\,h$	$\Delta t = 12\,h$	$\Delta t = 24\,h$
1927	31.0	33.0	33.0	38.0	56.0
1928	20.6	22.5	25.4	27.4	27.8
1929	21.0	33.5	36.5	61.5	98.0
1930	18.9	21.0	30.0	50.4	70.7
1931	62.2	64.2	65.2	65.2	72.6
1932	17.6	29.0	32.6	39.6	51.2
1933	32.4	36.0	36.0	40.0	60.0
1934	15.6	24.6	35.6	48.2	51.4
1935	31.0	52.0	54.8	68.2	68.4
1936	42.6	54.6	54.8	54.8	70.6
1937	31.0	33.0	33.0	35.4	51.6
1938	33.0	38.0	38.0	38.0	43.0
1939	24.2	24.2	25.4	32.2	44.6
1940	15.0	25.6	32.6	37.0	41.0
1941	15.6	23.0	25.0	41.0	66.0
1951	32.6	38.0	67.0	87.4	104.0
1952	30.4	30.4	30.4	44.0	65.6
1953	20.4	30.0	48.0	67.0	71.0
1954	53.6	69.0	70.6	70.6	72.8
1955	57.6	61.0	61.0	61.0	64.0
1956	30.8	45.6	45.6	64.6	81.6
1957	32.6	41.0	52.2	52.8	73.6
1958	23.8	23.8	27.8	46.0	73.0
1959	31.4	33.0	35.8	53.4	74.4
1960	36.0	48.0	70.0	110.0	140.0
1961	60.0	61.6	72.6	80.0	82.8
1962	58.4	61.2	61.2	78.0	135.0
1963	23.4	25.4	50.0	52.6	74.2
1964	27.2	52.0	64.0	73.6	93.0
1965	21.0	24.0	29.2	42.2	49.6
1966	28.0	28.0	28.0	28.0	28.0
1967	28.0	31.2	34.8	54.6	57.8
1968	25.0	43.0	43.0	43.0	43.0
1969	28.0	34.8	53.6	53.6	66.8

(*continued*)

Table 2.2. (*continued*)

Year	$\Delta t = 1$ h	$\Delta t = 3$ h	$\Delta t = 6$ h	$\Delta t = 12$ h	$\Delta t = 24$ h
1970	18.0	26.4	27.6	28.2	46.8
1971	26.2	27.0	38.0	40.2	43.0
1972	24.0	28.4	37.4	37.4	44.0
1973	36.0	40.0	48.6	57.0	66.0
1974	46.8	50.2	55.0	107.0	120.2
1975	19.8	30.8	31.0	42.2	74.4
1976	43.6	48.8	49.8	62.4	99.2
1977	22.6	47.2	66.8	85.0	85.0
1978	25.6	30.0	30.0	47.0	60.4
1979	51.0	55.2	55.2	55.2	55.2
1980	37.0	54.0	67.6	75.8	83.0
1981	20.8	26.2	39.6	52.0	99.0
1982	25.8	37.8	39.0	40.0	44.0
1983	28.2	31.2	34.2	47.4	60.6
1984	16.0	22.0	28.6	42.4	61.4
1985	32.2	38.2	44.0	45.8	52.6
1986	30.4	31.0	37.0	45.4	52.0
1987	35.6	48.0	54.0	56.8	80.8
1988	30.6	42.0	42.4	47.4	59.2
1989	23.6	35.4	39.4	64.2	100.2
1990	45.2	46.2	46.2	46.4	52.6
1991	36.4	56.0	87.2	88.6	99.4
1992	21.8	34.2	44.0	78.8	109.4
1993	32.6	51.8	53.2	53.2	74.4
1994	23.6	43.8	63.0	109.0	149.4
1995	28.4	32.8	32.8	44.1	45.9
1996	34.3	39.5	46.9	47.7	50.7
1997	49.6	57.1	57.1	61.5	66.5
1998	28.6	28.6	28.6	28.6	51.9
1999	25.8	41.2	41.6	41.6	46.8
2000	24.0	41.4	63.2	82.4	97.1
2001	21.2	24.6	26.1	26.9	28.1
2002	55.3	60.6	61.0	72.5	118.2
2003	29.5	29.5	31.2	37.8	51.3
2004	26.2	35.9	36.1	36.1	49.6
2005	65.2	74.2	77.7	79.5	80.5

Table 2.2. (*continued*)

Year	$\Delta t = 1\,h$	$\Delta t = 3\,h$	$\Delta t = 6\,h$	$\Delta t = 12\,h$	$\Delta t = 24\,h$
2006	31.0	34.8	46.7	64.4	106.9
2007	52.2	56.4	56.8	56.8	65.6
2008	20.6	23.8	34.2	56.6	86.4
2009	29.6	37.0	52.2	69.8	69.8
$\overline{X'}$	31.5	39.1	45.3	55.4	70.4
$\sigma_{X'}$	12.1	12.9	14.7	19.0	25.7
α	9.4	10.0	11.4	14.8	20.0
β	26.1	33.3	38.7	46.8	58.9
X^* ($T_{return} = 100$ yrs)	69.5	79.5	91.3	115.1	151.0

Note: These values are used to calculate the mean, standard deviation, and the parameters, α and β, of the Gumbel distribution, along with the rainfall values, X^*, with 100-year return period. All values are expressed in mm.

2.3 Interception

Not all precipitation reaching the land surface is available for streamflow or replenishing groundwater. Rather, a portion is temporarily stored by vegetation (**interception**), where it is subject to evaporation. If we consider the total amount of precipitation (gross precipitation) delivered to a point within a land area (p), such as might be measured by a tipping bucket gage placed in a clearing or above a forest canopy, some of that precipitation in a vegetated area will be intercepted by the plants, some will fall between plants to land on bare ground or ground covered by lower vegetation or leaf litter, and some will run down the stems and trunks of plants to the ground surface (Figure 2.9). Precipitation stored by the canopy or leaf litter is subject to evaporation. We can define the total interception (I_T) as the sum of the canopy interception (I_c) and litter interception (I_l). It is important to stress that, according to this definition, interception is associated with the evaporation of rainwater stored in canopy and litter. Interception should not be confused with the temporary retention of water in canopy or litter before it eventually drips to the ground.

The capacity of the canopy and litter to store water is limited, and the rain interception and snow interception capacities are not the same value. After a period of time during a storm, these "reservoirs" will begin to approach their limits, such that no additional water can be stored. Precipitation that is not intercepted by the canopy or which leaves the temporary storage of interception may be classified as either **stemflow** or **throughfall** (see Figure 2.9). The former is exactly what the name implies and the latter refers to water that reaches the ground directly or by dripping off leaves. This partitioning of precipitation between vegetated surfaces and the soil is critical in defining the reservoirs of water available for direct evaporation from surfaces and for transpiration by plants of water taken up from the soil through roots.

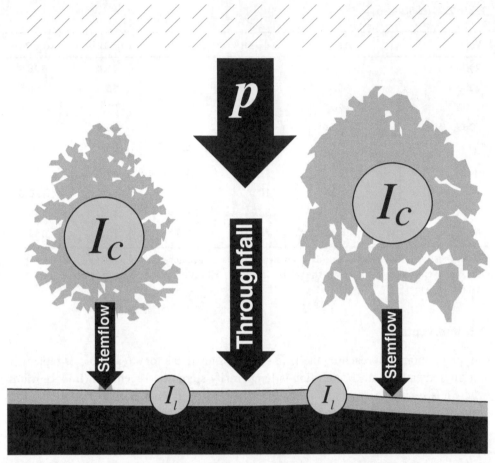

Figure 2.9 Depiction of fluxes (*arrows*) and interception stores (*circles*). Not all precipitation reaches the ground surface in vegetated areas, such that the combination of throughfall and stemflow is less than the total precipitation. Water stored in vegetation (I_c) or leaf litter (I_l) may return to the atmosphere through evaporation.

2.4 Evapotranspiration

The term **evapotranspiration** describes all the processes by which liquid water at or near the land surface becomes atmospheric water vapor. Evaporation is a physical process that occurs whenever a wet surface is exposed to unsaturated air. Transpiration is the vaporization and transport of plant water from leaf chloroplasts to the atmosphere through the **stomata**, small cavities existing on leaf surfaces. It is difficult in practice to separate evaporation (from wet surfaces) and transpiration (water evaporating inside plants) from each other. Hence, we focus on the combined quantity, evapotranspiration (*et*). As we saw in Chapter 1, looking at a global average, about two-thirds of the precipitation that falls on the continents is evapotranspired. Of this amount, 97% is *et* from land surfaces and 3% is open-water evaporation. It is through the process of evapotranspiration that the sun's energy is introduced to drive the hydrological cycle.

Both evaporation and transpiration require a diffusion mechanism (see 2.4.2) to sustain a water vapor flux that removes vaporized water molecules from the evapotranspiring surface. The two ingredients for vaporizing water are energy and water. Hence, evapotranspiration is where the surface-water balance and surface energy balance meet. The energy is derived from the solar radiation and the water is typically provided by local precipitation. Because both solar energy and available water are necessary to cause evaporation (and transpiration), energy will limit the rate of evapotranspiration at some times, and water availability will limit the rate at other times. Considering water availability, we see that evapotranspiration is a two-way street, where the amount of water present affects the rate of evapotranspiration, which in turn affects the amount of water present for subsequent allocation. When the near-surface soil is saturated or nearly saturated with water, evapotranspiration may proceed at a rate—known as the **potential evapotranspiration rate**—limited only by the availability of energy. When the soil becomes drier the actual evapotranspiration rate becomes reduced below that found for a wet surface.

Most plants have openings (stomata) on their leaves to allow them to take up carbon dioxide from the atmosphere. When the stomata are open, plants transpire water. Unlike evaporation, transpiration is not controlled solely by physical conditions because plants regulate the rate at which water is released in transpiration in a manner that varies by plant type. Of the water taken up by plant roots about 95% is transpired through the stomata. The remaining 5% or so is converted to biomass through photosynthesis. Hence, to first order, the water taken up by the roots is converted to vapor and lost to the atmosphere. When the availability of soil water is limited, plants conserve it by restricting flow to the atmosphere by contracting the stomata. However, the degree of restriction varies considerably across plant species and even throughout the year for a given species. This renders quantitative treatment possible only in an average sense, say with an entire stand of trees treated as a single "big leaf" for which the resistance to water flow is handled as a simple function of soil moisture content and time of year. Here hydrologic and ecologic conditions must be explored simultaneously, to incorporate a quantitative description of plant water use characteristics into the framework of hydrology.

The direct measurement of evapotranspiration is a difficult task and in most hydrological applications the rate of evapotranspiration is calculated as a function of quantities that are easier to measure. To that end, we can in principle use methods based on the water balance, the energy balance, or the diffusion of water vapor (mass-transfer methods). As it will be shown in the following subsection, water balance methods have poor accuracy. Thus, energy balance and mass-transfer approaches are most commonly used. To eliminate the dependence on some variables that are difficult to measure, these two methods are often combined (combination methods).

2.4.1 The water balance

Evapotranspiration may be estimated with a water balance approach, if the change in storage and all the inputs and outputs except *et* are known (see Equation 1.3):

$$\frac{dV}{dt} = p(t) + r_{si}(t) + r_{gi}(t) - r_{so}(t) - r_{go}(t) - et(t). \tag{2.7}$$

In this equation, the time rate of change in volume of water stored within the control volume (left side) is balanced by the difference between the inputs and the outputs. Each of the terms on the right side of the equation has units of volume per time $[L^3 T^{-1}]$, and all are functions of time. Equation 2.7 may be simplified for some problems, for example, by neglecting r_{gi} and r_{go} when considered unimportant. Writing Equation 2.7 for average annual quantities reduces it to Equation 1.3. Once all the terms except *et* have been measured, then *et* may be computed as the residual to force a balance that conserves mass in the system. An example of the use of this equation for an entire catchment is presented in Chapter 1. The result is a very reasonable estimate of average annual evapotranspiration, $\overline{et}$, for the catchment. In many cases, however, the water balance approach may suffer from the accumulation of errors in the measured variables. As an example, consider a lake with a very rapid throughflow rate, which over the course of a year does not see a net change in storage. Suppose that the measured average annual fluxes and associated errors for the lake are:

$$\overline{p} = 10^7 \pm 5 \times 10^5 \text{ m}^3 \text{ yr}^{-1} \ (\pm 5\%)$$

$$\overline{r}_{si} = 10^9 \pm 1.5 \times 10^8 \text{ m}^3 \text{ yr}^{-1} \ (\pm 15\%)$$

$$\overline{r}_{so} = 9.95 \times 10^8 \pm 1.5 \times 10^8 \text{ m}^3 \text{ yr}^{-1} \ (\pm 15\%).$$

If we neglect groundwater inflows and outflows, we can use these values and Equation 2.7 to solve for $\overline{et}$. The result, accumulating the errors as we go, is $1.5 \times 10^7 \pm 3 \times 10^8$ m³ yr⁻¹. (Bear in mind that for a lake such as this, the $\overline{et}$ is a relatively small part of the budget. You might want to compare this example with the example presented in Chapter 1, Section 1.3.2.)

It is unrealistic to expect to be able to quantify accurately all the terms in a water balance for a catchment to solve for *et*, especially over short time periods where storage changes are both significant and very difficult to measure or predict. Furthermore, this is a diagnostic rather than a predictive approach. In many cases we need to predict the rate of water loss to the atmosphere for certain anticipated conditions.

2.4.2 Mass-transfer methods

Water vapor fluxes from the soil surface or through the stomata typically occur by **diffusion**, the same process that is responsible for the spread of contaminants in streams and lakes or the transfer of heat through rock surrounding an igneous intrusion. In a diffusive process the flux takes place in the direction of decreasing concentration of the transported quantity and the rate of transport is proportional to the concentration gradient (i.e., the difference in concentration between two points in the flow direction divided by the distance between the two points). In the case of evapotranspiration the transported quantity is water vapor and its concentration (i.e., mass of water vapor per unit volume, or **water vapor density**) is proportional to the vapor pressure (see section 2.2). Therefore, evapotranspiration is a function of the vapor pressure difference between the evapotranspiring surface(s) and the overlying air (a) at a height z_a above the surface (Figure 2.10)

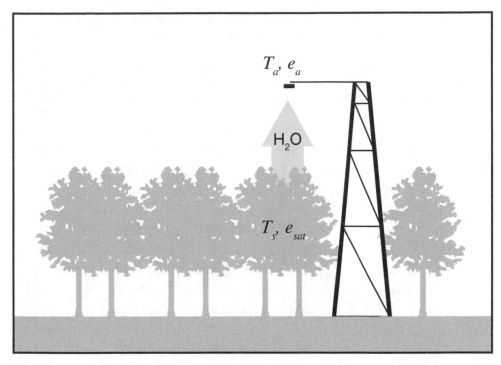

Figure 2.10 Water vapor fluxes from stomata and wet vegetation and ground surfaces to a measurement point a (e.g., top of the tower) at temperature T_a, vapor pressure e_a, and height z_a. The evapotranspiring surfaces are treated as a "big leaf" at temperature T_s and with saturation air humidity conditions.

$$et = K_E u_a (e_s - e_a), \tag{2.8}$$

where e_a and u_a are the vapor pressure and the wind speed at the height z_a, respectively, and e_s is the vapor pressure of the air in contact with the evapotranspiring surface. Because this air is saturated with water vapor (i.e., $e_s = e_{sat}$), e_{sat} can be calculated as a function of the surface temperature, T_s (see Figure 2.1 and Table 2.3).

The dependence of et on u_a reflects the fact that diffusion in a turbulent atmosphere is more effective with stronger wind speeds. (Turbulence is discussed in Chapter 3.) In general the diffusion of water vapor from the evaporating surface depends on the aerodynamic properties of the surface and is enhanced by surface roughness. The dependence on roughness is expressed through the K_E factor in Equation 2.8, which, in the case of evaporation from a wet surface (i.e., without accounting for the effect of stomatal regulation) and under neutral conditions (i.e., in the absence of convection or subsidence), can be expressed as

$$K_E = \frac{0.622 \rho_a k^2}{p_a \rho_w \left[\ln\left(\dfrac{z_a - z_d}{z_0} \right) \right]^2}, \tag{2.9}$$

Table 2.3. Values of air density, saturation vapor pressure, latent heat of vaporization, psychrometric constant, and slope of saturation vapor pressure curve as a function of air temperature

T (°C)	ρ_a (kg m^{-3})	e_{sat} (kPa)	λ_v (kJ kg^{-1})	γ (kPa °C^{-1})	S_a (kPa °C^{-1})
0	1.292	0.611	2.501×10^3	0.0654	0.044
1	1.284	0.657	2.499×10^3	0.0655	0.047
2	1.279	0.706	2.496×10^3	0.0656	0.051
3	1.274	0.758	2.494×10^3	0.0656	0.054
4	1.270	0.814	2.492×10^3	0.0657	0.057
5	1.269	0.873	2.489×10^3	0.0658	0.061
6	1.261	0.935	2.487×10^3	0.0659	0.065
7	1.256	1.002	2.484×10^3	0.0659	0.069
8	1.252	1.073	2.482×10^3	0.0660	0.073
9	1.247	1.148	2.480×10^3	0.0660	0.078
10	1.243	1.228	2.478×10^3	0.0661	0.082
11	1.238	1.313	2.475×10^3	0.0661	0.087
12	1.234	1.403	2.473×10^3	0.0662	0.093
13	1.230	1.498	2.470×10^3	0.0663	0.098
14	1.225	1.599	2.468×10^3	0.0663	0.104
15	1.221	1.706	2.466×10^3	0.0664	0.110
16	1.217	1.819	2.463×10^3	0.0665	0.116
17	1.213	1.938	2.461×10^3	0.0665	0.123
18	1.209	2.065	2.459×10^3	0.0666	0.130
19	1.205	2.198	2.456×10^3	0.0666	0.137
20	1.204	2.337	2.454×10^3	0.0667	0.145
21	1.196	2.488	2.451×10^3	0.0668	0.154
22	1.192	2.645	2.449×10^3	0.0668	0.161
23	1.188	2.810	2.447×10^3	0.0669	0.170
24	1.186	2.985	2.444×10^3	0.0670	0.179
25	1.184	3.169	2.442×10^3	0.0670	0.189
26	1.180	3.363	2.440×10^3	0.0671	0.199
27	1.176	3.567	2.437×10^3	0.0672	0.209
28	1.172	3.781	2.435×10^3	0.0672	0.220
29	1.169	4.007	2.433×10^3	0.0673	0.232
30	1.165	4.243	2.430×10^3	0.0674	0.243
31	1.161	4.494	2.428×10^3	0.0674	0.256
32	1.157	4.756	2.425×10^3	0.0675	0.269

Table 2.3. (*continued*)

T (°C)	ρ_a (kg m^{-3})	e_{sat} (kPa)	λ_v (kJ kg^{-1})	γ (kPa °C^{-1})	S_a (kPa °C^{-1})
33	1.153	5.032	2.423×10^3	0.0676	0.282
34	1.149	5.321	2.421×10^3	0.0676	0.296
35	1.149	5.625	2.418×10^3	0.0677	0.311
36	1.142	5.943	2.416×10^3	0.0678	0.326
37	1.138	6.277	2.414×10^3	0.0678	0.342
38	1.135	6.627	2.411×10^3	0.0679	0.358
39	1.131	6.994	2.409×10^3	0.0680	0.375

where ρ_a is air density (see Table 2.3), ρ_w is liquid water density, ln( ) indicates the natural logarithm, $k=0.4$ is the *von Karman*'s universal constant (dimensionless), p_a is the atmospheric pressure, z_a is the height at which u_a and e_a are measured (Figure 2.10), z_0 is the **roughness height**, a parameter that accounts for the roughness of the surface (Table 2.4); and z_d is a parameter known as **displacement height**, which accounts for the fact that in the presence of a forest canopy the surface over which the wind blows is not at the ground level but displaced by a height that is a function of the tree height, h (i.e., $z_d \approx 0.65h$) (Campbell and Norman, 1998). In the case of transpiration the diffusion of water vapor from the stomata is controlled by the plant, which can reduce the opening of the stomata to limit water vapor losses in conditions of limited soil water availability. Stomatal regulation can, therefore, reduce the value of K_E with respect to the values given by Equation 2.9.

As expected, looking at Equations 2.8 and 2.9, evapotranspiration rates are greater in the presence of dry air (i.e., $e_a < e_{sat}$) and stronger winds. Evapotranspiration is also enhanced by the roughness of the surface. This representation of evapotranspiration as a diffusive flux is due to the English scientist John Dalton (1802).

2.4.3 The energy balance

According to the first law of thermodynamics, the net radiant energy received at the land surface must be conserved. As in Chapter 1 for the conservation of mass, we consider a control volume and seek to balance energy inflows with outflows and net changes in energy stored within the system. For convenience, we establish an imaginary control volume near the land surface that includes a very thin top layer of soil (say, 10 mm thick), the vegetation, and the immediate surrounding air (Figure 2.11). Thermodynamic principles hold that the net radiant energy arriving across the boundary of this system must be exactly balanced by other energy fluxes across the boundary and the net change in energy held within the volume. The energy may change among its possible forms (radiant, thermal, kinetic, and potential), but it must be conserved.

All matter has internal energy, known as heat Q [M L^2 T^{-2}], which is due to the kinetic and potential energy associated with individual molecules. Heat is an extensive property (it depends on the amount of material) and is expressed in units of calories or joules. There are two types of heat: sensible and latent heat.

Table 2.4. Some example of surface roughness for a few types of ground surfaces and vegetation covers

Surface	z_0 (cm)
Ice	0.001
Dry lake bed	0.003
Calm, open sea	0.01
Closely mowed grass	0.1
Snow-covered farmland	0.2
Tilled bare soil	0.2–0.6
Thick grass (50 cm high)	9
Forest (on level ground)	70–120

Source: Data from Campbell and Norman (1998).

Sensible heat is proportional to temperature. As the name implies, this is the heat you would "sense" by contact or touch. The **specific heat capacity** c_p [$L^2 \Theta^{-1} T^{-2}$] provides a measure of how a substance's sensible heat changes with temperature. It is the amount of heat that is required for a unit increase of temperature in a unit mass of that substance. For instance, in the case of air c_p is about 1 kJ kg^{-1} K^{-1}, which means that 1 kJ of energy is required to increase by 1 K the temperature of 1 kg of air. Based on this definition, c_p can be expressed as

$$c_p = \frac{1}{m} \frac{\Delta Q}{\Delta T},$$

where m is mass and T (K) is absolute temperature (note that absolute temperature is represented in degrees kelvin, which is indicated as "K"). We can use the specific heat capacity to determine how the temperature of a given mass of water will change if we add energy (heat the water). The heat capacity of water at 20°C is approximately 4.2×10^3 J kg^{-1} K^{-1} or 1.0 cal g^{-1} K^{-1}. If we have 1.0 kg of water and add 12 kJ (kilojoules) of energy, then the temperature change can be calculated as:

$$\Delta T = \frac{\Delta Q}{mc_p} = \frac{12000 \, \text{J}}{1.0 \, \text{kg} \times 4.2 \times 10^3 \, \text{Jkg}^1 \, \text{K}^{-1}} = 2.9 \, \text{K}$$

or the water temperature will increase by approximately 3°C. We have assumed in this problem that we have not evaporated any of the water.

Latent heat is the portion of internal energy that *cannot* be sensed or felt. Instead, the latent energy is the amount of internal energy that is released or absorbed during a phase change, *at a constant temperature*. Evaporation involves a liquid to vapor conver-

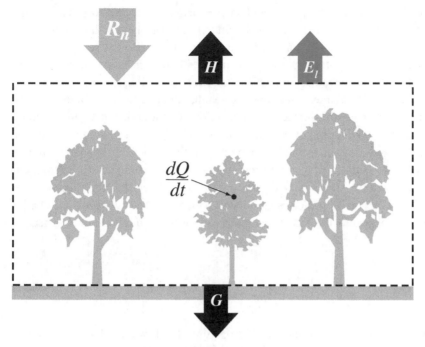

Figure 2.11 A control volume for energy conservation. Solar energy (R_n) entering the control volume must be balanced by fluxes of energy out of the volume and by the time rate of change in energy stored (dQ/dt).

sion, which requires energy to be added to the water. This energy is called the **latent heat of vaporization**:

$$\lambda_v = 2.45 \times 10^6 \text{ J kg}^{-1} \text{ (at 20°C)} \tag{2.10}$$

This tells us that we need to add about 2.5 million joules of energy to evaporate 1 kilogram of water. Other types of latent heat energy include

- latent heat of melting: $\lambda_m = 3.34 \times 10^5$ J kg^{-1} at 0°C

- latent heat of sublimation: $\lambda_s = 2.83 \times 10^6$ J kg^{-1} at 0°C

Note that evaporation of snow or ice involves a change in phase from solid to vapor. This may occur in either two separate steps, melting and then evaporation, or in one step as sublimation, which is the direct phase change from ice to water vapor. In either case latent heat is added to support the sum of the two unique phase changes, as evidenced by the fact that $\lambda_s = \lambda_v + \lambda_m$.

The net solar radiation (R_n, i.e., the fraction of solar radiation that is received at the surface and is not reflected back to the atmosphere) warms the exposed surfaces inside the control volume. When water is present some of this heat energy is absorbed by the water to support the phase change from liquid to vapor (evapotranspiration). Evapotranspiration will not typically absorb all the energy, and so the surface continues to warm. As the surface becomes warmer than the air and the underlying soil, heat will be conducted from the hot surface to the air (H) and from the surface down into the soil (G). During this process, both heat and water vapor have been added to the air inside the control volume. The moist, warm air becomes less dense than the surrounding air and tends to rise up and out of the control volume. Therefore, we see that the main energy outlets for the sun's energy that hits the Earth's surface are conduction into the soil, conversion of liquid water to water vapor, and heating of the overlying air. The energy used to evaporate the water is stored in the water vapor and is removed from the system as the water vapor is removed. Finally, the latent heat (evaporation energy) is converted back to thermal energy when and where the water recondenses. From this we see that the flux of water vapor induced by evapotranspiration is associated with a flux of latent heat, E_l. The **latent heat flux** is related to the rate of evapotranspiration through the latent heat of vaporization:

$$et = \frac{E_l}{\rho_w \lambda_v}, \tag{2.11}$$

where et = the evapotranspiration rate [L T^{-1}]; ρ_w = density of water [M L^{-3}]; λ_v = latent heat of vaporization at the temperature of interest [L^2 T^{-2}].

The energy balance approach to evaporation was proposed by Bowen (1926) and is based on the fact that evapotranspiration involves an energy flux (i.e., of latent heat energy to the atmosphere). The rate of evapotranspiration can be calculated using Equation 2.11 with the latent heat flux determined through the energy balance equation (Figure 2.11):

$$\frac{dQ}{dt} = R_n - G - H - E_l, \tag{2.12}$$

where R_n = the *net* (solar) radiation input, G = energy output through conduction to the ground, H = net output of sensible heat to the atmosphere, E_l = output of latent heat to the atmosphere (the latent heat flux) due to evaporation, and Q = the amount of heat energy stored in the control volume per unit area of surface. Each of the inputs and outputs on the right side of Equation 2.12 is in the form of an *energy flux* (i.e., energy per unit area per unit time). In SI units this would be J m^{-2} s^{-1}, or W m^{-2}. R_n is an input of energy to the control volume. Therefore, the convention adopted here is to express R_n as a positive quantity. Conversely, in the energy balance Equation 2.12, G, H, and E_l are assumed to be positive when directed outward from the control volume (i.e., when they are energy outflows). We can rearrange Equation 2.12 to solve for E_l, the latent heat flux:

$$E_l = R_n - G - H - \frac{dQ}{dt}. \tag{2.13}$$

We can substitute Equation 2.13 into Equation 2.11 and rearrange to solve for the evapotranspiration rate:

$$et = \frac{R_n - G - H - dQ/dt}{\rho_w \lambda_v}.$$
(2.14)

Let's consider a simple example of how we might apply the energy balance approach to calculate the daily evaporation from a forest on a sunny day ($R_n = 200 \ \mathrm{W \ m^{-2}}$) in Virginia, USA. If we can neglect G and H and assume that the heat (Q) stored within the forest remains approximately constant, then Equation 2.14 becomes:

$$et = \frac{R_n}{\rho_w \lambda_v} = \frac{200 \ \mathrm{W \ m^{-2}}}{(1000 \ \mathrm{kg \ m^{-3}})(2.5 \times 10^6 \ \mathrm{J \ kg^{-1}})} = 8.0 \times 10^{-8} \ \mathrm{m \ s^{-1}} = 0.7 \ \mathrm{cm \ day^{-1}}.$$

In practice, it is common to neglect changes in heat storage because of the relatively low heat capacity of air (the same assumption cannot be made in the case of water or soils; therefore, we choose a control volume that does not include a substantial mass of water or soil). However, typically we cannot neglect the sensible heat flux H. Also, for averaging times significantly less than a day we cannot safely neglect G either. Thus, to estimate the evapotranspiration rate using Equation 2.14 we need to determine H. Sensible heat is transported by diffusion across gradients of temperature. This indicates that H should be proportional to the temperature difference between the surface and the overlying air (Figure 2.10), and can be expressed using a formulation similar to the case of water vapor fluxes

$$H = K_H u_a (T_s - T_a)$$
(2.15)

with

$$K_H = \frac{c_p \rho_a k^2}{\left[\ln\left(\dfrac{z_a - z_d}{z_0} \right) \right]^2}.$$
(2.16)

2.4.4 Other methods

In the previous sections we have considered two methods that can be used to calculate the evapotranspiration rates. The mass-transfer method outlined in Section 2.4.2 uses Equations 2.8 and 2.9 to determine et as a function of wind speed (u_a), air vapor pressure (e_a), the saturation vapor pressure at the surface (e_{sat}, which is a function of surface temperature T_s), and a number of parameters that depend on air temperature and surface characteristics (Tables 2.3 and 2.4). The energy balance method outlined in Section 2.4.3 uses Equations 2.14 to 2.16 to calculate et as a function of the net radiation (R_n), heat conduction to the ground (G), wind speed (u_a), surface temperature (T_s), the air temperature (T_a) at the height (z_a), and the set of parameters reported in Tables 2.3 and 2.4.

It can be convenient to combine the two methods to eliminate one of these variables, preferably a quantity that is hard to measure. The **Bowen ratio** method eliminates the dependence on wind speed. It is based on the definition of Bowen ratio (B)—the ratio between sensible and latent heat fluxes—and can be expressed (using Equations 2.8, 2.9, 2.15, and 2.16) as

$$B = \frac{H}{E_l} = \gamma \left(\frac{T_s - T_a}{e_{sat} - e_a} \right), \tag{2.17}$$

where $\gamma = c_p \, p_a / (0.622 \, \lambda_v)$ is known as "psychrometric constant" and is calculated as a function of the air temperature, T_a (Table 2.3). If we express sensible heat in Equation 2.14 as $H = B \, E_1 = \rho_w \, \lambda_v \, et$ and solve for et we obtain

$$et = \frac{R_n - G}{\rho_w \lambda_v (1 + B)}. \tag{2.18}$$

The Bowen ratio method uses Equations 2.17 and 2.18 to calculate the evapotranspiration rate. Because Equation 2.17 is obtained using Equations 2.8 and 2.9, it does not account for the effect of stomatal regulation and can only be used in the case of evaporation from a lake, or other wet surfaces. We see from Equation 2.17 that as a surface becomes warmer (T_s increases) and drier (e_{sat} decreases), the Bowen ratio tends to increase. Consequently, sensible heat flux increases relative to latent heat flux. The Bowen ratio ranges from approximately 0.1 in very humid regions ($B \sim 0.1$ over the oceans) to values greater than 1 in arid regions. In deserts, B may be greater than 10. With estimates or measures of R_n, G, and B, Equation 2.18 provides a simple and direct estimate of et.

The **combination method** (or **Penman's method**) combines mass transfer and energy balance approaches to eliminate the dependence on the surface temperature (T_s), which is often difficult to measure. In this case the evaporation rate is expressed as

$$et = \frac{S_a[R_n - G]}{\rho_w \lambda_v (S_a + \gamma)} + \frac{K_H u_a [e_{sat}(T_a) - e_a]}{(S_a + \gamma)}, \tag{2.19}$$

where S_a is the slope of the saturation water pressure curve (Figure 2.1); its values are reported in Table 2.3. As in the case of the mass transfer method presented in Section 2.4.2, Equations 2.18 and 2.19 do not account for the effect of stomatal regulation. Thus, the Bowen ratio method and the Penman method can be used to estimate evaporation but not transpiration. The Penman Equation 2.19 can be modified to account for stomatal regulation (known as the Penman-Montieth method), though this introduces a number of additional parameters. Brutsaert (2005) provides a good explanation of the Penman-Montieth method.

2.4.5 Soil moisture controls on evapotranspiration

In general we need to consider the state (wetness) of a surface to understand and quantify how the solar energy reaching the earth surface is partitioned. When water is in limited supply, the surface becomes warmer than in the wet cases and more of the energy is removed from the control volume through conduction into the soil and heating of the air. Hence, in this case the surface properties rather than the atmospheric properties control the rate of evapotranspiration. Under wet surface conditions, the rate of evapotranspiration is governed by the supply of radiant energy, the relative dryness of the air, and the efficiency of the wind in removing the water vapor from the surface. This is evidenced by how quickly water evaporates from wet clothes on a dry day, as compared to a humid day. We know intuitively that for the same example, higher winds will increase the evaporation rate, and reduced solar input, say from heavy cloud cover, will reduce the evaporation rate.

Through field experiments we can explore the relationship between the surface wetness and the partitioning of the received energy between evaporation and heating of the air and soil. Figure 2.12 shows a set of experimental measurements of the terms in the

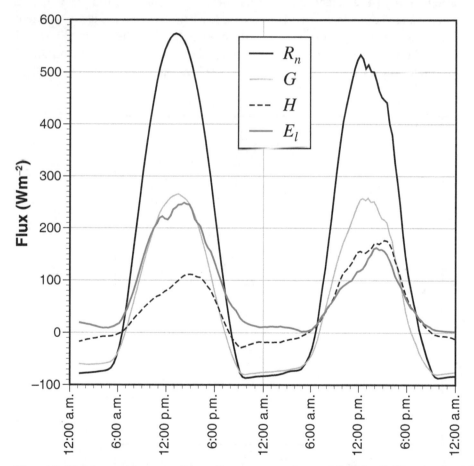

Figure 2.12 Measured energy fluxes from an experimental field in California. As the surface dries during the experiment, the latent heat flux (E_l) is reduced and the sensible heat flux (H) increased.

Data courtesy of John D. Albertson.

energy balance over an irrigated bare soil field in Davis, California. After wetting the soil with sprinkler irrigation the energy balance terms were monitored and graphed against time for a two-day period. The net radiation was very similar from day to day, except for a few passing clouds on the afternoon of day 2. The main difference between the two days was that the surface was wet on day 1 and partially dry on day 2. Notice how the measured latent heat flux (from evapotranspiration) is nearly twice as large as the heating of the air (H) on the wet day. On the drier day the two terms are nearly equal. If we were to plot additional days (following day 2), where the soil becomes increasingly dry, we would see H begin to exceed E_l as the evapotranspiration steadily declines.

The rate of *et* that occurs under the prevailing solar inputs and atmospheric properties, if the surface is fully wet, is commonly referred to as potential evapotranspiration (PET). By definition PET does not depend on surface properties and is not affected by stomatal regulation. For a catchment water balance, we are interested in the actual *et*, because that is the rate at which water is removed. The concept of PET is useful as a tool if we have some means to estimate actual *et* from knowledge of the potential rate and the nature of the surface conditions. By definition, when the surface is wet the ratio of *et* to PET will be unity. Conversely, when the surface is completely dry, the ratio will go

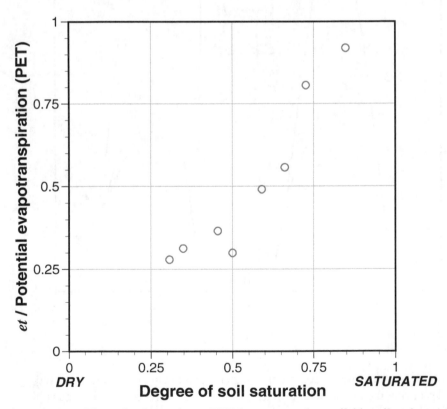

Figure 2.13 The ratio of actual *et* to PET decreases as the available soil moisture decreases.
Data courtesy of John D. Albertson.

to zero. The transition between these two extremes is shown in Figure 2.13 for the simple case of evaporation from a bare soil field. The actual *et* was determined from detailed measurements of wind velocity and vapor pressure in the atmosphere; PET was calculated from measured atmospheric properties.

Currently there are no widely used methods for estimating areal distributions of actual *et*, although satellite-derived surface data offer promise of delivering useful estimates of components of the surface energy balance from which actual *et* might be inferred (Moran and Jackson, 1991). In hydrology, computations using empirical evapotranspiration equations typically are used to estimate actual *et*. For a complete discussion of techniques to measure and estimate *et*, see Brutsaert (1982).

2.5 Concluding Remarks

This chapter has presented the concepts of precipitation, evapotranspiration, and interception, which are fundamental components of the hydrological cycle. A number of issues were raised in the introduction dealing with how we quantify these fluxes and stores. Our conclusion that it is very difficult to predict precipitation led us to investigate how hydrologists use statistical analyses to describe the spatial and temporal patterns of precipitation. Because we typically use point measurements of precipitation, there are significant problems with extrapolating these values to land areas of interest (e.g., catchments). Studies have shown that two precipitation gages (on a ridge exposed to the wind), separated by only a small distance (3 m), can produce results that differ consistently by as much as 50% (Court, 1960). Because of the temporal and spatial variability in precipitation fluxes, it is no wonder that water management in a basin as large as that drained by the Colorado River is a significant challenge. Despite problems with interpretation, it is apparent that radar and satellite data will be essential for resolving the measurement problem of areal precipitation.

It has been known for many years that changes in catchment vegetation can alter the water budget. One dramatic example is deforestation. Clearing land areas of trees typically decreases the amount of transpiration and interception (and consequently evaporation from interception stores); as a result, surface runoff and soil moisture increase. Data summarized in Dunne and Leopold (1978) indicate a linear relationship between the percentage of vegetation removed and the first-year increase in streamflow. More recent reviews (Stednick, 1996; Brown et al., 2005) indicate that a simple relationship between deforestation and increased streamflow is not so apparent, and requires further study (see Chapter 9). As vegetation regenerates, interception and evapotranspiration increase, and surface runoff declines. Other changes in catchment water dynamics accompanying deforestation have also been observed. Evaporation is increased from exposed soils. Infiltration is increased, enhancing leaching of soil solutes. The increased surface runoff accelerates soil erosion. It may take many years for reforestation to halt these effects (Stednick, 1996).

In introducing the concepts of the hydrological cycle and catchment water budgets in Chapter 1, several fluxes and reservoirs were discussed. This chapter has dealt with the connections between the land and the atmosphere. Ultimately, there is a close coupling between the atmosphere and the land surface, which controls the flux of water across this interface. Subsequent chapters will deal with the terrestrial portion of the hydrological cycle. To proceed to a physical treatment of water flows in streams and beneath the Earth's surface, we will need to gain a basic understanding of fluid mechanics. The physics of fluids will be the subject of Chapters 3 and 4.

2.6 Key Points

- The three primary steps in the generation of precipitable water in the atmosphere are: creation of saturated conditions in the atmosphere, condensation of water vapor into liquid water, and growth of small droplets by collision and coalescence until they become large enough to precipitate.

- Precipitation onto the land surface is measured with relative ease at individual points or stations by using precipitation gages. A record of precipitation versus time is called a hyetograph. Areal distribution of precipitation can be estimated using weather radar. {Section 2.2.1}

- Mean areal precipitation can be estimated using data from multiple stations within a catchment using the isohyetal method. {Section 2.2.2}

- Hydrologists use frequency analysis to describe the temporal variability of historical observations from individual stations. Annual precipitation data often are described well by a normal distribution {Section 2.2.3}, while rainfall events are typically distributed according to a Gumbel distribution. {Section 2.2.4}

- A portion of the gross or total precipitation may be temporarily stored above the land surface by vegetation and then evaporated. This is referred to as interception. {Section 2.3}

- Evapotranspiration from a catchment includes a variety of different vaporization processes, including evaporation from open water, soils, and vegetation surfaces, and transpiration from plants. {Section 2.4}

- By making appropriate assumptions and by collecting appropriate field data, a long-term average rate of evapotranspiration can be estimated using the mass balance approach. {Section 2.4.1}

- The mass transfer approach is based on a representation of the water vapor fluxes associated with evapotranspiration as a diffusion process along gradients of decreasing water vapor pressure {Section 2.4.2}

- The energy balance approach is based on the principle of conservation of energy in which the net solar radiation (R_n) is partitioned between latent heat flux (E_l) to the atmosphere, sensible heat flux (H) to the atmosphere, heat flux to the ground (G), and the change in energy stored in the control volume (dQ/dt). {Section 2.4.3}

- Some of the most commonly used methods for the calculation of evapotranspiration (e.g., the Bowen ratio and Penman methods) are based on a combination of mass transfer and energy balance approaches. {Section 2.4.4}

- Over land surfaces water may be limited in availability (e.g., in dry soils), and rates of evapotranspiration are then reduced from those over a fully wet surface. Actual evapotranspiration (*et*) is commonly much less than potential evapotranspiration in catchments where moisture limitations occur, but *et* can approach potential evapotranspiration in extremely wet climates where energy may become limiting. {Section 2.4.5}

2.7 Example Problems

Problem 1. Two tipping bucket rain gages are used to collect the following rainfall data:

Time	Cumulative precipitation (mm) Station #1	Cumulative precipitation (mm) Station #2
4:00 a.m.	0.0	0.0
6:00 a.m.	0.0	0.0
8:00 a.m.	1.0	1.0
10:00 a.m.	4.0	3.0
12:00 noon	13	11
2:00 p.m.	17	15
4:00 p.m.	19	16
6:00 p.m.	19	17
8:00 p.m.	19	17
10:00 p.m.	19	17
12:00 midnight	19	17
2:00 a.m.	19	17
4:00 a.m.	19	17

A. Calculate the mean daily rainfall intensity for each station (mm hr^{-1}).

B. Calculate the maximum 2-hour rainfall intensity for each station (mm hr^{-1}).

C. Calculate the maximum 6-hour rainfall intensity for each station (mm hr^{-1}).

D. Using the arithmetic average method and knowing that the drainage basin area is 176 mi^2, calculate the total volume of rainfall (m^3) delivered to the basin during the event.

Problem 2. Measurement of changes in volume of water in an evaporation pan is a standard technique for estimating potential evapotranspiration. United States Class A evaporation pans are cylindrical with the following dimensions: depth = 10.0 inches and diameter = 47.5 inches. An evaporation pan can be considered a hydrological system with an inflow, outflow, and storage volume. Evaporation from pans is not the same as evaporation from natural surfaces for a variety of reasons. For example, water temperatures in shallow pans will be much more variable than temperatures in a nearby lake. Evaporation measured in pans is adjusted by a factor called a pan coefficient to convert to an estimate of potential evapotranspiration (see Brutsaert, 1982).

A. Calculate the cross-sectional area (m²) of a United States Class A evaporation pan through which inflows and outflows of water can pass. Also, calculate the total storage volume of the pan (m³).

B. Initially, the pan contains 10.0 U.S. gallons of water. Calculate the depth of water in the pan (mm).

C. Assuming a water density of 997.07 kg m⁻³ (25°C), calculate the mass (kg) of water in the pan.

D. After 24 hours in an open field (no precipitation), the pan is checked and the volume of water left in the pan is determined to be 9.25 gallons. Calculate the average evaporation rate (mm hr⁻¹) from the pan.

E. The pan is emptied and refilled with 10.0 gallons of water and left in an open field for another 24 hours. During this period, rain fell for a 3-hour period at a constant intensity of 2.5 mm hr⁻¹; after 24 hours, the volume of water in the pan was 11.50 gallons. Calculate the average evaporation rate (mm hr⁻¹) from the pan during this period

F. If the evaporation rate calculated in E remains constant and no additional precipitation occurs, then estimate the time (days) for the pan to empty as a result of evaporation.

Problem 3. For Problem 2 part D, calculate the flux of latent heat from the water in the pan to the atmosphere (W m⁻²). Use a water density, $\rho_w = 1000.0$ kg m⁻³.

Problem 4. A small (area = 300 ha) catchment in Iowa absorbs a mean $R_n = 330$ W m⁻² during the month of June. In this problem, apply the energy balance approach to estimate evapotranspiration from the catchment during the month of June.

A. Write a complete energy balance equation (i.e., including all terms) for the catchment for the month of June.

B. Neglecting conduction to the ground (G) and the change in energy stored ($dQ/dt = 0$), simplify your energy balance equation for the catchment so that it can be solved for the latent heat flux, E_l. Also, replace the term H (the sensible heat flux) with $B \times E$, (where B is the Bowen ratio).

C. Using a mean Bowen ratio of 0.20 for the catchment, calculate the mean daily flux of latent heat to the atmosphere (W m^{-2}) and the mean evapotranspiration rate (mm day^{-1}) from the catchment. Use a water density, $\rho_w = 1000.0$ kg m^{-3}.

D. Calculate the total evapotranspiration from the catchment during the month of June (mm).

2.8 Suggested Readings

Campbell, G.S., and J. M. Norman. 1998. *An Introduction to Environmental Biophysics*, New York: Springer.
Oke, T.R. 1987. *Boundary Layer Climates*, 2nd ed. New York: Routledge. Chapter 1, pp. 3–32.

3 The Basis for Analysis in Physical Hydrology: Principles of Fluid Dynamics

3.1 Introduction

Hydrology is concerned with the occurrence and movement of waters on and beneath the surface of the Earth. Water moves, or flows, in response to forces arising from gravity and pressure, for example, that act on water above and below the ground. Much can be learned about the movement of water on and in the Earth by considering the forces causing fluid motion and the response of fluids to those forces. The study of the physical processes governing fluid motion is called fluid dynamics. Flow of water in streams and subsurface aquifers, infiltration of precipitation into soils, evaporation, design of flood control measures, and transport of groundwater contaminants are just some of the problems hydrologists study that depend on a knowledge of fluid dynamics. In hydrology we are obviously most concerned with understanding the behavior of water, but the principles discussed in this chapter apply to a broad range of fluids.

The notions of force balance and Newton's second law (force equals mass times acceleration), familiar from basic solid mechanics, can be extended to fluids, including water. The forces that cause fluids to move are due to gravity, pressure differences, and surface stresses. In this chapter we analyze relatively simple cases such as flow through a pipe or hose, but the same principles apply to flow in streams and through soil and rocks.

These applications are explored in subsequent chapters. The basic principles developed in this chapter also apply to the flow of air, lava, glaciers, and the Earth's mantle. While all of these can be described as fluids, they differ dramatically in their viscosity. The viscosity of lava is almost a billion times that of air. We will find that fluid viscosity plays an important role in the nature of fluid flow. Flows in which viscous forces dictate the nature of flow are called **laminar flows**; flows in which viscous forces are relatively unimportant are often **turbulent**. Most groundwater flow is laminar, while most surface-water flow is turbulent.

3.2 Definitions and Properties

We usually classify matter as either solid, liquid, or gas, based on macroscopic properties. Typical classifications are made on the basis of the most easily specified differences among the three types of matter: a gas takes on the shape and volume of a container, a liquid takes the shape of the portion of the container that it fills but retains a fixed volume and a solid has its own defined shape as well as volume. Liquids and gases are called **fluids**.

When describing fluid motion, a more fundamental definition of fluid is needed. The word fluid comes from the Latin *fluere* (to flow) and a definition found in a dictionary might be "a substance capable of flowing." The scientific use of the term is not much different: a fluid is a substance that continuously deforms when subjected to a shear stress. **Shear stress** is a tangential force (i.e., a force parallel to a surface) per unit area acting on a surface. Our experience tells us that fluids move in the presence of a shear stress; for example a puddle of water will be disturbed if a wind exerts a tangential stress on the water surface.

The above definition is valid for any fluid. The rate at which deformation occurs, however, is not independent of the fluid itself. The property of a fluid that describes the resistance to motion under an applied shear stress is termed **viscosity**. The viscosity of honey is greater than the viscosity of water, which in turn is greater than the viscosity of air. This common sense notion of viscosity corresponds with the scientific notion. To formulate a precise definition of viscosity we will consider a simple conceptual experiment. Imagine a very large, thin sheet of material (area $= A$) floating on a very shallow layer of water (depth $= d$) lying above a horizontal bottom surface. Suppose further that the sheet is pulled (with force F) across the surface of the water. This experiment can be conceptualized in more concrete terms as attaching a rope to the edge of a square sheet of Styrofoam 2 meters on a side, floating this on a layer of water that is 5 mm deep at the bottom of a large, flat-bottomed pool, and pulling slowly on the rope to cause the Styrofoam to move *tangentially* (parallel to the water surface) across the surface of the water. We would find that the fluid adjacent to the bottom of the pool would "stick" to the bottom and not move and that the fluid adjacent to the underside of the floating sheet of Styrofoam would stick to that surface and, therefore, move with the same velocity as the sheet itself, say u_{plate} (Figure 3.1). After a short time we would find that the velocity of the water between the sheet and the pool bottom is proportional to the distance above the bottom, as long as the product $u_{plate} \times d$ is less than about 0.002 m^2 s^{-1}; the reason for this limit is explored in Example Problem 3 (Section 3.10). For example, at a distance of $d/2$ units above the bottom—halfway between the bottom surface and the floating plate—the water velocity would equal $u_{plate}/2$. We would also find that the applied force divided by the area of the plate is proportional to the surface (plate) velocity divided by the water depth or

$$\frac{F}{A} = \mu \frac{u_{plate}}{d}. \tag{3.1}$$

The constant of proportionality, μ, is the viscosity of the fluid.

Equation 3.1 is a specific case of an important relationship for understanding the flow of water, air, and many other fluids. More generally, F/A is the tangential force per unit area (shear stress) acting on the fluid surface and u_{plate}/d is the slope of the curve (a straight line in this case) relating fluid velocity to distance above the bottom boundary. A curve relating velocity to distance above the bottom boundary is called a **velocity profile**; the slope of such a curve is called the velocity gradient. We can rewrite Equation 3.1 to relate the shear stress to the velocity gradient:

$$\tau = \mu \frac{du}{dz}, \tag{3.2}$$

where τ = shear stress [M L^{-1} T^{-2}]; du/dz = the velocity gradient [T^{-1}]; u = velocity in the x-direction [L T^{-1}]; and μ = dynamic viscosity [M L^{-1} T^{-1}]. (Note that equivalent dimensions for viscosity are [F L^{-2} T], where F is force. Thus viscosity is often reported in SI units of pascal·seconds.) Equation 3.2 is known as Newton's law of shear and provides a definition for viscosity: the viscosity of a fluid is the ratio of the shear stress at a point in the fluid to the velocity gradient at that point.

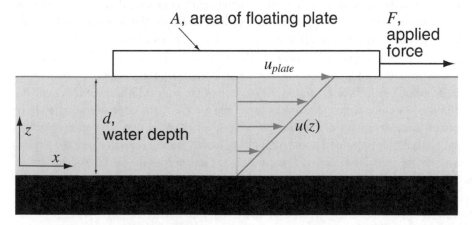

Figure 3.1 Schematic diagram of hypothetical experiment to determine the viscosity of water. Viscosity is given by the ratio of the applied force per unit area (F/A) to the slope of the velocity profile (u_{plate}/d).

The phrase "at a point" in the above definition deserves further comment. We want to use the idea of a point in the mathematical sense because the use of calculus is necessary for quantitative descriptions of fluid systems. However, we know, because of the molecular nature of matter, that if we picked a point at random in a fluid, that point might be interior to an atomic particle or alternatively might be in the space between particles. The fluid "properties" associated with such a point would obviously depend on the location of the point. We avoid the problems associated with an approach that recognizes the discrete nature of matter by making the continuum assumption: the fluid is idealized macroscopically as being continuous throughout its entirety—the molecules are pictured as being "smeared" or "averaged" to eliminate spaces between atomic particles. Thus, we will use the terminology "at a point" so that we have the mathematical tools that are necessary, but it should be borne in mind that the "point" actually represents average conditions over a small volume. For our purposes—explanation in physical hydrology—this assumption is valid. (There are situations involving rarefied gasses important in other branches of science for which the continuum assumption may not be acceptable.)

Once the continuum assumption has been made, we can define the density of fluid at a point. **Density** is simply mass per unit volume. If density varies spatially, the ratio of mass to volume will vary with the sample size. Hence, the definition of density at a point is the limit of the ratio of mass to volume, as the volume under consideration shrinks to an infinitesimal:

$$\rho = \lim_{\Delta V \to 0} \frac{\Delta M}{\Delta V}, \tag{3.3}$$

where ρ = density [M L^{-3}] and ΔV = is a small volume of fluid [L^3] with a mass ΔM [M]; the "Δ" before V and M just reminds us that we are considering small volumes and masses. Of course, the continuum assumption means that density "at a point" is really the average density of a volume that is small compared with the macroscopic scale of the fluid motions being studied but large compared to intermolecular distances. If the density of a fluid does not vary spatially, the fluid is said to be **homogeneous**. A fluid property related to density is unit weight. The **unit weight** (γ) of fluid is its weight per unit volume and is equal to the product of density and the acceleration of gravity, that is $\gamma = \rho g$.

The viscosity of water at 20°C is approximately 1.0×10^{-3} pascal·seconds (Pa · s) and the density of water at that temperature is slightly less than 1000 kg m^{-3} (or 1.0 g cm^{-3}); a pascal is 1 newton m^{-2} (see Appendix 1). Both viscosity and density are temperature dependent. Viscosity decreases with increasing temperature from 1.787×10^{-3} Pa · s at 0°C to 0.7975×10^{-3} Pa · s at 30°C (Table 3.1), a range that includes much of the variation expected in the natural environment. Density varies with temperature in a somewhat more complex fashion. Water exhibits a density maximum at about 4°C; the density decreases slightly both above and below that temperature. More extensive information on the density and viscosity of water is provided in Appendix 2.

We should note that density also depends on pressure. Pressure will be defined in the next section, but, roughly speaking, it is the force per unit area that tends to compress a fluid. Values of density given in Table 3.1 are for a pressure of 1 atmosphere. The deviations

Table 3.1. Properties of water as functions of temperature

Temperature, °C	Viscosity, μ, Pa·s	Density, ρ, kg m^{-3}
0	1.787×10^{-3}	999.87
5	1.519×10^{-3}	999.99
10	1.307×10^{-3}	999.73
15	1.139×10^{-3}	999.13
20	1.002×10^{-3}	998.23
25	8.904×10^{-4}	997.07
30	7.975×10^{-4}	995.67

in density caused by changes in pressure are often negligible. In these cases, the fluid is said to be **incompressible**. However, there are cases of importance in environmental sciences where the compressibility of water must be considered and, thus, the variation of density with pressure taken into account. These cases include the variation of water density with depth in the oceans and with the variation of confining pressure in deep aquifers.

3.3 Forces on Fluids

In discussing the flow of water (or any other fluid) it is necessary to consider the forces acting on a fluid particle (a very small volume of fluid). These forces are usually divided into two classes, body forces and surface forces. **Body forces** are those that do not require direct contact with the fluid; they "act at a distance." One body force that plays an important role in hydrological systems is the force due to gravity, the weight of the fluid. Another example of a body force is an electromagnetic force. **Surface forces** are those caused by direct contact between two fluid particles or between fluid and solid. The tangential force in the conceptual experiment we employed in defining viscosity is an example of a surface force—the dragging of the water along by the floating plate required the plate to be in contact with the fluid surface. A normal force is one oriented perpendicular to a surface.

Instead of speaking of surface forces directly, we usually employ the concept of force per unit area or **stress**. There are two types of stresses that we consider in fluid dynamics— **normal stresses** and tangential stresses. The latter, as we have already seen, are termed shear stresses. The inward-directed (compressive) normal stress, when applied to a fluid medium, is referred to as **pressure**.

3.4 Fluid Statics

A special case of fluid motion is that of a fluid at rest, i.e., the case of no motion. If a fluid is not moving, there are no shear stresses present (by definition) and only pressure need be considered when analyzing surface forces. The special case of no motion has important applications and the study involving such problems is referred to as fluid statics.

The variation of pressure with depth in a fluid at rest is one of the fundamental relationships in fluid dynamics. The equation describing this variation in pressure is known as the **hydrostatic equation**. We will derive this equation by considering a thin slice of fluid at rest and considering the forces acting on it (Figure 3.2). The slice has a surface area A and a thickness Δz. The horizontal pressure forces acting on the sides on the slice must be equal and opposite so that they sum to zero. If not, there would be a net pressure force causing the slice of fluid to move sideways, at odds with our assumption that the fluid is not moving. The upward force on this slice, pA, is due to the pressure p on the bottom face of the slice. There are two downward forces, the weight of the slice itself and the force due to pressure on the top face of the slice. The weight of the slice is given by the product of density ρ times gravitational acceleration g times the volume of the slice V ($= A\Delta z$). The pressure on the top face of the slice will differ from the pressure on the bottom by some small amount Δp, so that the downward-directed pressure force on the top face is $(p + \Delta p)A$. Because the fluid is not moving, the upward and downward forces acting on the slice must balance each other, giving:

$$(p + \Delta p)A + \rho g A \Delta z = pA,$$

which we can solve for the change in pressure Δp from the top to the bottom of the slice:

$$\Delta p = -\rho g \Delta z$$

or

$$\frac{\Delta p}{\Delta z} = -\rho g$$

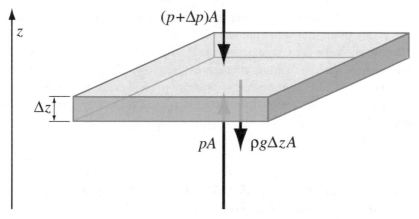

Figure 3.2 A thin slice within a static fluid. Because the fluid is at rest, the upward and downward forces acting on the slice must balance.

As the thickness of the slice shrinks to an infinitesimal,

$$\frac{dp}{dz} = -\rho g. \tag{3.4}$$

Equation 3.4 is the hydrostatic equation. It indicates that the rate of decrease of pressure with distance as we proceed upward (relative to gravity) is ρg, the unit weight of the fluid. The pressure is caused by the weight of the overlying fluid. In many instances involving water we can consider density to be constant and we can then integrate Equation 3.4 to give pressure as a function of depth. Suppose the slice in Figure 3.2 is at a level d units below the surface of the fluid and that $z = 0$ at the surface. We will integrate Equation 3.4 from $z = -d$ to $z = 0$, i.e., from the slice to the surface. Rewriting Equation 3.4 in differential form,

$$dp = -\rho g dz \tag{3.5}$$

we can integrate to obtain:

$$\int_p^{p_s} dp = -\rho g \int_{-d}^0 dz,$$

where p_s is the pressure at the surface, typically atmospheric pressure. Evaluating the integral gives

$$p_s - p = -\rho g [0 - (-d)]$$

or

$$p - p_s = \rho g d. \tag{3.6}$$

Equation 3.6 is a form of the hydrostatic equation that can be used to determine the absolute pressure at any point in a static fluid of constant density. Commonly, we take the pressure at the surface to be zero **gage pressure** ($p_s = 0$) so that what we actually are calculating is the pressure relative to atmospheric pressure. To convert gage pressure to absolute pressure, we would add the atmospheric pressure to gage pressure. Throughout the remainder of this text, we will follow the convention that p is the gage pressure. In this case, Equation 3.6 becomes:

$$p = \rho g d. \tag{3.7}$$

Thus, the pressure in a constant density, static fluid is the product of the unit weight and the depth. Pressure increases linearly with depth.

As an example, let us calculate the pressure on the bottom of the 1-meter high pond wall shown in Figure 3.3a. The unit weight of water is 9.8 kN m^{-3} and the depth is 1 meter.

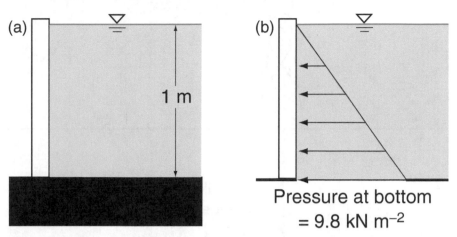

Figure 3.3 Water pressure on a vertical wall (*a*) increases linearly with depth at a rate of 9.8 kN m⁻² per meter (*b*).

Therefore, the pressure at the bottom of the pond, and hence at the base of the wall, is 9.8 kN m⁻². The pressure on the wall varies linearly with depth such that the pressure at 0.5 m is 4.9 kN m⁻² and the (gage) pressure at the surface is 0 kN m⁻² (Figure 3.3b). The total force on the wall is the integral of pressure over area. If we consider a 1-meter width of the wall (perpendicular to the plane of the page in Figure 3.3), then the force on the wall is determined by:

$$\text{Force per meter of width} = \int_1^0 \left(-9.8\frac{\text{kN}}{\text{m}^3}\right) z\,dz = 4.9\frac{\text{kN}}{\text{m}}.$$

One application of the hydrostatic equation is in calculations for drilling fluid when an oil well is drilled into a pressurized formation (Figure 3.4). If the pressure in the oil-bearing formation is not controlled, then a blowout can occur. These occurrences are now rare, although the 2010 Deepwater Horizon event in the Gulf of Mexico is a reminder of the consequences of blowouts. One method for preventing blowouts as a well is drilled is to fill the borehole with a drilling fluid (or drilling mud). The hydrostatic pressure created by the column of mud counteracts the formation pressure of the oil reservoir. If a well is drilled into an oil reservoir with formation pressure equal to 1.8×10^4 kN m⁻² at a depth of 2000 m, then the height of a column of drilling mud required to balance the formation pressure can be calculated using the hydrostatic equation

$$p_{required} = formation\ pressure = \rho_{mud}gd,$$

where d is the height of the mud column above the formation. For a drilling mud with unit weight ($\rho_{mud}\,g$) = 10.5 kN m⁻³,

$$p_{required} = 1.8\times10^4\ \text{kN m}^{-2} = (10.5\ \text{kN m}^{-3})\times d.$$

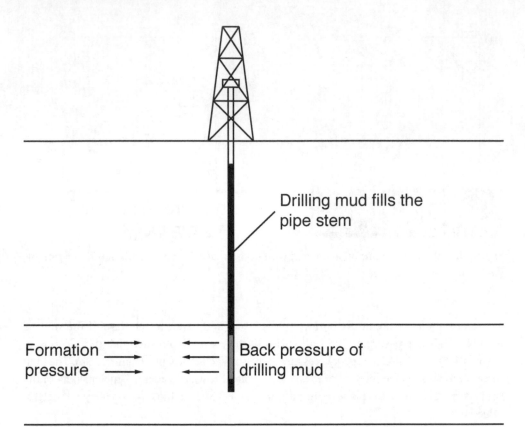

Figure 3.4 Drilling mud, with a density typically 10 to 20% greater than that of water, is used to stabilize wells by providing a hydrostatic pressure at depth to counterbalance the formation fluid pressure.

So the required height of the drilling mud in the well is about 1700 m. The depth of the formation is 2000 m so there the balancing pressure of the drilling mud is more than adequate to prevent a blowout.

3.5 Fluid Dynamics

As indicated previously, a description of fluid dynamics requires specification of the relationship between forces and acceleration. Fluid motions (other than the special case of no motion that we have just discussed) are described using Newton's second law, $F = ma$. The force F is the sum of all the forces that cause fluid motion. Acceleration a is the rate of change of velocity with time. In this section, we will investigate the nature of fluid acceleration. Acceleration will be related to the forces acting on a fluid to derive the Bernoulli equation, one of the fundamental equations of fluid dynamics. Several examples of the application of the Bernoulli equation are also considered.

3.5.1 Fluid acceleration

Acceleration of a solid object, such as an automobile, can be found by measuring its velocity as a function of time $u(t)$ and calculating $a = du/dt$. It is difficult to take the same approach to determining fluid acceleration because a fluid does not move as a body with a single velocity that can directly be measured. Instead it is a collection of many small fluid "elements," not all of which are moving with the same velocity. Rather than attempt to measure the velocity or acceleration of each fluid element as it moves downstream, we typically use measurements of velocity at fixed points to calculate the acceleration.

The velocity of a flow can vary in space and time. The change in velocity with time at a single location is the **local acceleration**. A change in velocity from point to point in a fluid at a single time gives rise to a **convective acceleration**. The local and convective components of acceleration in a region of a flow can be determined from velocity measurements made over a period of time at several locations along the flow path in the region of interest. The total acceleration experienced by individual fluid elements moving through the region is the sum of these two components.

An example may serve to clarify this concept. Suppose you leave Washington, D.C., at 8 a.m. on a summer day and drive due south. That morning, the temperature throughout the mid-Atlantic region of the U.S. increases at a rate of 2°C per hour. This is the local rate of change of temperature. The temperature in Washington, D.C., for example, may increase from 20°C at 8 a.m. on a summer morning to 26°C at 11 a.m. In addition to the local change we might expect a warming trend in the southerly direction—the 8 a.m. temperature at Raleigh, North Carolina, may be 23°C—so that one would encounter higher temperatures on traveling south. Suppose this warming occurs at a rate of 0.8°C per hundred kilometers. Now, if you measure the temperature as you drive south from Washington at an average speed of 75 km hr^{-1}, what will be the total rate of change of temperature with time that you record? It will be the sum of the local rate of change (2°C/hr) and the convective rate of change (velocity times rate of change with distance $= 75$ km hr$^{-1} \times 0.8$°C per 100 km $= 0.6$°C hr^{-1}) or 2.6°C hr^{-1}.

To express mathematically the local and convective components of the rate of change of temperature we employ the notation for partial derivatives. We write the local rate of change in temperature as the partial derivative of temperature with time, $\partial T/\partial t$. This is just a notational convenience for expressing the change of temperature with time at a fixed location (e.g., Washington). Likewise, $\partial T/\partial y$ is the partial derivative of temperature with distance in the y direction or the rate of change of temperature with distance at a fixed time (e.g., 8 a.m.). The total rate of change of temperature with time is written as dT/dt and is equal to $\partial T/\partial t + v\,(\partial T/\partial y)$ where v is the velocity in the y direction.

Acceleration is treated in an entirely analogous manner to temperature in the above example. Total acceleration is the sum of local acceleration and convective acceleration, and acceleration in the x-direction can be written:

$$\frac{du}{dt} = \frac{\partial u}{\partial t} + u\frac{\partial u}{\partial x}, \tag{3.8}$$

where $u =$ velocity in the x-direction.

In this text we will often restrict our discussion to **steady flow**. The term *steady* applied to fluid flow means that conditions at a point are not changing in time. For example, if the velocity of water at some point in a stream does not change over a certain time period, then the flow has been steady over that period. Steady flow implies that the local acceleration, $\partial u/\partial t$, is zero. Velocity can still vary from point to point in the fluid, however, and if it does, fluid acceleration will be non-zero due to the convective term. For example, an upland stream may be characterized by a sequence of pools and riffles, rapids following a slow-flowing pool following another set of rapids, and so forth. Although the average velocity at any fixed point in the stream is not changing with time, water accelerates as it flows from a pool to a riffle, or vice versa. In contrast, the flow in a straight canal with a constant cross-sectional geometry would not vary from point to point along the flow path. Such a flow is said to be **uniform**, whereas the flow through a series of riffles and pools is nonuniform. A flow that is both steady and uniform has no acceleration; the local acceleration is zero and the convective acceleration is zero.

3.5.2 The Bernoulli equation

To be able to predict the behavior of a moving fluid, for example water flow in a river or through an irrigated field, it is necessary to develop a model, or more specifically, a mathematical equation. This equation must express the correct relationship between the basic physical laws and our observations of fluid motion. We want a relatively simple equation, easily solvable, but one that nevertheless is general enough to describe practical flow situations. An aim of hydrology is to determine what simplifying assumptions are acceptable for giving useful predictions of the behavior of real hydrological systems.

We will consider the problem of the flow of water through a garden hose as the starting point for developing an equation for fluid motion. The results, as we will see, apply to a far wider range of problems than just a garden hose. The hose is connected to a house spigot with sufficient pressure to cause an ample stream to leave the open end, perhaps 10 m away. Clearly the pressure at the spigot will affect the flow in the hose. We must also consider what other factors might be influencing the water in the hose. As in the hydrostatic case (i.e., no motion), we approach the problem by making an assessment of the body and surface forces acting on the fluid.

We will consider an arbitrary short section of the hose of length ds oriented in the s (an arbitrary) direction to begin constructing the force balance for the water movement through this small volume (Figure 3.5). Water is flowing into this small piece from the house, or upstream, end and out of it at the downstream end.

Pressure causes a surface force to act on the small volume. There is both an upstream pressure, p_1, and a downstream pressure, p_2. Thus there is a force in the downstream direction, $p_1 A$, and a force in the upstream direction, $p_2 A$. Another force that acts on the volume of water in the short section of hose is the weight of the water itself, that is, its mass multiplied by the acceleration of gravity:

$$F_g = (\rho dV)g, \tag{3.9}$$

where F_g = body force or weight [M L T^{-2}] and dV = volume = $A ds$ [L^3].

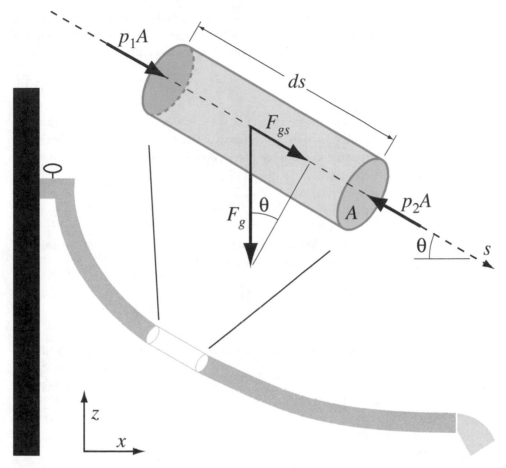

Figure 3.5 A section of a garden hose used to analyze the pressure and gravitational force acting on a small volume of water.

The forces from pressure and the weight of the water are the only two forces we will include in our simple model. It is easy to imagine a more complex situation where other forces would be important. If, for example, the fluid were beryllium, a relatively magnetic fluid, instead of water, then it might be appropriate to consider magnetic forces. A more obvious force we have not listed is the frictional force between the water and the hose or within the flowing water. Frictional forces are tangential surface forces due to the viscosity of the water and the fact that fluids in contact with a surface "stick" to that surface. Although friction is certainly a real surface force, we wish to keep our model simple, and so will neglect frictional forces for the moment. We must keep in mind, however, that this assumption of negligible friction is not correct. We recognize from the start that the final result will be useful only in cases when this assumption does not cause too much error. In many cases in hydrology, the equation resulting from this simple, frictionless case has to be modified, as we will see.

To develop our flow model, knowledge of physical laws and observations of the flow itself are used. According to Newton's second law, the product of the mass and acceleration of the fluid within the hose must be equal to the sum of the forces acting on the fluid volume:

$$F_g + F_p = ma, \tag{3.10}$$

where F_g is the body force due to gravity and F_p is the surface force due to pressure, the net force resulting from the sum of the force due to upstream pressure, p_1A, acting in the $+s$-direction (positive force) and the force due to downstream pressure, p_2A, acting in the $-s$-direction (negative force).

$$F_p = p_1A - p_2A = (p_1 - p_2)A. \tag{3.11}$$

Over the small section of hose that we are considering, the pressure at the downstream end, p_2, differs from the upstream pressure, p_1, by a small amount dp,

$$p_2 = p_1 + \frac{dp}{ds}ds.$$

The term dp/ds is the pressure gradient in the s direction. Substituting this into Equation 3.11,

$$F_p = \left[p_1 - \left(p_1 + \frac{dp}{ds}ds \right) \right]A$$

or

$$F_p = -\frac{dp}{ds}dsA. \tag{3.12}$$

The body force F_g acts in the vertical direction. We are interested only in the component of this force acting along the axis of the hose, the s direction. F_{gs} can be computed from a simple trigonometric relationship (see Figure 3.5):

$$F_{gs} = F_g \sin\theta. \tag{3.13}$$

where θ is the angle the hose is inclined from the horizontal. We can further simplify this expression if we note that

$$\sin\theta = -\frac{dz}{ds}. \tag{3.14}$$

Equation 3.13 can be written

$$F_{gs} = (\rho g dV)\sin\theta = -\rho g dV \frac{dz}{ds}$$

or

$$F_{gs} = -g\frac{dz}{ds}\rho A ds. \tag{3.15}$$

We now have all of the elements of the force balance in useful terms. The mass of the fluid in the control volume is the density times the volume. The forces are as indicated in Equations 3.12 and 3.15. Assembling these in the form of Newton's second law ($F = ma$) gives:

$$(\rho A ds)a = -g\frac{dz}{ds}\rho A ds - \frac{dp}{ds}A ds,$$

where a is the acceleration in the s-direction.

If we assume that the flow is steady in time, then $\partial u/\partial t$ is zero, and Equation 3.8 becomes

$$a = u\frac{du}{ds}, \tag{3.16}$$

where u is the velocity in the s-direction. (Because u is now a function of s only, we write the convective acceleration term as a regular rather than a partial derivative.) This assumption would imply that the supply pressure at the house is steady in time. Finally, we can divide through by $\rho A ds$,

$$u\frac{du}{ds} + g\frac{dz}{ds} + \frac{1}{\rho}\frac{dp}{ds} = 0. \tag{3.17}$$

To review briefly how Equation 3.17 was developed, recall that it is a simple statement of Newton's second law of mechanics. The acceleration of a fluid element was equated to the sum of the forces per unit mass exerted on the element. We have assumed that the frictional force is small enough to neglect and that the flow is steady in time.

Now that we have derived an equation describing flow in a garden hose, we can use it to predict some characteristics or properties of this flow. For example, if we know what the pressure gradient is, can we compute the water velocity or the volume discharge rate? How is pressure related to slope and velocity? To answer these questions, it is first necessary to solve Equation 3.17. Our solution will be simplified if we recall from the rules for differentiation that

$$u\frac{du}{ds} = \frac{1}{2}\left(2u\frac{du}{ds}\right) = \frac{1}{2}\frac{d(u^2)}{ds}. \tag{3.18}$$

Using Equation 3.18, Equation 3.17 can be written:

$$\frac{d}{ds}\left(\frac{u^2}{2g}+z+\frac{p}{\rho g}\right)=0. \tag{3.19}$$

Equation 3.19 can be solved by integration along the s axis. However, to do so we must make another assumption: that the fluid is both homogeneous and incompressible. Making this assumption, we can integrate with respect to s, and write,

$$\frac{u^2}{2g}+z+\frac{p}{\rho g}=\text{constant}. \tag{3.20}$$

Equation 3.20 is known as the **Bernoulli equation**, after Daniel Bernoulli (1700–1782), a Swiss physician, mathematician, and physicist, who was one of the principal founders of the field of fluid mechanics. The constant on the right side of Equation 3.20 is **total head**, H.

The terms in the Bernoulli equation derived above contain elevation, pressure, and velocity as variables. When written as in Equation 3.20, each term in the Bernoulli equation has units of length, and can be thought of as a component of the total head:

Velocity head: $\dfrac{u^2}{2g}\underset{\mathrm{dim}}{=}\dfrac{L^2}{T^2}\dfrac{T^2}{L}=L$

Elevation head: $z\underset{\mathrm{dim}}{=}L$

Pressure head: $\dfrac{p}{\rho g}\underset{\mathrm{dim}}{=}\dfrac{ML}{T^2L^2}\dfrac{L^3}{M}\dfrac{T^2}{L}=L$

The notation "$\underset{\mathrm{dim}}{=}$" is used to denote dimensional equality, for example, $z\underset{\mathrm{dim}}{=}L$ means that elevation has dimensions of length. Each of the total head components can also be thought of as having units of energy per unit weight. Multiplying the Bernoulli equation by the weight of a volume of fluid (ρgV) gives the equivalent equation in units of energy:

$$pV+mgz+\frac{mu^2}{2}=\text{constant}.$$

We see that pV is the "flow work" or the work due to pressure, mgz is the potential energy, and $mu^2/2$ is the kinetic energy. Thus, the Bernoulli equation can also be thought of as a statement of the conservation of energy.

At this point it is desirable to recall all of the assumptions needed to derive the Bernoulli equation, namely:

a. no friction,
b. incompressible fluid,

c. homogeneous fluid,

d. flow steady with time.

In addition, the solution is valid only along the path of integration, the s axis, a line everywhere parallel to the flow field that is referred to as a **streamline**.

3.5.3 Applications of the Bernoulli equation

The use of the Bernoulli equation can best be illustrated by considering a simple example. A tank is filled with water and drains at a small spigot at the bottom (Figure 3.6). By allowing for overflow at the top, and a continual supply of water, the level in the tank remains constant. Our problem is to compute the velocity of the water flowing out of the spigot. To apply the Bernoulli equation, all of our assumptions must be valid, and we must choose an appropriate streamline. The fluid is water, and we can reasonably assume it is both homogeneous and incompressible. Since the level in the tank does not change with time, our problem satisfies the steady assumption. We must assume that friction is unimportant, an assumption whose validity can be tested only by experiment. The streamline to be followed is a question of convenience. A possible streamline from point 1 to point 2 is postulated and the Bernoulli equation is applied. Restating the Bernoulli equation,

$$\frac{u^2}{2g} + z + \frac{p}{\rho g} = \text{constant} = H.$$

Because the total head is constant, the sum of the velocity, elevation, and pressure heads will be the same at every point on a streamline. Using subscripts to identify points 1 and 2 on the streamline, the Bernoulli equation can therefore be written in the useful form:

$$\frac{u_1^2}{2g} + z_1 + \frac{p_1}{\rho g} = \frac{u_2^2}{2g} + z_2 + \frac{p_2}{\rho g} \tag{3.21}$$

If a spot of dye were placed carefully on the surface of the tank, its speed of motion would be seen to be very small with respect to the speed at which the water exits the tank because the surface area of the tank is so much larger than the cross-sectional area of the spigot at the outlet. Because u_1 is so small relative to u_2 ($u_1 \ll u_2$), we can assume that u_1 is equal to zero in our analysis. The pressure on the surface would be the local atmospheric pressure, as would be the pressure at the outlet of the spigot, point 2. The small difference in atmospheric pressure due to the elevation difference between 1 and 2 can be ignored. With these considerations, we can now write,

$$z_1 + \frac{p_{atmos}}{\rho g} = \frac{u_2^2}{2g} + z_2 + \frac{p_{atmos}}{\rho g}$$

or

$$\frac{u_2^2}{2g} = z_1 - z_2.$$

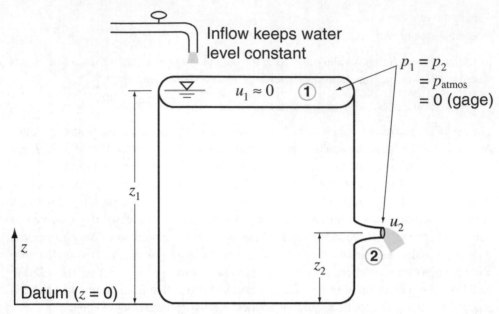

Figure 3.6 Tank with steady flow of water. The Bernoulli equation provides a relationship between the depth of water in the tank and the velocity of water exiting the tank at point 2.

If we let $d = z_1 - z_2$ (the depth of water between points 1 and 2), then,

$$u_2 = (2gd)^{1/2}. \tag{3.22}$$

Thus, the velocity of the water leaving the tank depends on the depth of the water in the tank.

If a plug were placed in the spigot, shutting off the flow, we could use this same approach to compute the pressure at point 2. In this case, both u_1 and u_2 are zero but p_2 is no longer equal to atmospheric pressure:

$$z_1 = z_2 + \frac{p_2}{\rho g}$$

or

$$p_2 = \rho g d. \tag{3.23}$$

This equation, for the *static* situation, is identical to Equation 3.7, the hydrostatic equation. Thus, we have shown that the hydrostatic equation is a special case of the Bernoulli equation.

Let us return to our general form of the Bernoulli equation (Equation 3.20),

$$\frac{u^2}{2g} + z + \frac{p}{\rho g} = \text{constant,}$$

and consider the relationship between pressure and velocity for the garden hose. If the velocity along a horizontal segment of a streamline (i.e., one for which z did not change) were to increase, say because of a constriction in the hose, we would conclude that the pressure would decrease. That is, pressure and velocity are inversely related, assuming no change in elevation, z.

Often we are more interested in the average or mean velocity of flow through a hose or pipe than we are in the velocity along any particular streamline. The **mean velocity** U at any cross section is the discharge Q divided by the cross-sectional area A, that is, $U = Q/A$. The mean velocity also can be thought of as the average value of the velocities at each point in the cross-section. For the frictionless flow described by the Bernoulli equation, the velocity at any point in a cross section is equal to the mean velocity, so $u = U$. This is not true for flows in which friction is important.

One of the fundamental constraints on flow through pipes or channels is conservation of mass: rate of inflow minus rate of outflow equals rate of change of storage (Eq. 1.1). In the case of steady flow in a full hose, there are no changes in the amount of water in any segment of the hose at any time (no change in storage) so inflow rate must equal outflow rate. Because we are taking density to be constant, the *volumetric* inflow rate must equal the *volumetric* outflow rate. The conservation of mass (or volume, in this case) equation is often referred to as the **continuity equation**, simply written,

$$Q = UA = \text{constant.} \tag{3.24}$$

The continuity equation and the Bernoulli equation are two of the fundamental relationships of fluid mechanics.

As an example of the way conservation of mass can be used to provide information about flow, consider a section of hose where the diameter changes from D_1 to D_2, where $D_2 = 2D_1$ (Figure 3.7). Because we have assumed the flow to be steady, the same quantity of water flowing through the narrow section must be flowing through the wide section:

$$Q_1 = Q_2 \tag{3.25}$$

and thus,

$$U_1 A_1 = U_2 A_2 \tag{3.26}$$

where the cross-sectional area of the hose is $A = \pi D^2/4$ and D is the hose diameter. Substituting the expression for cross-sectional area into Equation 3.26 gives,

$$U_1 \frac{\pi D_1^2}{4} = U_2 \frac{\pi D_2^2}{4}$$

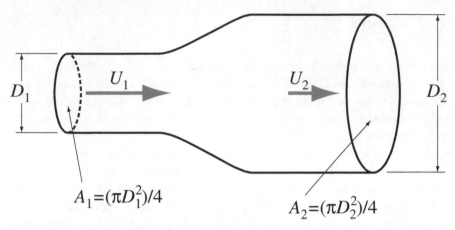

Figure 3.7 An expansion joint in a hose. Discharge is the same at points 1 and 2 (upstream and downstream ends of expansion joint) for steady flow through the hose. As a result, if cross-sectional area increases, velocity must decrease.

or

$$U_1 D_1^2 = U_2 D_2^2$$

Substituting the relationship between D_1 and D_2, we find

$$U_1 D_1^2 = U_2 (2D_1)^2,$$

which may be rearranged to give:

$$\frac{U_1}{U_2} = 4. \tag{3.27}$$

Thus, the velocity in the narrow section of the hose is four times the velocity in the wider section.

If the hose is level ($z_1 = z_2$), the Bernoulli equation is given by:

$$\frac{U_1^2}{2g} + \frac{p_1}{\rho g} = \frac{U_2^2}{2g} + \frac{p_2}{\rho g}.$$

(Recall that for frictionless flow, $U = u$, and therefore the Bernoulli equation can be written in terms of mean velocity, U.) Because U_1 is greater than U_2 (Equation 3.27), we see that the pressure at point 1 must be smaller than the pressure at point 2. Thus, we have shown that an increase in hose diameter results in a *decrease* in mean velocity and a corresponding *increase* in pressure.

3.6 Energy Loss

In deriving the Bernoulli equation, we assumed that fluid friction was negligible. Laboratory experiments can be conducted to test the validity of this hypothesis for simple flow situations involving, for example, flow through glass tubes. There are many flow situations in which fluid friction can be neglected, especially if we are concerned with flows over short distances. For this reason, the Bernoulli equation has been used as the basis of many flow measurement devices used in pipes and streams. An accurate description of flow, however, often requires that we account for the energy loss due to friction, which is associated with the viscous properties of fluids. Reconsider the example of the garden hose. If we were to punch very small holes in the hose every meter along its length, with the water running, it would cause a number of small fountains, as in a lawn soaker (Figure 3.8). The height of each fountain is a measure of the internal pressure at that point in the hose,

$$h_i = \frac{p_i}{\rho g},$$
(3.28)

where the subscript i refers to position along the hose (1, 2, and 3 in Figure 3.8).

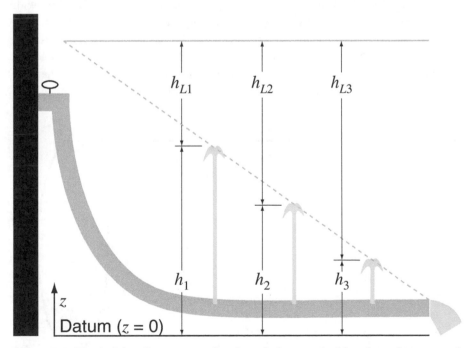

Figure 3.8 The height of water spouting from holes punched in a hose decreases along the length of the hose because of frictional losses of energy (head loss) as the flow travels downstream in the hose.

For a horizontal section of the hose with no friction, the Bernoulli equation would predict that pressure and velocity along the hose would be constant provided the hose has a constant diameter. This implies that the h_i's in Figure 3.8 would all be the same. However, observations of the heights of sprinklers in a real hose with punctures show that $h_1 > h_2 > h_3$; that is, adjacent fountain elevations become successively smaller as the open end of the hose is approached. Because the hose is level in the section we are interested in and the discharge and diameter are constant (and therefore U is not changing along the hose), the pressure must be continuously decreasing along the hose. But, this conclusion contradicts the conservation of energy equation (Bernoulli) that we obtained using the assumption of zero friction:

$$\frac{p}{\rho g} + z + \frac{U^2}{2g} = H = \text{constant}.$$

In fact, we observe a loss in total head along the hose, h_L, which if added to the other terms in our equation would give a total head equal to the initial sum H. That is, the Bernoulli equation must be modified to account for the energy loss to friction:

$$\frac{p}{\rho g} + z + \frac{U^2}{2g} + h_L = \text{constant}, \tag{3.29}$$

where h_L is called the **head loss**. Head loss is simply an empirical way of dealing with the fluid friction that is dissipating energy (converting mechanical energy to thermal energy) inside the hose over a specified length of hose. For a fluid with a non-zero viscosity, like water, friction causes fluid adjacent to a solid surface to "stick" to that surface. This means the velocity of the water in contact with (stuck to) the wall of the hose must be zero because the hose is not moving. As a result, there will always be a velocity

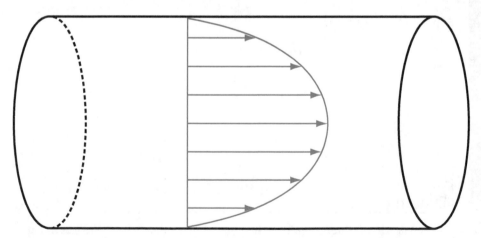

Figure 3.9 Water "sticks" to the walls of the hose and viscosity causes a frictional energy (or head) loss. Velocity is zero at the walls of the hose and a maximum at the centerline.

gradient away from the wall within a frictional flow (Figure 3.9). Energy dissipation is a result of this velocity gradient.

3.6.1 Friction factor

For flow through a circular pipe or hose, head loss between the two ends of the hose depends on the viscosity of the fluid, the velocity of the flow, hose diameter and length. Because head loss also is a function of the roughness of the wall material itself, we must include a **friction factor** in its definition:

$$h_L \equiv f \frac{L}{D} \frac{U^2}{2g}, \tag{3.30}$$

where f = friction factor [dimensionless], L = length [L], and D = diameter [L].

If we consider the special case of a horizontal hose of constant diameter (and therefore constant cross-sectional area), velocity and elevation don't change along the hose. We can write the Bernoulli equation including head loss as

$$\frac{p_1}{\rho g} + z_1 + \frac{U_1^2}{2g} = \frac{p_2}{\rho g} + z_2 + \frac{U_2^2}{2g} + h_L, \tag{3.31}$$

where h_L is the head loss between cross-sections 1 and 2 of the hose. If velocity and elevation don't change along the hose, then $U_1 = U_2$, $z_1 = z_2$, and therefore $h_L = (p_1 - p_2)/\rho g$. If we express head loss as head loss per unit length, then

$$\frac{h_L}{L} = \frac{1}{\rho g} \frac{p_1 - p_2}{L} = -\frac{1}{\rho g} \frac{\Delta p}{L} = -\frac{1}{\rho g} \frac{dp}{dx}, \tag{3.32}$$

where L is the length of the hose between cross-sections 1 and 2. If we combine Equations 3.30 and 3.32, we can obtain the following expressions for velocity:

$$U = \left(\frac{h_L}{L} \frac{2gD}{f} \right)^{1/2} = \left(\frac{p_1 - p_2}{L} \frac{2D}{\rho f} \right)^{1/2} = \left(-\frac{dp}{dx} \frac{2D}{\rho f} \right)^{1/2}. \tag{3.33}$$

Note that dp/dx is negative for flow in the positive x-direction. In other words, there is a pressure drop in the direction of flow. Equation 3.33 suggests that if we can measure the pressure drop between two points along a hose or pipe a distance L apart and we know the friction factor, we can calculate the velocity or discharge of flow through the pipe.

The problem of calculating or predicting the velocity at which water or other fluids would flow through a given pipe, channel, or other conduit is a problem that has great practical importance in the design and management of water supply and sewer systems, waterways, and industrial pipelines. Calculating velocity using Equation 3.33 requires knowledge of the correct friction factor. Friction factor is a quantity that must be determined experimentally. Because of its importance, many measurements of friction factor f

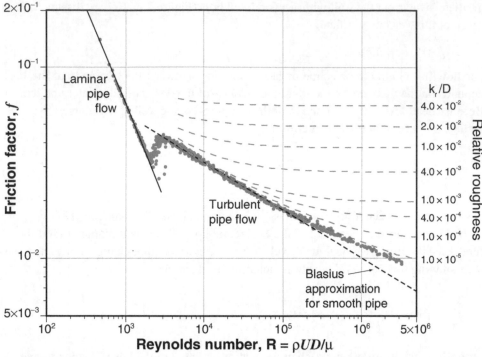

Figure 3.10 Friction factor diagram for pipes. Laboratory measurements of friction factor f versus Reynolds number **R** for smooth pipes over a range of Reynolds numbers are shown by the blue dots. The break in the measurements at Reynolds numbers between 2000 and 4000 marks the transition from laminar to turbulent flow. The Blasius approximation (Equation 3.39) is a linear approximation to the turbulent flow measurements valid for $4000 < $ **R** $ < 100,000$. Friction factors for turbulent flow in rough pipes are indicated by the gray dashed curves, where relative roughness, k_r/D, is the ratio of the roughness length, k_r (see Figure 3.12), to pipe diameter.

have been made under controlled laboratory conditions. A number of these measurements of friction factor are plotted in Figure 3.10 against the term $\rho UD/\mu$. This dimensionless term, the Reynolds number **R**, is discussed in the next section. When plotted in this manner, the data for all pipe diameters, discharges, and fluids describe a well-defined relationship for pipes that are smooth inside. We refer to this as a friction factor diagram.

3.7 Laminar and Turbulent Flows

It is obvious looking at the friction factor diagram (Figure 3.10) that something happens to the flow when the Reynolds number exceeds about 2000. It turns out that the frictional losses in a fluid flow depend on whether the flow is laminar or turbulent. In order to understand the difference between these flows, let us review an experiment first conducted by Sir Osborne Reynolds (1843–1912). Reynolds was an English civil engineer noted for both his theoretical and applied studies in fluid mechanics. Reynolds's experiment used a simple apparatus to visualize a flow field (Figure 3.11). In this apparatus, dye is injected along the centerline of a pipe in which fluid is flowing. The dye serves to show the nature of the flow as the fluid moves along the tube. At low velocities the dye remains approximately on the centerline, traveling downstream with, and at the velocity of, the flow. Because of molecular diffusion there will be a small flux of dye away from the centerline, but this will be so small that no departure from the line will be observed with the naked eye. Reynolds characterized this flow situation by describing the fluid as moving in distinct *laminae* (Latin for layers), and today we refer to such flow as **laminar flow**.

Now consider what happens as the velocity of fluid flow in the tube is increased. At some value of increased velocity, the dye is observed to become mixed within small, randomly located spots within the flow; with higher velocities the dye rapidly becomes completely mixed with the surrounding fluid in a short distance of flow. Reynolds referred to this flow as sinuous or disturbed flow. Today we refer to flows exhibiting such violent mixing as **turbulent flows**. In this case the mixing takes place much more rapidly than realized by molecular diffusion, and in fact we call this mixing *turbulent diffusion*. We can conceptualize the turbulent mixing as small parcels of the fluid "jumping" away from a streamline to a different portion of the flow field. Most flows we observe in nature, for example, clouds, rivers, ocean waves, are turbulent flows. In this regard, we note that "laminar" and "turbulent" refer to properties of the flow, and not properties of the fluid. For example, the water in the garden hose could be in either laminar or turbulent flow without changing its properties, that is, density or viscosity.

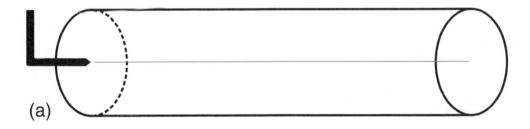

(a)

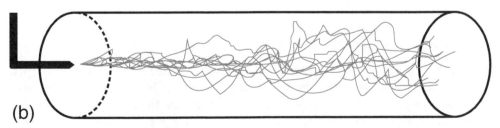

(b)

Figure 3.11 Reynolds's experiment. In laminar flow (*a*), the dye remains a thin line, moving with the fluid. The dye trace becomes convoluted in turbulent flow (*b*).

To explain the last statement, we return to Reynolds's experiment. By carefully changing the properties of the fluids in his apparatus, as well as the tube diameter and flow velocity, Reynolds empirically defined a dimensionless number **R** which describes the flow properties:

$$\mathbf{R} = \frac{UL\rho}{\mu} \tag{3.34}$$

where U = a characteristic velocity [L T^{-1}]; L = a characteristic length [L]; ρ = fluid density [M L^{-3}]; μ = viscosity [M L^{-1} T^{-1}]. **R** is called the **Reynolds number**. According to Figure 3.10, if **R** is small (with U taken as the average velocity and L taken as the diameter of the tube), say less than 2000, then the flow in a pipe will be laminar. If **R** is large, say greater than 4000, then the flow will be turbulent. At values between these limits, called the transition region, the flow is not easily characterized as one or the other. The choice of the velocity and length is a matter of accepted convention. For circular pipes such as the garden hose, the characteristic velocity is taken as the mean value over the cross section (the discharge divided by the cross-sectional area), and the characteristic length as the hose diameter. If water flows at 15°C at 2 m s^{-1}, through a 30-mm diameter hose, the Reynolds number would be calculated as follows:

U = 2 m s^{-1} (mean velocity);

L = 0.03 m (pipe diameter);

ρ = 10^3 kg m^{-3};

μ = 1.139 × 10^{-3} Pa · s;

$$\mathbf{R} = \frac{(2)(0.03)(10^3)}{(1.139 \times 10^{-3})} \approx 53{,}000,$$

which indicates that the flow is turbulent. The flow could be made laminar by somehow reducing **R** to some value less than 2000. We could accomplish this by:

a. reducing the velocity,
b. reducing the diameter,
c. reducing the density of the fluid, or
d. increasing the viscosity.

In order to accomplish (c) or (d), we would have to change the fluid temperature, or perhaps the fluid itself. This latter alternative suggests another important conclusion from Reynolds's experiments. Two *different* fluids with the same Reynolds number will have similar flows, that is, laminar or turbulent. Thus, we can expect the same degree of mixing or turbulence inside the hose for both water and glycerin, if they both have the same value for **R**. Because the viscosity of glycerin is much greater than that of water, it would be difficult to achieve this similarity in practice.

Now that we have some feeling for the meaning of the Reynolds number, let us reconsider the differences between laminar and turbulent flow as evidenced by Reynolds's experiment. For the laminar flow, any time the dye began to stray from the centerline of the tube, it was hindered, or its motion was restored to the centerline. The attempted deviation or acceleration is an apparent inertial force, and the restoring force is due to the viscous forces within the fluid. Thus, for laminar flow, the mixing is restrained as the viscous force must dominate over the inertial force. If the inertial force becomes greater than the viscous force, then the dye would have a tendency to deviate from the centerline, and the flow could become turbulent. We can therefore take the ratio of the inertial to viscous force as an index of when flow would be expected to be laminar or turbulent. If the viscous forces $\gg$ inertial forces, the flow is laminar, and if the inertial forces $\gg$ viscous forces, the flow is turbulent. We know that the inertial force is proportional to mass $\times$ acceleration, which could be expressed as $(\rho)(L^2)(\text{velocity}^2)$. Similarly, the viscous force is proportional to the shear stress $\times$ area, which could be expressed as $(\mu)(L)(\text{velocity})$. Thus,

$$\frac{\text{inertial force}}{\text{viscous force}} \propto \frac{\rho L^2 V^2}{\mu L V} = \frac{\rho L V}{\mu} = \mathbf{R}. \tag{3.35}$$

The Reynolds number is equal to the ratio of the inertial to viscous forces within the fluid.

3.7.1 Laminar pipe flow

Let us return to our example of water flowing through the garden hose. If the hose is narrow enough, the flow slow enough, or the viscosity high enough, the flow could be laminar. In this case, the experimental relationship between friction factor and Reynolds number indicated in Figure 3.10 is linear and is given by:

$$f = \frac{64}{\mathbf{R}} = \frac{64\mu}{\rho U D}. \tag{3.36}$$

If we substitute this into Equation 3.33, we obtain:

$$U = \left(-\frac{dp}{dx} \frac{2D}{\rho} \frac{\rho U D}{64\mu} \right)^{1/2} = \left(-\frac{dp}{dx} \frac{U D^2}{32\mu} \right)^{1/2}$$

or, if we square both sides and divide by U:

$$U = -\frac{dp}{dx} \frac{D^2}{32\mu}. \tag{3.37}$$

The velocity given by Equation 3.37 is the mean velocity for laminar flow through the pipe; this velocity times the pipe cross-sectional area gives the discharge, $Q = UA$. Steady laminar flow through a uniform pipe is referred to as Poiseuille flow and Equation 3.37 for the mean velocity is sometimes referred to as **Poiseuille's law**. As depicted in Figure 3.9, the actual velocity at any point in the pipe varies with distance between the wall and the center of the pipe. If we integrated this velocity profile over the pipe cross section, we would obtain the discharge Q; the mean velocity U is this discharge divided by pipe cross-sectional area A.

In deriving Poiseuille's law from the Bernoulli equation we assumed that the elevation along the hose or pipe is constant, that is, $z_1 = z_2$. We can generalize the equation by relaxing this assumption and allowing the hose to be inclined at an angle as in Figure 3.5. In this case a term related to the elevation difference is added to Poiseuille's law giving:

$$U = -\left(\frac{dp}{ds} + \rho g \frac{dz}{ds}\right)\frac{D^2}{32\mu}$$

or

$$U = -\frac{d}{ds}\left(\frac{p}{\rho g} + z\right)\frac{D^2 \rho g}{32\mu}. \tag{3.38}$$

The term in parentheses in Equation 3.38 is one that is used frequently in hydrology and is referred to as hydraulic head h.

3.7.2 Turbulent pipe flow

If the flow velocity, pipe (or hose) diameter, and fluid viscosity combine to produce a Reynolds number in excess of 4000, as is often the case, then the flow will be turbulent and Poiseuille's law does not apply. Instead, we must return to Equation 3.33 and use Figure 3.10 to determine the friction factor. For values of Reynolds number between 4000 and 100,000, the friction factor for smooth pipes can be approximated by the expression:

$$f = 0.316\mathbf{R}^{-1/4} \tag{3.39}$$

This approximation, proposed by Blasius, is indicated in Figure 3.10.

When the inside of the pipe is rough, the friction factor for turbulent flow increases. In this case flow experiences resistance not only from the pipe walls themselves (termed skin friction) but also from the **roughness** elements protruding up into the flow as suggested in Figure 3.12. This roughness resistance is due to drag (called form drag) produced as flow is forced to go around roughness elements on the flow boundary. Examples of roughness elements that contribute to bottom friction in streams include rocks, boulders, and other bed forms such as sand ripples and dunes. The rougher the surface is, the greater the form drag, and the higher the friction factor for turbulent flow, as shown in Figure 3.10. Thus, in turbulent flow, unlike laminar flow, it is necessary to characterize the roughness of the walls bounding the flow in addition to the other factors. We will

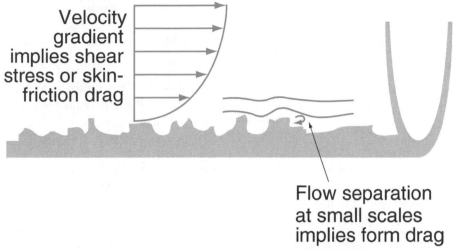

Figure 3.12 Flow through a rough-walled hose. The roughness length, k_r, is a measure of the size of the irregularities or unevenness of the surface.

see in the next chapter on channel flow that characterization of channel roughness is an important but difficult part of estimating flow velocity in streams and other turbulent surface flows.

3.8 Concluding Remarks

The material in this chapter was meant to introduce you to the basics of fluid mechanics. We should emphasize again that our objective is to apply these fundamental principles of fluid flow to hydrological problems. It may not yet be clear to you how a description of pressure distribution, flow and fluid drag in a garden hose is relevant to our objective. The applications to the "real world" are discussed in the succeeding chapters and we will see that the garden hose example will assist us in problems ranging from the relationship between river velocity and bed roughness to the interpretation of water levels in wells.

"Real world" applications of the principles of fluid mechanics are not restricted to the flow of water. The winds that blow over the surface of the Earth are flows of air in response to differences in atmospheric pressure or other forces. Flows of air near the Earth's surface are characterized by high Reynolds numbers because air has a relatively low viscosity and the flows are always "deep." In contrast, lava has a viscosity more than a thousand times that of water (the value depends strongly on temperature) while its density is only 2 to 3 times larger. As a result, flows of lava are low Reynolds number, laminar flows that can be described by a form of Poiseuille's equation. One of the many things that makes water interesting is that there are commonly occurring examples of both laminar and turbulent flow.

3.9 Key Points

- A fluid is a substance that deforms, or flows, when subjected to a shear stress, no matter how small. {Section 3.2}

- Viscosity is a temperature-dependent property of a fluid that characterizes its ability to resist deformation. The viscosity μ of a fluid is given by the ratio of the shear stress at a point in the fluid to the velocity gradient at that point (Newton's law of shear): $\mu = \tau/(du/dz)$. The viscosity of water at 20°C is 1.0×10^{-3} Pa · s. {Section 3.2}

- When defining properties of a fluid at a point we make the continuum assumption that, at the macroscopic level, the fluid is continuous throughout its entirety, that is, that the molecular structure of the fluid is smeared out to eliminate spaces between atomic particles. {Section 3.2}

- The density of a fluid is the mass per unit volume at each point in the fluid. A homogeneous fluid is one in which density is constant throughout the fluid. The density of water is approximately 1000 kg m^{-3}. The unit weight of a fluid is its weight per unit volume, ρg [N m^{-3}]. {Section 3.2}

- Two classes of forces act on fluids: body forces and surface forces. Body forces, such as gravity, act uniformly on each fluid element. Surface forces, such as pressure and friction, act on the surfaces of fluid elements. In fluid mechanics we typically represent surface forces as forces per unit area, or stresses. There are two types of stresses: normal stresses and tangential stresses termed shear stresses. The inward-directed (compressive) normal stress, when applied to a fluid medium, is referred to as pressure. {Section 3.3}

- In a fluid at rest, pressure increases with depth at a rate of ρg, the unit weight of the fluid. This is expressed by the hydrostatic equation, $dp/dz = -\rho g$. The integrated form of the hydrostatic equation for a fluid of constant density indicates that pressure increases linearly with depth below the surface, $p = \rho g d$. The pressure given by this relationship is gage pressure, the pressure relative to atmospheric pressure. To obtain absolute pressure, the atmospheric pressure acting on the fluid surface must be added to p. {Section 3.4}

- The acceleration of a fluid can be divided into two components. The local acceleration is the change in velocity with time at a point, $\partial u/\partial t$. The convective acceleration is related to changes in velocity from one place to another in the flow, e.g., $u\,(\partial u/\partial x)$ for flow in the x-direction. The total acceleration is the sum of these two components. A flow is steady if the local acceleration is zero. A flow is uniform if the convective acceleration is zero. {Section 3.5.1}

- The Bernoulli equation states that, for a frictionless flow, the sum of the velocity head $u^2/(2g)$ [L], elevation head z [L], and pressure head $p/(\rho g)$ [L] along a streamline is a constant termed total head, H [L]: $[u^2/(2g)] + z + [p/(\rho g)] = H$. The assumptions made in deriving the Bernoulli equation are 1) no friction; 2) incompressible fluid; 3) homogeneous fluid; and 4) steady flow. For frictionless flow, the velocity u on any streamline in a cross section is equal to the mean velocity U, so that U can be substituted for u in the Bernoulli equation to obtain an equation in terms of mean velocity. {Section 3.5.2}

- The continuity (conservation of mass) equation for *steady* flow through a pipe or channel states that the rate of inflow is equal to the rate of outflow, as long as there is no water entering or leaving the flow through the sides. If the density of the fluid is constant, this means that the discharge must be the same at each cross section of the flow: $Q = UA = constant$. The continuity equation is often combined with the Bernoulli equation when solving flow problems. {Section 3.5.3}

- Frictional losses of energy in a flow result in a loss in total head with distance downstream. When friction is important, head loss h_L must be added to the terms in the Bernoulli equation to keep the sum of the terms constant: $[u_1^2/(2g)] + z_1 + [p_1/(\rho g)] = [u_2^2/(2g)] + z_2 + [p_2/(\rho g)] + h_L$. For flow through a horizontal pipe, head loss is related to the pressure drop along the pipe: $h_L = (p_1 - p_2)/(\rho g)$. {Section 3.6}

- Head loss for flow through a pipe or tube is related to mean velocity U, pipe diameter D, and pipe length L by the empirical equation $h_L = f[(LU^2)/(2Dg)]$, where f is the friction factor [dimensionless]. Values of f are determined experimentally and are typically represented in friction factor diagrams of f as a function of pipe Reynolds number $\mathbf{R} = \rho UD/\mu$ (Figure 3.10). {Section 3.6.1}

- The Reynolds number $\mathbf{R} = \rho UD/\mu$ is a measure of the relative importance of viscous forces in a flow. When the Reynolds number is less than 2000 for pipe flow, viscous forces are large enough to damp any disturbance in the flow resulting in laminar flow. Viscous forces are less effective at higher Reynolds numbers (above about 4000 for pipes) so that disturbances to the flow can grow, causing the flow to become turbulent. Two different fluids with the same Reynolds number will have similar flows. {Section 3.7}

- The friction factor for laminar pipe flow is given by $f = 64/\mathbf{R}$. Combining this with the head loss equation gives an equation for the mean velocity for laminar flow in a pipe (Poiseuille's law): $U = -(dp/dx)(D^2/32\mu)$. {Section 3.7.1}

- For turbulent pipe flow, friction factor must be obtained from the friction factor diagram or determined experimentally. The friction factor for turbulent flow increases with increasing roughness of the pipe walls. {Section 3.7.2}

3.10 Example Problems

Problem 1. The following questions make use of the hydrostatic equation.

A. What is the gage pressure (Pa) at a depth of 10.0 m in a lake with a water temperature of 15°C?

B. Would the pressure change significantly if the water temperature was 22°C instead?

C. At what depth (m) is the gage pressure 300 kPa?

D. What depth (m) of mercury, with a unit weight of 133 kN m⁻³, would be required to produce a pressure of 300 kPa?

Problem 2. A plate is pulled over a horizontal layer of water that is 10.0 mm deep (Figure 3.1). The temperature of the water is 20°C. If the plate exerts a shear stress of 0.01 N m⁻² on the upper surface of the water, what is the speed (m s⁻¹) of the plate?

Problem 3. Observations show that flow in a circular pipe of diameter D remains laminar up to a Reynolds number $\mathbf{R}_{pipe} = \rho UD/\mu = 2000$ (Figure 3.10). What about flows in other flow in channels or pipes with different geometries? Consider the example of flow between two flat plates shown in Figure 3.1. In this case, the appropriate length scale for the Reynolds number is $L = d$. It also makes sense to use plate speed rather than mean flow velocity speed as the characteristic velocity in $\mathbf{R}$, giving $\mathbf{R}_{plate} = \rho u_{plate} d/\mu$. Changes in length and velocity scales can alter the upper limit for laminar flow, but for the case of flow between two parallel plates, flow is again laminar up to $\mathbf{R}_{plate} = 2000$. (A parameter called the hydraulic radius [Chapter 4.5] can be used to find values of $\mathbf{R}$ corresponding to the laminar–turbulent transition for different flow cross-sections. This is explored further in an example problem in Chapter 4.)

A. For 20°C water between 2 plates separated by a distance of 4.0 mm (i.e., $d = 4.0$ mm; Figure 3.1), what is the maximum speed that the upper plate can move and still maintain laminar flow?

B. Is the flow in Problem 2 laminar? (Note: if not, Equation 3.1 is no longer correct.)

Problem 4. Surface temperature in a river is measured by a thermometer drifting with the water at a rate of 1 km hr⁻¹. The water in the river as a whole is warming at a rate of 0.2°C hr⁻¹, and the temperature along the stream increases by 0.1°C every kilometer in the downstream direction. What change in temperature (°C) does the thermometer record in 6 hours?

Problem 5. A tank like the one pictured in Figure 3.6 is filled to a constant level of 0.70 m. The center of the outflow opening near the bottom is 0.10 m above the bottom of the tank. What is the velocity (m s⁻¹) of flow exiting from the outflow opening?

Problem 6. The pressure drop through a well-designed constriction can be used to measure the velocity of flow through a pipe. If the pressure drop from a 0.1-m diameter cross section to a 0.05-m diameter cross section is 7.5 kPa, what is the velocity (m s⁻¹) in the 0.1-m diameter section of the pipe? Hint: Use the conservation of mass equation to relate the velocity at the smaller cross section to that at the larger cross section.

Problem 7. A steady discharge of 2.0×10^{-4} m³ s⁻¹ is flowing through a 20-mm diameter hose. The viscosity of the water is 1.0×10^{-3} Pa·s, and the density of the water is 1000.0 kg m⁻³.

A. Calculate the Reynolds number. Is the flow laminar or turbulent?

B. What is the friction factor and the head loss per unit length for this flow (both are dimensionless)?

C. What is the change in pressure (Pa) over a 10-m length of the hose?

Problem 8. Lava, with a density of $2700 \, kg \, m^{-3}$ and viscosity of 1.0×10^3 Pa-s, flows through a conduit that is circular in cross section. The diameter of the conduit is 1.0 m. Flow of lava through the conduit is driven by a pressure gradient of $-2.0 \, kPa \, m^{-1}$. Assuming the flow is laminar, what is the discharge of lava ($m^3 \, s^{-1}$)? Is the assumption of laminar flow valid?

3.11 Suggested Readings

Fox, R.W., A.T. McDonald, and P.J. Pritchard. 2011. *Introduction to Fluid Mechanics*, 8th ed. New York: John Wiley & Sons.

Middleton, G.V., and P.R. Wilcock. 1994. *Mechanics in the Earth and Environmental Sciences*. Cambridge: Cambridge University Press. Chapters 9 and 11, pp. 296–336, 365–394.

4 Open Channel Hydraulics

4.1 Introduction

In this chapter, the principles of fluid mechanics are applied to surface-water flow in channels, for example, rivers, streams, and canals. Observations of flow in channels suggest many interesting and important questions. Why does the flow in rivers vary from deep and tranquil to shallow and torrential? What controls the depth of water in a river? How are water depth and discharge in a stream related? How does the velocity in a stream change as the amount of water carried by the stream increases?

Consider the last question: How are velocity and discharge in a stream related? It turns out that the answer to this question is a key to determining how fast a flood will move down a river valley. We also need to know about the speed of water movement if we want to know how fast a contaminant that is spilled into a river moves toward the water intake of a city's water supply. An instance of the latter occurred in the early 1990s when a train car containing a hazardous chemical derailed and spilled into the river upstream of a city in California. It was imperative in this case to know when the city water intake had to be shut down to avoid drawing the contaminant into the city's water-supply system. It turns out that, for river channels, there is generally a useful relationship between mean velocity of flow in a channel and the discharge in the channel (Figure 4.1).

The relationship between water depth and discharge is one of very broad importance in hydrology. We often need to know how the depth of water in a channel is related to the quantity of water the channel is carrying. Unlike the pipes and hoses that we considered in the last chapter, channels do not have "lids." The water depth varies along with the other hydraulic quantities. In the following discussion of hydraulics in stream channels, the relationship between water depth and discharge will be seen to be important for

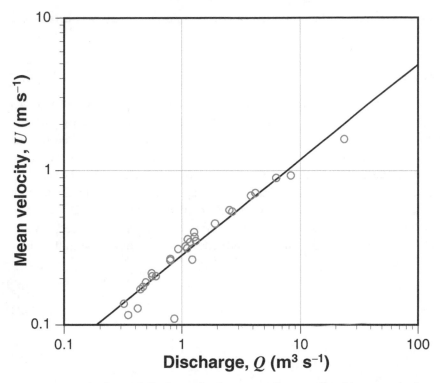

Figure 4.1 Velocity and discharge in rivers usually are related by a power function, which plots as a straight line on logarithmic axes. The equation of the straight line shown in the graph is $U = 0.28Q^{0.62}$.

Data from Miller et al. (1971).

flow measurement. Our primary goal in this chapter is to explore the physical basis for relationships among the hydraulic variables that we can measure in channels. In Chapter 5 these principles will be applied to stream hydrology.

Obviously, all of the important questions relating to flow in channels cannot be answered fully in one brief discussion on channel hydraulics. Many important flow phenomena can be described in a reasonably straightforward manner, however, by considering the equation of continuity, the Bernoulli equation, and the friction factor relationships developed in the previous chapter.

In developing some of the principles of fluid mechanics in Chapter 3, we started with the assumption of frictionless flow. Of course, there are no fluids that have zero viscosity or any pipes that have no roughness at their surfaces. Nevertheless, the Bernoulli equation assuming no friction is a useful starting point for a discussion of the physics of fluid flow in pipes and hoses and can be applied directly in many problems. A similar approach is useful for channel flow. Some concepts about how water depths vary with discharge are appreciated best by considering frictionless flow. It again turns out that the concepts derived from an assumption of frictionless conditions are useful for measuring flow in channels. Therefore, we will start our discussion by exploring specific energy in a frictionless channel and will then proceed to consider flow in "real" channels where friction has a dominant effect.

4.2 Specific Energy

To begin, consider how velocity and water depth are related in open channel flow. An open channel flow differs from flow in a pipe in that the channel flow is only partially enclosed by a solid boundary. The upper "boundary" of an open channel flow, between the water and the atmosphere, is called a free surface. To keep the analysis simple we will consider the flow to be steady and frictionless so that the Bernoulli equation can be used.

Our first step will be to adapt this equation to a form particularly suited for application to channel flow. First consider flow in a short section of channel (Figure 4.2). In a very short channel section, or **reach**, frictional losses often can be disregarded, as can any change in the elevation of the bed. The Bernoulli equation for a streamline in this short segment of channel is:

$$\frac{u^2}{2g} + \frac{p}{\rho g} + z = H \tag{4.1}$$

or

$$\frac{u^2}{2g} + \frac{p}{\rho g} + (z_b + h - d) = H,$$

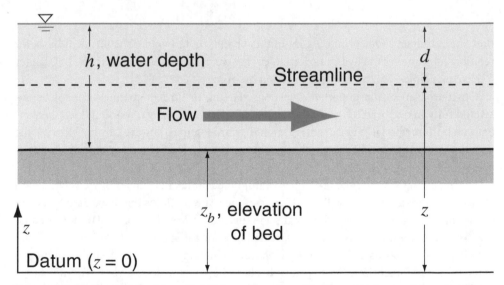

Figure 4.2 Horizontal flow in a short channel section. The flow can be considered uniform and frictionless over this section. The total energy per unit weight of the water along the streamline is $U^2/2g + h + z_b = E + z_h$, where E is the specific energy.

where z_b is the elevation of the stream bed above datum [L], h is total water depth [L], d is the depth of the streamline below the free surface [L], and H is the total energy per unit weight or total head [L]. The "datum" is just a convenient reference elevation; often mean sea level is used. As discussed in the derivation of the Bernoulli equation in Chapter 3, H is a constant. The term $z_b + h - d$ is the elevation head. If the flow is essentially horizontal, then the fluid acceleration in the vertical direction is negligible and the pressure will vary with depth beneath the free surface according to the hydrostatic equation. Thus, $p = \rho g d$ and Equation 4.1 can be written as:

$$\frac{u^2}{2g} + d + (z_b + h - d) = H.$$

At this point a useful modification can be introduced: $d + (z_b + h - d)$ is equal to $z_b + h$, where h is the total water depth and z_b is the elevation of the stream bed above datum. This would be true for any streamline because it is independent of d (e.g., the gage pressure at the water surface ($d = 0$) is zero and the total elevation head is $z_b + h$).

For frictionless flow the velocity is assumed constant over the depth. In this case, the velocity on each streamline is the same and is equal to the channel mean velocity U [L T^{-1}], so that Equation 4.1 can be written:

$$\frac{U^2}{2g} + h + z_b = H. \tag{4.2}$$

The first two terms of this equation, $U^2/2g + h$, can be interpreted as the energy per unit weight of the flowing water *relative to the stream bottom*. This combined term is defined as the **specific energy**, E [L]:

$$E = \frac{U^2}{2g} + h. \tag{4.3}$$

Equation 4.2 can be rewritten:

$$E + z_b = H. \tag{4.4}$$

From Equation 4.4 we see that if the elevation of the channel bottom remains constant, then because H is constant, E must also be constant.

The continuity equation for the steady, uniform flow of a constant-density fluid states that the discharge Q [L^3 T^{-1}] remains constant so that $U_1 A_1 = U_2 A_2$, where A is the cross-sectional area of flow [L^2]. To simplify the analysis even further, let the channel have a rectangular cross section. Then the cross-sectional area is equal to the channel width, w [L], times the water depth, h:

$$Q = Uwh. \tag{4.5}$$

If, for the moment, we also assume that the width of the channel is constant, Equation 4.5 implies that:

$$q_w = \frac{Q}{w} = Uh = \text{constant},$$ (4.6)

where q_w is called the **specific discharge** [$L^2 \, T^{-1}$]. From Equation 4.6, $U = q_w/h$, which can be substituted into Equation 4.3 to give:

$$E = \frac{q_w^2}{2gh^2} + h.$$ (4.7)

Equation 4.7 provides a means for determining the depth of a frictionless channel flow of known discharge and specific energy. That is, given numerical values for E and for q_w, h is the single unknown quantity in Equation 4.7. Although it is possible to solve Equation 4.7 directly, a graphical solution is more convenient. A **specific energy diagram** is a graph of specific energy, E, versus depth, h, for a given value of specific discharge, q_w (Figure 4.3). From the specific energy diagram, we see that for any specific discharge there are three depths for each value of specific energy. (Equation 4.7 is a cubic equation in h, so there are three roots). Only two of these possible depths, h_1 and h_2, are positive; the third depth, h_3, is negative and can be disregarded in that it has no physical meaning. The two positive depths that are possible for a single value of the specific energy and the given specific discharge are called **alternate depths**.

The specific energy diagram (Figure 4.3) exhibits a minimum in E, which corresponds to the condition (called the **critical condition**) in which the solutions h_1 and h_2 of Equation 4.7 coincide. Thus, under critical conditions, a flow with specific discharge, q_w, has the minimum specific energy. The corresponding depth is known as the **critical depth**, h_c. Critical flow is a unique condition in which, for any given value of specific discharge, there is only one possible water depth (i.e., the critical depth). For values of E with two possible (alternate) depths, one will be larger than the critical depth and the other will be smaller (Figure 4.3). The flow depth h_1 corresponds to a flow that is deeper and slower than the critical flow and is called **subcritical flow**. The flow with depth h_2 is shallower and faster than the critical flow, and is termed **supercritical flow**.

The original problem posed at the beginning of this discussion was to determine the water velocity and depth for a given discharge and stream cross section. Application of the Bernoulli equation and the continuity equation led to the specific energy diagram. This suggests that an answer to the question demands additional knowledge (the specific energy of the flow) and that two combinations of depth and velocity are possible. Thus, we will have to examine the question further to determine how specific energy and depth are related.

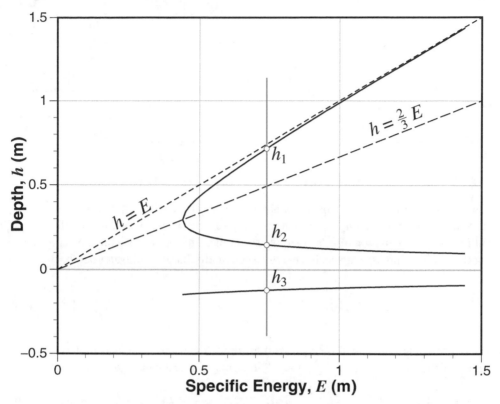

Figure 4.3 A specific energy diagram for $q_w = 0.5$ m^2 s^{-1}. The three depths, h_1, h_2, and h_3, are those allowed by theory for a given specific discharge and a given specific energy (0.75 m in this figure). The three depths are the solutions to the cubic equation relating E, h, and q_w. Because negative depths have no physical meaning, the upper part of the diagram, with positive values of h, is used in practice.

4.2.1 Flow over a vertical step

A comparison between flow through a hose with a contraction and the free surface analog, a stream with a shallow rise, or step, in the bottom illustrates the role of specific energy in understanding problems of open channel flow (Figure 4.4). From the equation of continuity and Bernoulli equation, it is easy to compute the velocity or pressure in the narrow section of the hose if we know these values for the larger section. As found in Chapter 3, $U_2 > U_1$ and $p_2 < p_1$ for a steady, homogeneous, frictionless flow through a pipe contraction such as that illustrated in Figure 4.4. Unlike the flow through the hose contraction, the stream would seem to have a "choice" as it flows over the step. To constrain the flow we will stipulate that the width of the stream is constant and that flow remains within the channel. What will the surface do as the flow crosses the step: rise, fall, or remain unchanged? Assuming the transition is essentially frictionless, the Bernoulli equation (4.2) can be written:

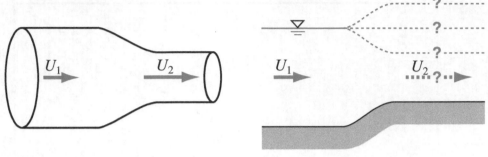

(a) Pipe with constriction   (b) Channel with step

Figure 4.4 For the case of a pipe (*a*) or an open channel (*b*), can U_2 and p_2 or h_2 be calculated if U_1 and p_1 or h_1 are known? The answer is a straightforward "yes" for (*a*), but for (*b*) the calculation of h_2 is not quite so simple. The answer is gained by considering the relationship among specific energy, specific discharge, and water depth.

$$\frac{U_1^2}{2g} + h_1 = \frac{U_2^2}{2g} + h_2 + \Delta z, \tag{4.8}$$

where the stream bottom at the upstream location is chosen as the datum and Δz is the height of the step. Restating Equation 4.8 in terms of specific energy,

$$E_1 = E_2 + \Delta z = H. \tag{4.9}$$

Because this is a frictionless flow, the total energy H must be constant. However, Equation 4.9 shows that specific energy will not be constant if the elevation of the channel bottom changes.

We begin by considering the case in which the flow upstream of the step is subcritical. Presuming that its flow velocity and depth are known, we can compute the upstream specific energy,

$$E_1 = h_1 + \frac{U_1^2}{2g} \tag{4.10}$$

and we can locate E_1 on a specific energy diagram because the specific discharge also is known (Figure 4.5):

$$q_w = U_1 h_1. \tag{4.11}$$

From Equation 4.9 we next can find E_2, the specific energy over the step; it is equal to E_1 minus the height of the step, Δz. As the flow moves over the step it gains potential energy at the expense of specific energy. We can make this subtraction directly on the energy axis of the diagram. We now have the specific energy over the step, E_2, but we

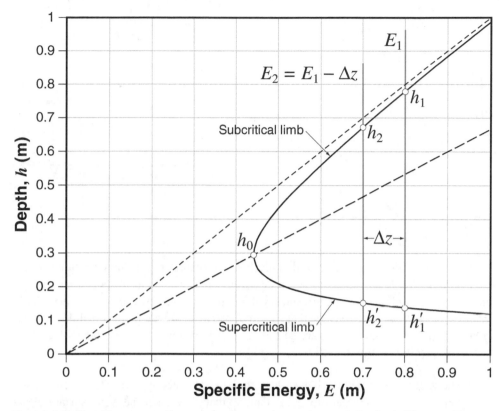

Figure 4.5 Specific energy diagram for flow over a step; $q_w = 0.5$ m^2 s^{-1}. The specific energy at station 1, 0.8 m, is represented by the vertical line to the right. The specific energy at station 2 is 0.1 m less than that at station 1 because a step up in the bottom has increased the potential energy, thereby decreasing the specific energy. E_2 is shown as a vertical line on the diagram, 0.1 m to the left of (less than) E_1.

still have to determine which of the two depths possible for the single values of E_2, h_2 or h_2' is correct.

The proper choice of depth is best argued from a consideration of the specific energy diagram, Figure 4.5. Regardless of the depth the flow establishes over the step, the change must proceed along the same specific discharge curve, because q_w is constant. Thus, as the flow moves up the step with decreasing specific energy, the depth decreases, first to h_2, and then perhaps to h_2'.

To go from h_1 to h_2 requires only a rise in the bed of elevation z. For the flow finally to reach h_2', it must pass through the minimum specific energy at depth h_0, and then proceed to h_2'. But to go from h_0 to h_2' would require an upward step of a certain critical height followed by a downward step such that the final height of the step is less than the initial height (Figure 4.6). It is obvious then, because our step does not include this hump, that the *only* depth accessible to the flow is the first choice, h_2.

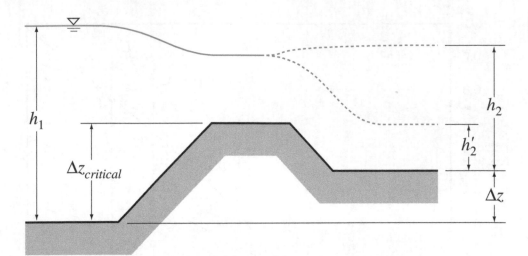

Figure 4.6 A critical step is required to reach an alternate depth. If the upstream flow (*shown to the left in the diagram*) is subcritical, the alternate depth h_2' can be reached only if the flow follows the specific energy diagram around the "nose" of the curve. This means that the flow must be critical at some point. The only way that critical flow can be created under the conditions we are considering is by having a step sufficiently high to cause the specific energy to drop to E_0.

We now know the water depth over the step, h_2, and the height of the step, z, but we still don't know if $h_2 + \Delta z$ is larger or smaller than h_1; that is, whether the water surface elevation increases or decreases over the step (see Figure 4.4). To resolve this question, Equation 4.8 can be rearranged in the form:

$$h_1 - (h_2 + \Delta z) = \frac{U_2^2}{2g} - \frac{U_1^2}{2g}.$$

We already have found from Figure 4.5 that h_2 is less than h_1; therefore, U_2 must be greater than U_1 if the discharge is constant, and so,

$$\frac{U_2^2}{2g} - \frac{U_1^2}{2g} > 0.$$

Thus, $h_1 - (h_2 + \Delta z) > 0$, or

$$h_1 > h_2 + \Delta z. \tag{4.12}$$

So, in a subcritical flow the water surface *drops* and velocity *increases* as water flows over the step. This conclusion explains why canoeists are wary of sudden changes in stream elevation, as they often mean that large rocks are below.

Note that if the upstream flow is supercritical with specific energy E_1 and depth h_1' the specific energy will still decrease as the flow crosses the step. However, in this case the decrease in E is accompanied by an increase in depth from h_1' to h_2', and, therefore, a decrease in velocity because the discharge remains constant (see Figure 4.5).

It is also important to stress that if the step height, Δz, is greater than the difference between E_1 and the specific energy of the critical flow (Figure 4.5), E_1 would not be sufficient to allow for the flow of the specific discharge q_w across the step. In these conditions it is observed that the step modifies the flow and induces an increase in E in the surroundings of the step, thereby allowing the flow to occur across the step with minimum specific energy (i.e., in critical conditions). In this case the step functions as a broad crested weir, as explained in Section 4.3.

4.2.2 Flow criticality

From the example of the flow over the step we have shown that for a single value of the specific discharge, q_w, and specific energy, E, there generally are two possible depths. It is also clear from the specific energy diagram, however, that there is a minimum specific energy, E_0, which has only a single corresponding depth, h_0 (Figure 4.5). This implies that for any value of specific discharge, there is a depth at which specific energy $E = h + U^2/(2g)$ is a minimum. Note, however, that this does not mean that either the depth or the velocity is at a minimum value. We can derive a relationship between the velocity and depth at this condition of minimum specific energy as follows:

$$E = h_0 + \frac{U_0^2}{2g} = h_0 + \frac{q_w^2}{2gh_0^2}.$$

(4.13)

At E equal to a minimum,

$$\frac{dE}{dh} = 1 - \frac{q_w^2}{gh_0^3} = 1 - \frac{U_0^2}{gh_0} = 0$$

(4.14)

or

$$U_0^2 = gh_0$$

(4.14)

and

$$\frac{U_0}{\sqrt{gh_0}} = 1.$$

(4.15)

This last equation states that when the mean velocity is equal to $\sqrt{gh_0}$, the specific energy is a minimum.

The line defining critical flow conditions divides the specific energy diagram into two regimes (Figure 4.5). As noted in the previous section, for any value of specific discharge, the regime above this line has slow, deep flow relative to the flow below the line. This upper regime is called **subcritical flow** and the lower regime is referred to as **supercritical flow**. **Critical flow** occurs along the dividing line of these two regimes.

For any value of q_w, the specific energy will be at a minimum when the flow is critical, that is, when,

$$\mathbf{F} = \frac{U}{\sqrt{gh}} = 1. \tag{4.16}$$

The dimensionless parameter $\mathbf{F}$ is called the **Froude number**. If $\mathbf{F} > 1$, the flow is supercritical, and if $\mathbf{F} < 1$, the flow is subcritical. When $\mathbf{F} = 1$, the flow is critical.

4.3 Discharge Measurements Using Control Structures

How does a flow come to be either subcritical or supercritical? One of the obvious ways that flow can be controlled is by using some type of structure. We intuitively expect that a dam will cause a river to back up and that the resulting flow within the reservoir will be deep and subcritical. On the other hand, the rapid, shallow flow of water down the face of a steep spillway must be supercritical. For the case of frictionless flow considered so far, subcritical flow can be thought of as controlled by a structure downstream (e.g., the dam is downstream of the subcritical region) and supercritical flow controlled by a structure upstream.

The control that a structure, such as a dam, has on flow through a channel can be exploited for making measurements of channel discharge. Upstream of the dam the flow is subcritical. The flow over the spillway is supercritical. Therefore, somewhere over the dam crest the flow must reach the critical condition where $U = \sqrt{gh}$, that is, the Froude number equals 1. In fact, from the discussion of critical flow in the previous section, we know that we can force any channel flow to become critical by placing in the flow a properly designed step across the channel. An artificial obstruction like a step or dam over which all the water in a channel must flow is referred to as a **weir**. There are various types of weirs, including sharp-crested, broad-crested, and notched weirs. Here we will focus on broad-crested weirs.

Consider the flow over the broad-crested weir shown in Figure 4.7. Because we are interested in only a short reach of flow we can expect that frictional energy losses will be small enough that we can use the Bernoulli equation. We also will assume that the flow is steady on the time scale required for making a measurement. Using the Bernoulli equation to relate the flow over the weir crest (position 0) to the flow upstream of the weir (position 1), we obtain:

$$z_1 + h_1 + \frac{U_1^2}{2g} = z_0 + h_0 + \frac{U_0^2}{2g} = H. \tag{4.17}$$

The upstream flow is subcritical, so $U_1^2/gh_1 < 1$. In fact, generally the flow just upstream of the weir will be relatively slow and deep, in which case we can make the assumption that the Froude number is small enough that $U_1^2/2g \ll h$. In this case, we can neglect the

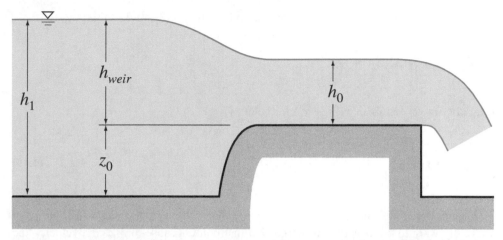

Figure 4.7 A broad-crested weir. The structure causes water in the channel to "back up" until the flow over the weir is critical. Because we know how velocity and water depth are related for critical flow, measurement of the height of the water in the pool behind the weir can be used to determine the discharge in the channel.

velocity term on the left side of Equation 4.17. If we choose our datum to be at z_1, the level of the channel bottom upstream of the weir, Equation 4.17 reduces to

$$h_1 = z_0 + h_0 + \frac{U_0^2}{2g} = H$$

or, if we define h_{weir} as the height of water upstream of the weir relative to the weir crest, i.e., $h_{weir} = h_1 - z_0$, then

$$h_{weir} = h_0 + \frac{U_0^2}{2g}. \qquad\qquad (4.18)$$

At some point over the weir crest, the flow must be critical. At this point $U_0^2 = gh_0$, or $U_0^2/2g = h_0/2$. Substituting this into Equation 4.18 gives:

$$h_{weir} = h_0 + \frac{h_0}{2} = \frac{3}{2}h_0,$$

or

$$h_0 = \frac{2}{3}h_{weir}.$$

If we then substitute this expression for h_0 into Equation 4.18, we find that:

$$h_{weir} = \frac{2}{3} h_{weir} + \frac{U_0^2}{2g}.$$

Rearranging to obtain an expression for velocity,

$$U_0 = \left(\frac{2}{3} g h_{weir} \right)^{1/2}. \qquad \qquad (4.19)$$

Equation 4.19 tells us the velocity at a particular point over the weir crest, but what we really want to know is the discharge Q at that point because, unlike the velocity, the discharge over the weir will be the same as that upstream and downstream of the weir owing to conservation of mass. To obtain discharge from velocity, we need to multiply the velocity by the cross-sectional area of the flow. Over the weir where the flow is critical, the depth is h_0 and the across-channel width of the weir crest is a constant we will call w_c. Thus,

$$Q = U_0 w_c h_0 = \left(\frac{2}{3} g h_{weir} \right)^{1/2} w_c \frac{2}{3} h_{weir} = \left(\frac{8}{27} g \right)^{1/2} h_{weir}^{3/2} w_c \qquad (4.20)$$

Equation 4.20 is the discharge equation or weir formula for a broad-crested weir. In general the flow through a weir can be expressed as in Equation 4.20 in the form

$$Q = C_q \sqrt{2g}\, h^{3/2} w_c,$$

where C_q is a (dimensionless) discharge coefficient. Using Equation 4.20 we find that in the case of the broad-crested weir $C_q = 0.385$. Other kinds of weirs have similar discharge equations but with a different discharge coefficient. For example, in the case of a sharp-crested weir C_q is bigger (0.41), and for the typical dam spillway C_q can be as high as 0.49–0.50. The important thing to note is that we can use a measurement of h_{weir} (a simple task) to obtain a measure of Q (an otherwise relatively difficult task).

A broad-crested weir is one example of how a control structure can be used to measure discharge. Other measuring devices that work on a similar principle include the Parshall flume in which the flow is forced to go through a width constriction that causes the flow to become critical. Downstream of a control structure, such as the spillway of a dam (or waterfall), the flow in a channel will once again become subcritical (if the channel bed has a relatively gentle slope), this time without the benefit of a control structure. To explain how the channel itself exerts control requires that the influence of friction be considered.

4.4 The Effect of Bed Roughness

Until now, we have considered specific energy in channels with no slope. In "real" channels, the bottom elevation drops in the downstream direction. First consider the implication of assuming frictionless flow in a channel with a sloping bottom (Figure 4.8). If the total head, H, is constant and the elevation of the stream bed above datum, z_b, is decreasing, then the specific energy must increase continually since $H = E + z_b$. This cannot be a general conclusion because cases of very nearly uniform flows (E, in particular, not changing in the downstream direction) are observed in channels with sloping beds. The alternative is to consider that H is not constant but decreases in the downstream direction. That is:

$$\frac{dH}{dx} = -S_f, \qquad (4.21)$$

where S_f is the slope of the total energy line as indicated in Figure 4.8. S_f is known as the **friction slope** because it is frictional head loss that causes H to decrease. Note that Equation 4.21 can be written as:

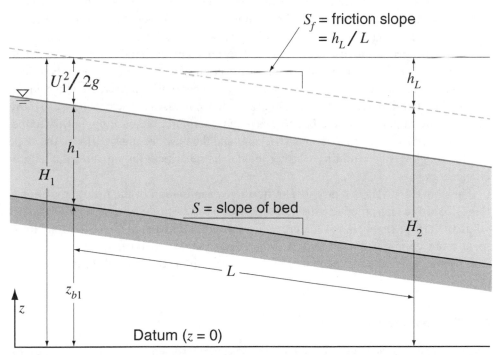

Figure 4.8 Friction along the bed causes head loss. The schematic diagram shows the case of uniform flow in a channel, a condition often approximated in streams. If water depth and water velocity remain constant in the downstream direction and the elevation of the bed decreases, the Bernoulli equation indicates that energy is lost to friction as the water flows along the channel.

$$\frac{dE}{dx} + \frac{dz}{dx} = -S_f. \tag{4.22}$$

As dz/dx is the negative of the bottom slope, S, Equation 4.22 can be written:

$$\frac{dE}{dx} = S - S_f. \tag{4.23}$$

For the case of uniform flow (u, h, and hence E unchanging downstream), Equation 4.23 implies that the friction slope is equal to the bottom slope. Does the constant specific energy of uniform flows mean that these flows do not dissipate energy? Even though E is constant, the total energy $H = E + z$ decreases in the flow direction. The fact that E is constant and that friction slope is equal to the bed slope means that uniform flows dissipate as much energy as is provided by the bed slope. In other words, if the flow is uniform the energy dissipation in a channel is the same as the energy added by the change in bed elevation between the upstream and downstream sections of the channel. Another way to look at uniform flows is that they exhibit a balance between friction forces and the component of gravity in the flow direction.

Equation 4.23 can be used to infer how flow is controlled in a river channel. To picture this, imagine a laboratory experiment in which water is caused to flow over a bed of sand and create a channel. The inflow might be controlled by a reservoir of water and the upstream specific energy would then be determined by the height of the water in that reservoir. We might suppose that initially the friction slope would exceed the bed slope and thus that $dE/dx < 0$. This means that the specific energy would decrease in the downstream direction through bottom friction. If the flow possesses sufficient energy to move the sand, we might expect the bed to erode and, therefore, increase the bottom slope, S. As S approaches S_f, dE/dx approaches zero and the flow becomes uniform, the slope and the roughness determining whether the flow is subcritical for a given value of specific energy.

The above analysis is really just an intuitive examination of the meaning of Equation 4.23. In actual practice, this equation can be used to determine how depth and velocity vary in channels but the procedures are rather involved (see Chanson, 2004, for a more complete discussion). A simpler approach is to investigate the relationship between velocity and head loss in a manner similar to that used for flow through pipes.

4.5 Channel Flow Equations

Friction slope is the head loss per unit length and can be represented as h_L/L (see Figure 4.8). In Chapter 3, an equation for the head loss in a pipe was presented:

$$h_L \equiv f \frac{L}{D} \frac{U^2}{2g}, \tag{3.30}$$

where D is the pipe diameter and f is the friction factor, a (dimensionless) function of the Reynolds number and surface roughness. A similar head loss equation can be written for a free surface flow, but because there is no "diameter" of the channel, an alternate measure of size must be used. This measure is the **hydraulic radius** and the head loss equation becomes:

$$\frac{h_L}{L} = f \frac{1}{4R_H} \frac{U^2}{2g},$$

(4.24)

where R_H is the hydraulic radius and is defined as the ratio of the cross-sectional area of the flow to the channel wetted perimeter. The numerical factor 4 is introduced because for a pipe with a circular cross section, the hydraulic radius is equal to the diameter divided by 4. For a rectangular channel (Figure 4.9) the cross-sectional area is wh, and the wetted perimeter is $2h+w$. Thus,

$$R_H = \frac{wh}{2h+w}.$$

(4.25)

If the width is much greater than the depth, $w \gg h$, then

$$R_H \approx \frac{wh}{w} = h,$$

(4.26)

and so for wide streams and rivers it often is appropriate to approximate the hydraulic radius by average channel depth. (For example, if $w = 40h$, the error in using $R_H \approx h$ is 5%.)

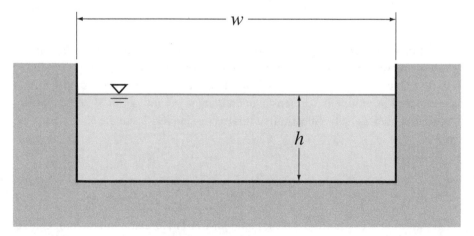

Figure 4.9 Cross section of a rectangular channel. The hydraulic radius is the ratio of the area of flow (wh) to the wetted perimeter ($w + 2h$). For a very wide rectangular channel, the wetted perimeter is very close to the width of the channel so the hydraulic radius approximately equals the water depth.

Note that if we use hydraulic radius to define the flow Reynolds number $\mathbf{R} = \rho U R_H / \mu$, the critical Reynolds numbers below which the flow is laminar and above which the flow is turbulent are now 500 and 1000, respectively. It follows that almost all flows in channels are turbulent! For example, a "shallow, slow" flow intuitively might be thought to be "laminar" when, in fact, it is not, a mistake made quite often. As a concrete example, suppose water at 20°C flows in a wide channel with a depth of 0.20 m and a mean velocity of 0.30 m s^{-1}. This flow has a Reynolds number of about 6000, well into the turbulent range.

Rearranging Equation 4.24, we can solve for the velocity as a function of the slope, hydraulic radius and friction factor,

$$U = \sqrt{8g/f}\sqrt{SR_H} = C\sqrt{SR_H}, \tag{4.27}$$

where we have assumed uniform flow so that $h_L/L = S$. However, if the flow is not uniform, this equation and its subsequent variations (e.g., Equation 4.31) still can be applied provided that the friction slope rather than the bottom slope is used. Equation 4.27 was originally proposed by Antoine Chézy in the late 1700s and the equation is referred to as the **Chézy equation**. The constant, C, called the **Chézy number**, has the same units as $g^{1/2}$.

Assuming that the value of C is known, the Chézy equation permits a very simple computation of the mean velocity (average over the cross section) for steady, uniform flow. Unfortunately, the "constant" C depends on the hydraulic radius and the bottom roughness. Several equations were proposed after Chézy's initial effort, many of which sought to replace the C with a parameter that is a function of roughness only. In 1891, another Frenchman, A. Flamant, proposed,

$$C = \frac{kR_H^{1/6}}{n}, \tag{4.28}$$

where n is a function of only roughness and is today called the **Manning coefficient**; the same expression was separately proposed by others and, in fact, in Europe n is often known as the Gauckler-Manning coefficient. Here we adopt the convention that n is dimensionless and introduce k, a dimensional constant ($k = 1$ m$^{1/3}$ s^{-1} in SI units) to maintain dimensional homogeneity. Substituting this expression for C into the Chézy equation we have,

$$U = \frac{k}{n}R_H^{2/3}S^{1/2}. \tag{4.29}$$

Typical values of n range from 0.02 for smooth-bottom streams to 0.075 for coarse, overgrown beds (Table 4.1). For a wide range of natural stream and river channels in the United States, Barnes (1967) reports values of n from 0.024 to 0.075. In upland streams, the effective roughness is controlled by the riffle-and-pool structure and not by the size of the bed material. In these cases Manning's n can be very large (Beven et al., 1979).

Table 4.1. Roughness coefficients for channels

Channel form	Manning's n
Laboratory flume (glass)	0.01
Smooth concrete	0.012
Unlined earth canal in good repair	0.02
Natural streams	0.024–0.075
Very steep upland channels	0.075–?? (>1)

Note: For natural streams, the report by Barnes (1967) is an excellent reference.

As an example of how bottom roughness affects stream velocity, consider two streams that are identical except that for one n equals 0.025, and for the other n equals 0.075. Assume that the slope is 0.0006 m/m, and the hydraulic radius is 3.0 m. In this example,

$$U_1 = \frac{1.0}{0.025}(3.0)^{2/3}(0.0006)^{1/2} = 2.35 \text{ m s}^{-1}$$

$$U_2 = \frac{1.0}{0.075}(3.0)^{2/3}(0.0006)^{1/2} = 0.79 \text{ m s}^{-1} \tag{4.30}$$

This variation in roughness, equivalent to the difference between a stream with a sand bed and a stream with a coarse, weedy bed, can have a significant effect on velocity, and thus discharge. As noted above, for a wide rectangular channel $R_H \approx h$, and Equation 4.29 can be written:

$$U = \frac{k}{n}h^{2/3}S^{1/2}. \tag{4.31}$$

Multiplying both sides of Equation 4.31 by depth gives an expression for specific discharge:

$$q_w = Uh = \frac{k}{n}h^{5/3}S^{1/2}. \tag{4.32}$$

This equation suggests that channel characteristics such as bed roughness and channel slope control the flow in the sense that the relationship between q_w and h is established once n and S are specified.

4.5.1 Velocity distribution in open channels

Manning's equation provides a method for estimating channel mean velocity U given depth or hydraulic radius, channel slope, and an estimate of Manning's roughness coefficient. As we found for flow through pipes (Section 3.6), once we include friction in the description of the flow, the velocity is not uniform throughout the cross section. Friction with the channel bottom requires that the fluid velocity adjacent to the bottom be zero. Clearly the velocity is non-zero away from the bottom, so velocity must vary with distance from the channel bottom. We found that for laminar pipe flow, the shape of the velocity profile is parabolic (see Figure 3.9). The velocity profile for a laminar channel flow also would be parabolic, with the maximum velocity at the channel surface (rather like cutting the pipe in half). The presence of turbulence in the flow greatly enhances the mixing of fluid in a flow compared to the mixing resulting from molecular viscosity. As a result, the velocity distribution in a turbulent channel flow is logarithmic rather than parabolic (Figure 4.10).

The logarithmic velocity distribution of a turbulent channel flow is given by the Karman-Prandtl equation:

$$u(z) = \sqrt{gR_H S}\left[2.5\ln\left(\frac{z}{k_r}\right) + 8.5 \right],$$
(4.33)

where k_r is a channel roughness parameter. In writing the equation in this form we have assumed that the channel bottom is hydraulically rough, a good assumption for most stream beds. For channel beds of nearly uniformly-sized sand or gravel, k_r is approximately equal to the diameter of the bed material. For more complicated stream beds the problem of determining k_r is a difficult one and often requires measurements of flow velocity in the stream reach of interest.

4.6 Measuring Flow in Natural Channels

Often we don't know enough about a stream to make even good estimates of flow velocity or discharge without making some direct measurements. Channel cross section and slope can be measured using meter sticks and surveying equipment. Channel roughness cannot be measured directly. Instead, it must be determined by measuring all of the other variables in Manning's equation or the Karman-Prandtl equation and rearranging the equations to provide an expression for n or k_r. To do this requires measurements of velocity in the channel using a current meter. Current meters measure velocity at a point. How can we most efficiently use point measurements to estimate discharge?

Mean velocity in a channel is the sum, or integral, of the velocities at each point in a cross section divided by the cross-sectional area:

$$U = \frac{1}{A}\int_0^w \int_0^h u(y,z)\,dz\,dy.$$
(4.34)

The Karman-Prandtl equation tells us how velocity varies in the vertical direction (z), but there are no generalizations for the variation of velocity in the cross-stream direction (y). If the channel is wide and rectangular, then we can reasonably assume that the velocity is the same at each point across the channel. In this case integrating across the channel is equivalent to multiplying by channel width, so that Equation 4.34 simplifies to:

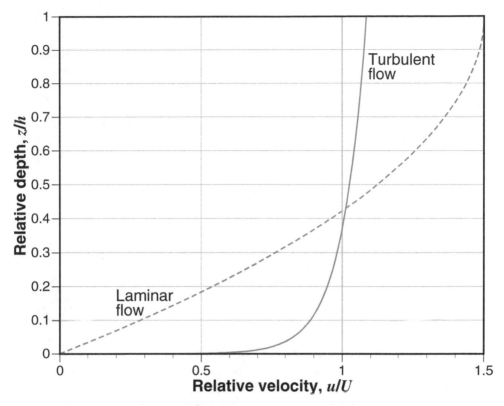

Figure 4.10 Velocity distribution in a channel. For laminar flow (almost never observed in natural channels), the velocity profile is parabolic, similar to the case examined for pipe flow in Chapter 3. For turbulent flow, the velocity near the bed changes much more rapidly than it does in laminar flow. The velocity profile for turbulent flow generally is described using a logarithmic equation.

$$U = \frac{1}{wh} w \int_0^h u(z)\,dz = \frac{1}{h} \int_0^h u(z)\,dz.$$

If we substitute the Karman-Prandtl equation for $u(z)$ and carry out the integration, we obtain:

$$u(z) = \sqrt{ghS}\left[2.5\ln\left(\frac{h}{k_r}\right) + 6.0\right], \tag{4.35}$$

where, because we are assuming the channel is wide and rectangular, we can take $R_H \approx h$. We can deduce an interesting and useful result by comparing Equations 4.33 and 4.35. The velocity in a channel varies from zero at the channel bed to a maximum value at the channel surface (Figure 4.10). The mean velocity is some intermediate value and thus there will, in general, be some level z_U at which the velocity $u(z_U)$ is equal to U. Setting

Equations 4.33 and 4.35 equal to each other and solving for the value of z_U gives $z_U = 0.37h$ (Figure 4.10). This means that all we have to do if we want to estimate mean velocity is to measure the velocity at a level roughly $0.4h$ above the bottom or $0.6h$ below the surface.

Of course, most natural channels are not rectangular in cross section so that estimating U with Equation 4.33 letting h=average depth cannot be expected to be very accurate. However, we still can make use of the results for rectangular cross sections by dividing a more complicated channel cross section into a number of rectangular subsections as illustrated in Figure 4.11. For each subsection, the width and depth can be measured with a tape and meter stick and the mean velocity can be approximated as the measured velocity at a depth $0.6h$ below the surface. Taking this approach, we can re-write Equation 4.34 as:

$$U = \frac{1}{A}\sum_{i=1}^{N} w_i h_i U_i,$$
(4.36)

where N is the number of subsections, $U_i = u(z = 0.4h_i)$, and

$$Q = UA = \sum_{i=1}^{N} w_i h_i U_i.$$
(4.37)

Once we know mean velocity, slope, and depth (or hydraulic radius), we can determine the value of Manning's n using Manning's equation. If we did this for a number of different discharges in a given stream reach we could estimate a characteristic value of n for the reach that could then be used to predict the mean velocity and discharge for depths other than those for which direct measurements were made. Leopold and Wolman

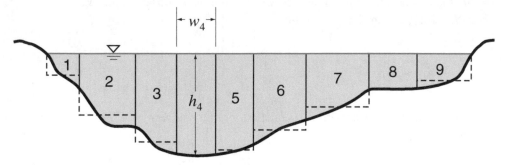

Figure 4.11 Method for determining discharge in a stream. The stream is subdivided into a number of rectangular elements. A current meter is used to measure the speed of the flow at a distance $0.4h$ from the stream bottom in each rectangle. The water velocity at this depth is approximately the average velocity for that segment, assuming that the logarithmic velocity profile for turbulent flow is valid. Discharge is calculated by multiplying the average velocity for each rectangle by the area of that rectangle and summing across the stream. (For additional detail on this method, see www.rcamnl.wr .usgs.gov/sws/SWTraining/WRIR004036/VAMethod/VelocityAreaMethod.swf)

(1957) derived an empirical relationship between the friction factor, water depth, and D_{84}, the size of bed material of which 84% of bed material is finer (see also Leopold, 1994, for other examples of data supporting the empirical relationship). In terms of Manning's n, the empirical relationship is (assuming that $R_H = h$)

$$\frac{1}{n} = \frac{\sqrt{8g}}{h^{1/6}} \left[2.0 \log_{10} \left(\frac{h}{D_{84}} \right) + 1.0 \right].$$ (4.38)

Measured and calculated (using Equation 4.38) values of n are shown in Figure 4.12 for a site on Brandywine Creek in southeast Pennsylvania. Manning's n generally decreases with increasing discharge at a site, at least for flows in which the river stays within its banks. This is at least in part because the bottom roughness, indicated by some measure of size of the bottom material, e.g., D_{84}, protrudes farther up into the flow for shallow flows than for deep flows, and thereby increases frictional energy loss.

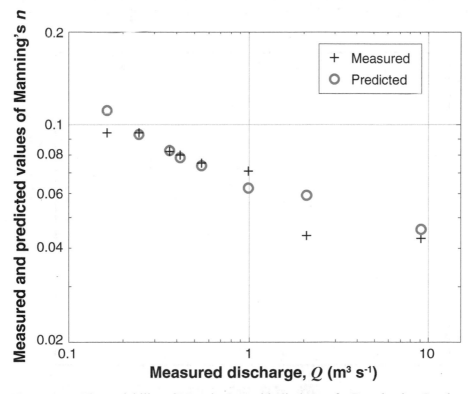

Figure 4.12 The variability of Manning's n with discharge for Brandywine Creek at Cornog, Pennsylvania, can be determined from the values of measured flow, water depth, and other factors reported by Wolman (1955). Values calculated with Equation 4.38 used $D_{84} = 0.15\,\text{m}$ and assumed that $R_H = h$.

Manning's equation, in conjunction with an estimated value of n, can be used to infer stream velocity and discharge in the absence of direct hydraulic measurements. For example, Manning's equation provides a means of estimating velocity and discharge under very high flow conditions when it is not possible to make flow measurements directly.

4.7 Concluding Remarks

There are many important applications of the theory developed above and some of these will be explored in subsequent chapters. Here we briefly consider examples of how the relationships derived above can be useful in interpreting information about flows in channels.

Recall that our consideration of Manning's equation suggested that discharge should be related to the 5/3 power of stream depth for a stream with a rectangular cross section (Equation 4.32). Streams rarely have strictly rectangular cross sections, of course, but our analysis could be used to suggest that a power relationship, perhaps with an exponent different from 5/3, might be useful for describing stream data. In fact this often is the case. For Brandywine Creek near Cornog, an equation $Q = 28.26h^{2.26}$ fits the obser-

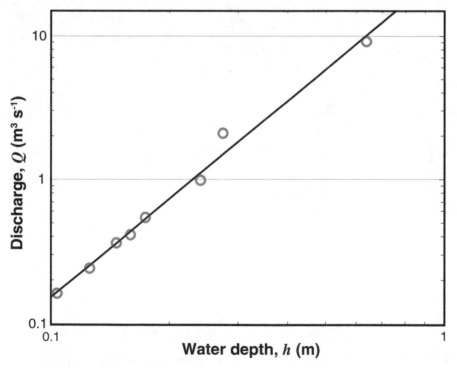

Figure 4.13 Stream discharge is related to a power of stream-water depth for many streams and rivers. For Brandywine Creek at Cornog, the exponent in the empirical relationship between Q and h is 2.26.

vations of discharge and depth reasonably well (Figure 4.13). As we will see in the next chapter, such power relationships are used routinely in making continuous measurements of stream discharge, measurements that are essential for solving many practical and theoretical hydrological problems.

Fluvial geomorphologists use power relationships to describe the interrelationships among many streamflow variables of interest to them, not only Q and h. Consider some general observations on hydraulic variations along the longitudinal profile of a river, that is, on the changes in slope, width, depth, and velocity in the downstream direction. Leopold et al. (1964) and Leopold (1994) present data of this type for a number of rivers. Relationships tend to be straight lines on logarithmic axes (i.e., to be power relationships) for most river systems; the Brandywine Creek basin is an example (Figure 4.13). Of course, slope decreases as a river progresses from the headwaters to the sea and discharge generally increases because of the steady increase in drainage basin area. The observed increase in width and depth also seems intuitively plausible. Velocity also increases in the downstream direction.

There are many interrelated changes that occur among hydraulic variables in streams and these must be considered collectively in explaining the observed changes in mean velocity of a river system. Smith (1974) assumed that Manning's equation held for a

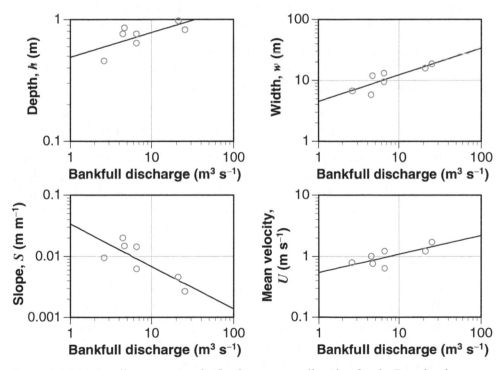

Figure 4.14 Hydraulic parameters in the downstream direction for the Brandywine Creek basin.

Data from Miller et al. (1971).

hypothetical river system with constant roughness (Manning's n constant in the downstream direction), that discharge increased directly with drainage area, and that the basin was eroding at a constant rate. He used *bankfull discharge*, the discharge in the stream when it is filled right to the top of the channel, in his analysis. Using empirical equations for sediment transport in the river, he then showed that slope should be proportional to discharge to the negative 0.18 power, that velocity should be proportional to the 0.09 power of discharge, that depth should be proportional to the 0.27 power of discharge, and that channel width should be proportional to the 0.64 power of discharge. The corresponding powers for the Brandywine basin (the slopes of the lines in Figure 4.14) are −0.69, 0.30, 0.20, and 0.43. The magnitudes of the powers for this particular basin do not correspond all that closely with Smith's theory, but again the form of the relationship is similar and the hydraulic interpretations help provide a consistent conceptual basis for explaining natural phenomena. These ideas also are useful in assessing the potential impact of modifications on a stream channel (e.g., how stream *channelization* affects all of the pertinent variables in a channel as a result of changes in slope).

4.8 Key Points

- Many important flow phenomena in rivers can be described in a reasonably straightforward manner by considering the equation of continuity, the Bernoulli equation, and the friction factor relationships developed in Chapter 3. {Section 4.1}

- The Bernoulli equation for flow in a frictionless channel can be written $U^2/2g + h + z_b = H$, where U is mean velocity [L T^{-1}], h is water depth [L], and z_b is the elevation of the channel bottom above datum [L]. {Section 4.2}.

- Specific energy E [L] is the energy per unit weight of water in the stream relative to the stream bottom: $E = U^2/2g + h$. Total energy, H [L], is therefore $E + z_b$. {Section 4.2}

- For channels with a rectangular cross section, a useful relationship among specific energy, specific discharge, and water depth follows from the definition of specific energy: $E = q_w^2/2gh^2 + h$, where q_w is the specific discharge, the discharge per unit width of channel [L^2 T^{-1}]. For given values of E and q_w, this equation is cubic in h giving three mathematically allowable depths of water for the given conditions, only two of which are positive. These two physically meaningful depths are called the alternate depths. {Section 4.2}

- For any specific discharge, there is some minimum value of specific energy possible. This minimum occurs at critical flow, with the defining condition being that the Froude number **F** [dimensionless] is one: $U_0/\sqrt{gh_0} = 1$. For Froude numbers below one, the flow is subcritical, slow and tranquil. For Froude numbers greater than one, the flow is supercritical, rapid and torrential. {Section 4.2.2}

- Flow measurement devices for water in channels can be devised on the basis of the known relationship between velocity and water depth at critical flow. An example is a broad-crested weir for which the principles developed in this chapter yield a way to estimate discharge from a measurement of the height of water above the weir crest: $Q = (8g/27)^{1/2} h_{weir}^{3/2} w_c$, where h_{weir} is the distance from the top of the weir to the water surface in the pool behind the weir and w_c is the width of the weir crest. {Section 4.3}

- For steady, uniform open channel flows, the friction slope is equal to the bottom slope, and depth, velocity, and specific energy are constant. {Section 4.4}

- The hydraulic radius R_H [L] of a stream channel is defined as the ratio of the cross-sectional area of flow to the wetted perimeter. {Section 4.5}

- Using the friction factor discussed in Chapter 3, the Chézy equation can be derived: $U = \sqrt{8g/f} \sqrt{SR_H} = C\sqrt{SR_H}$, where C is the Chézy number [$L^{1/2}\,T^{-1}$], S is channel slope [dimensionless], and R_H is the hydraulic radius. {Section 4.5}

- Manning's equation, $U = kR_H^{2/3}S^{1/2}/n$, is an expression relating mean velocity to slope and hydraulic radius that is very widely used. Values of the roughness parameter, n, have been tabulated for a range of channel conditions. {Section 4.5}

- The vertical profile of velocity in a river channel often is assumed to be logarithmic by virtue of a derivation due to Prandtl and von Karman. The mean velocity for flows with a logarithmic velocity profile occurs at a distance of $0.4h$ above the bottom. The common method of gaging streams using current meter measurement relies on this approximation. {Sections 4.5.1 and 4.6}

- Manning's n generally decreases as a function of discharge in streams, at least for flows within the stream banks. This decreasing roughness can be thought of intuitively as due to the relatively larger size of the bottom roughness (e.g., pebbles in an upland stream) with respect to stream depth at low flows versus the relatively smaller size at high flows. {Section 4.6}

4.9 Example Problems

Problem 1. Water flows in a 3.00-m wide rectangular channel. The water depth is 1.50 m and the discharge is 1.50 m³ s⁻¹. The channel bottom drops smoothly by 0.100 m over a short distance (a step down in the bottom) with no head loss or change in the width of the channel.

A. Calculate the specific discharge (m² s⁻¹) and specific energy (m) at the upstream station.

B. Calculate the specific discharge (m² s⁻¹) and specific energy (m) at the downstream station.

C. Calculate the water depth (m) at the downstream station.

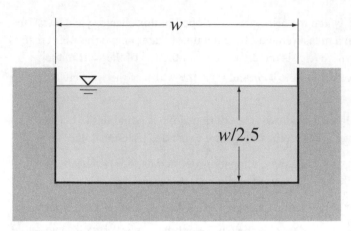

Figure 4.15 Canal cross section for Problem 2.

Problem 2. A discharge of 2.0 m³ s⁻¹ is carried in a canal with the cross section shown in Figure 4.15. The canal is 1400 m long and drops 0.50 m in elevation over that distance. Manning's n for the channel is estimated to be 0.020. What is the value of w (m) for this canal?

Problem 3. As noted in Problem 3 of Chapter 3, channels and pipes of varying geometry will be characterized by different formulations of the Reynolds number. For example, $\mathbf{R}_{pipe} = \rho UD/\mu$, where D is pipe diameter, whereas a more appropriate form for channels might be $\mathbf{R} = \rho UR_H/\mu$, where R_H is hydraulic radius. Use of a different length scale in $\mathbf{R}$ will generally alter the upper limit of $\mathbf{R}$ for laminar flow relative that that found for pipes. In this problem, we reformulate $\mathbf{R}$ using hydraulic radius as the length scale to obtain a more general form of $\mathbf{R}$ that we can use to relate the laminar-turbulent transition in pipes to other flow geometries.

A. Find an expression for the hydraulic radius R_H in terms of pipe diameter D for a pipe with a circular cross-section.

B. Rearrange the expression found in 3A to get a relationship for D in terms of R_H. Substitute this into $\mathbf{R}_{pipe} = \rho UD/\mu$ to get an expression for $\mathbf{R}$ in terms of R_H.

C. Set the expression found in 3B equal to 2000, the critical Reynolds number for pipes. Rearrange this to find the equivalent critical value for $\mathbf{R}_{RH} = \rho UR_H/\mu$.

D. Use the equation you developed in 3C to find the critical Reynolds number for the transition from laminar to turbulent flow in a relatively wide, rectangular channel (such that $R_H \approx h$).

Problem 4. Use Equation 4.32 to estimate the depth and mean velocity of a flow in a channel with a slope $S = 0.003$, width $w = 15$ m, discharge $Q = 1.0$ m³ s⁻¹, and Manning's $n = 0.075$. Does the assumption that $R_H \approx h$ seem reasonable for this flow?

4.10 Suggested Readings

Dunne, T., and L.B. Leopold. 1978. *Water in environmental planning.* San Francisco: W. H. Freeman. Chapter 16, pp. 590–660.

Leopold, L.B. 1994. *A view of the river.* Cambridge: Harvard University Press. Chapter 9, pp. 148–167.

5 Catchment Hydrology: Streams and Floods

5.1 Introduction

The principles governing flow in channels, discussed Chapter 4, provide the basis for describing the flow of water in rivers and streams. In this chapter we emphasize the response of a river to precipitation and the nature and dynamics of the resulting flood. The problems associated with floods are many and varied, and solving them is an important application of hydrology.

Flood control is an issue that is important throughout the world. Every year we read in the newspapers about damage and death caused by river flooding. Flood insurance, levee construction, and future land-use zoning are common topics of discussion in the wake of floods. Such discussions point out the range of options that must be considered in flood protection schemes. Structural approaches to flood control, such as levees, dams, and channel widening, and non-structural approaches, such as floodplain zoning and preservation of natural wetlands, must be weighed in terms of costs and benefits. The issues often cause contentious debate and rarely are settled to everyone's satisfaction.

As an example, consider the situation for Sacramento, California. The American River basin lies to the east of Sacramento. The basin area is about 5000 km^2 with headwaters in the Sierra Nevada. Three forks of the river flow through scenic canyons in the mountains, valued by white-water sports enthusiasts and hikers, and merge just above Folsom Dam. Below the dam, the American River flows between levees through downtown Sacramento. More than one million people live in the Sacramento metropolitan area, within the floodplains of the American and Sacramento rivers.

Extreme precipitation and snowmelt cause high discharges in the American River from time to time, and the city of Sacramento occasionally has suffered flood damage. After a major flood in 1850, the city of Sacramento decided to rebuild on the floodplain. A series of flood-control measures have been implemented, but the city nevertheless has remained susceptible to devastating floods (NRC, 1995a).

One of the chief protection measures for Sacramento is Folsom Dam, on the American River. Folsom Reservoir is a multipurpose facility; it is regulated for water supply, recreation, and flood control. Different purposes often dictate conflicting strategies for reservoir operation. To provide an adequate water supply through a dry period, one would want to keep a reservoir nearly full; however, to protect against floods one would want to keep a reservoir empty so floodwaters could be stored. When Folsom Dam was planned, it was supposed to protect against a flood that would be expected to occur only once in 500 years on average. A series of floods from 1955 through 1986 caused a reevaluation of the protection levels offered by the dam, reducing the expected protection offered by Folsom Dam to floods that occur once in 70 years on average. New construction begun in 2007 and scheduled to be complete in 2017 is designed to double the flood safety rating of Folsom Dam and to increase the level of protection for the Sacramento area to a 200 year flood.

The question of what options should be considered in planning for flood protection for Sacramento is not a purely technical one. Socioeconomic factors are very important and ultimately the question is one that must be answered as part of the political process. Ideally, the political decision is informed by solid technical knowledge. Thus, hydrological questions will arise. For example, the flows of water into and out of a reservoir must be understood when considering the construction of a new dam or when evaluating the operating policy of an existing dam.

How can an estimate be made of the level of protection offered by a dam? A flood-control reservoir works by storing water during high inflows to the reservoir and later releasing this water at a rate much lower than the highest flood rates. The problem of deducing how outflow from a reservoir is related to its inflows is known as **flood routing**. As will be described in this chapter, flood routing depends on some of the principles of fluid flow that were covered in earlier chapters.

In this chapter we discuss some selected elements of surface hydrology and show how the formulation and solution of surface-water problems are developed from fundamental hydraulics. We consider the propagation of flood waves through reservoirs, flood routing, and the frequency with which a flood of a given magnitude would be expected to occur. There are many facets of surface-water hydrology that are not included in this chapter. The material presented here is intended to introduce the subject, particularly with regard to river floods.

5.2 The Hydrograph

In the analysis of floods, water supply, and other subjects included in surface-water hydrology, the basic quantity to be dealt with is river discharge, the rate of volume transport of water. Both river discharge and depth (or **stage**) change with time, and an understanding of this temporal variability is a prerequisite for approaching hydrological analyses. A graph of river stage or discharge versus time at a point is called a **hydrograph** (Figure 5.1). The former is referred to as a stage hydrograph and the latter a discharge hydrograph.

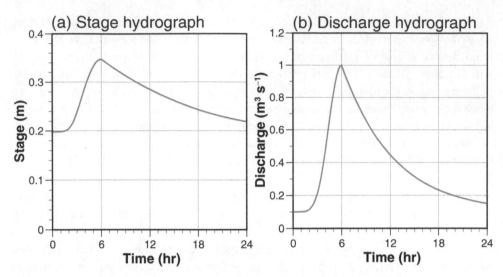

Figure 5.1 The hydrograph. River stage (*a*) and discharge (*b*) as functions of time.

The discharge hydrograph is not measured directly, but is inferred from the stage hydrograph. In the United States, the United States Geological Survey is responsible for the measurement of river stage at over 7000 gaging stations. At many of these locations, data are continuously recorded by a float gage installed inside a stilling well (Figure 5.2). As the river rises and falls, the float moves with the water level and the motion of the float is recorded. The stilling well serves both to protect the float mechanism and to dampen the rapid fluctuations of the water surface due to local disturbances and turbulence. In recent years, the float mechanism has been replaced by a pressure transducer, a device that records the pressure at the bottom of the stilling well. The hydrostatic Equation 3.6 is used to calculate water depth from the measured pressure. Radar is also used to measure stage in some rivers.

The record of water depth versus time measured at a **gaging station** is the stage hydrograph (Figure 5.1a). To construct a discharge hydrograph from these data, determinations must be made of the river discharge for various values of flow depth or stage. By measuring stream velocity at numerous positions across the width of the stream, the discharge at a given stage can be determined as described in Section 4.6. These measurements are generally made with a current meter either attached to a wading rod or suspended from a cable, depending upon the depth and discharge of the river.

After discharge has been measured for a range of flow depths, a **rating curve** can be constructed that relates stage to discharge. Rating curves typically are nonlinear and often are approximated using a power function (Figure 5.3). To be useful, a stream gage should be located at a point where the stage-discharge relationship will not change from year to year; such a location is called a control section. An example of a control section might be a place where the flow is constricted by a natural rock formation or by bridge abutments. A station where the cross section changes almost continually because of streambed erosion or deposition would not provide good control.

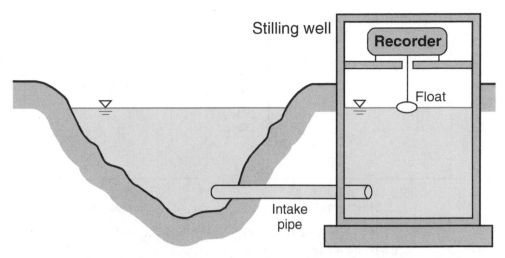

Figure 5.2 Schematic diagram of a stream gaging station used to record flow depth (stage) as a function of time. The resulting stage hydrograph can be converted to a discharge hydrograph using a rating curve.

After the rating curve for a gaging station is determined, it can be used to convert each value of stage to discharge. Consider, for example, the stage hydrograph shown in Figure 5.1a. The stage peaks at 0.35 m (at time = 6 hr). Using the rating curve shown in Figure 5.3, the corresponding peak discharge is $Q = 76.5(0.35 \text{ m})^{4.1} = 1.0$ m^3 s^{-1} (Figure 5.1b). In this way, a continuous measurement of river stage is used, in conjunction with an established rating curve, to determine discharge as a function of time. Almost all discharge hydrographs are determined in this manner.

A typical stream discharge hydrograph is shown in Figure 5.4. This hydrograph of Holiday Creek near Andersonville, Virginia, for a period in 1974, indicates that discharge is highly variable. During and after rainfall and snowmelt events, water moves through the catchment to the stream channel and the discharge increases. The resulting peak in the hydrograph is termed a **flood**, regardless of whether the river actually leaves its banks and causes damage. Background discharge between floods is termed **baseflow** and is supplied by inflow of groundwater. There were a number of floods on Holiday Creek within the 3-month period of the hydrograph shown in Figure 5.4. Each of these floods has a characteristic hydrograph, with a steeply rising segment followed by a less steeply falling segment as the flood passes the station. The exact nature of each hydrograph depends on watershed and storm characteristics.

The difference among hydrographs is well illustrated by the examples in Figure 5.5. Poplar Creek is a perennial stream in Tennessee. Floods occur throughout the year, but the magnitude of the discharge peaks varies seasonally in response to changing weather conditions. In contrast, the Frio River is an ephemeral stream in Texas that is dry for long periods with occasional large floods. Little Blackfoot River lies in a snowmelt-dominated catchment. The hydrograph from a stream like this is characterized by a single, extended period of high discharge in the late spring and early summer when the snow melts. Even

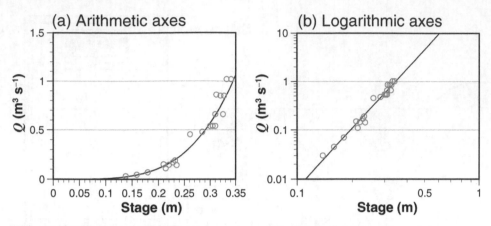

Figure 5.3 Rating curve (stage-discharge curve) for the Snake River above Montezuma, Colorado. The equation for the lines is $Q = 76.5 \, (\text{stage})^{4.1}$.
Data from Boyer (1994).

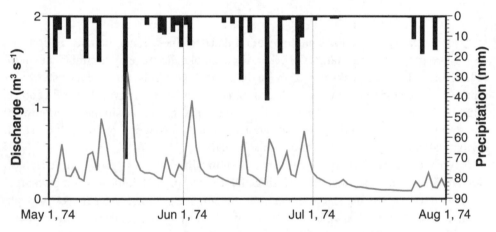

Figure 5.4 Hyetograph (*black bars, scale at right*) and hydrograph (*blue line, scale at left*) for Holiday Creek near Andersonville, Virginia.

within a single climatic zone, differences in catchment shape, geology, and vegetation will produce differing hydrographs in response to similar precipitation events.

5.3 Movement of Flood Waves

The floods shown in Figure 5.4 may be thought of as waves that propagate downstream. The way in which flood waves propagate—their speed, peak height, and peak discharge—is an important area of study in hydrology. Planning for the response of rivers to both normal and extreme rainfall depends on this understanding.

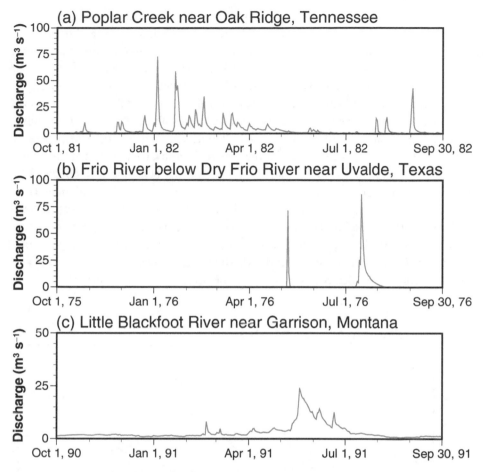

Figure 5.5 Annual hydrographs for three rivers in catchments in three climate zones illustrating the differences among streamflow in perennial (*a*), ephemeral (*b*), and snowmelt (*c*) streams.

In an "ideal" channel with a frictionless fluid, we might conceive of a flood wave traveling with no change in form from its point of origin. However, mechanical energy is lost (dissipated) in a natural channel because of the frictional resistance of the rough channel bed. Water also becomes stored in pools and backwaters and is subsequently released. As a result of friction and storage, we might expect the magnitude of a flood wave to be reduced, or attenuated, as it travels downstream. This attenuation will depend on several factors: the ability of the channel system to store and release water, the channel roughness, the amount of vegetation in the channel and its floodplain, the relative straightness of the channel, and the number of bridges and other obstructions in the path of the flood wave.

The picture is complicated further because the lateral inflow of water from tributaries and groundwater will increase the volume of water in the channel as one moves downstream. In general, discharge increases downstream as drainage area increases. An example may serve to illustrate the relative effects of drainage area and flood wave attenuation in a river or stream network. Hydrographs for the South Fork of the South River at Waynesboro, the Shenandoah River at Front Royal, and the Potomac River at Washington, D.C., for the same event are quite different (Figure 5.6). The South River is a tributary of the Shenandoah and the Shenandoah flows into the Potomac well above Washington. Perhaps the most noticeable difference among the hydrographs is the highest peak flow at Washington, the lowest peak at Waynesboro, and the intermediate peak at Front Royal. This seems to contradict the "attenuation by friction and storage" hypothesis stated above, but as we have already indicated, the behavior is complicated by the increasing drainage area in the downstream direction. The effect of increasing drainage area can be taken into account by dividing discharge by drainage area to "normalize" the discharges before they are compared. The normalized hydrographs indicate that the flood wave is indeed attenuated as it moves through the channel (Figure 5.6). The normalized hydrographs show clearly the time delay between the flood peaks and the reduction in peak discharge per unit area in the downstream direction.

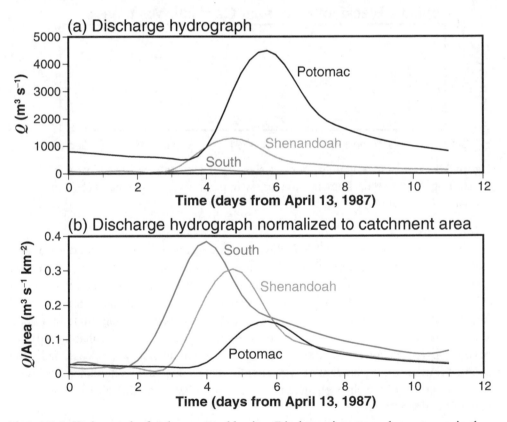

Figure 5.6 Hydrographs for three nested basins. Discharge increases downstream in the stream network due to increased drainage area (*a*). Dividing discharge by catchment area (*b*) reveals both attenuation and delay of the flood wave as it moves downstream.

5.4 Flood Routing

Flood warning and flood mitigation depend on understanding how quickly a flood crest travels downstream and how the height of the crest changes as it does so. Essentially, the problem is to be able to predict what a downstream hydrograph would be if we know the upstream hydrograph. This is known as flood routing. In addition to the upstream hydrograph, the prediction of the downstream hydrograph requires an estimate of how fast the flood wave is moving, how much water is being added by inflow, the influence of friction, and in general a complete understanding of the hydrology and hydraulics of the drainage basin. We can simplify the problem by identifying depth and velocity of the flood wave as the two most important variables. As in the channel transition problem in Section 4.3, the equations of continuity and Newton's second law provide the two equations we need to solve for the two unknowns, depth and velocity.

If a section of river channel, or reach, is taken as the control volume, then the inflow is given by the upstream hydrograph and the outflow by the downstream hydrograph. We will assume that the lateral inflow of water is small relative to the change in storage associated with the flood wave and, therefore, can be neglected. This assumption may not be valid if the reach is too long. Analyzing the movement of a flood wave is different from analyzing the steady flows considered in Chapter 4. Depth and velocity change quickly with time during a flood and, therefore, we must solve this as an unsteady flow problem.

In simplest terms, neglecting any lateral inflow, the continuity equation for unsteady flow can be written as

$$\frac{dV}{dt} = I - O, \tag{5.1}$$

where I is the known upstream (inflow) hydrograph, O is the downstream (outflow) hydrograph that we want to predict, and V is the volume of water stored in the reach. Under steady-state conditions, the storage volume is constant and $I = O$. During a flood, inflow increases and stage increases. As a result, the volume of water stored in the reach increases. The greater the change in storage in a given interval of time (dV/dt), the greater the difference between the inflow and outflow hydrographs.

Solving for the two unknowns V and O in Equation 5.1 requires a second equation. Equation 5.1 is a statement of conservation of mass. Therefore we expect the second equation to follow from Newton's second law (conservation of momentum). All hydrological problems must obey these physical laws. An example of this second type of equation is the weir equation that relates discharge to depth. Depth, in turn, is related to the volume of water in a stream reach or reservoir. Thus we see that an equation such as a weir equation can also provide a relationship between volume V and outflow O, that is, $O = f(V)$. Substituting this expression for O into Equation 5.1 produces an equation in which the rate of change of volume (dV/dt) depends on volume. Such an equation is called a differential equation.

Typically, we cannot solve the differential equation for V directly. Instead, a **numerical method** is used to solve the equation. Many problems in hydrology must be solved numerically. Basically, this involves transforming the differential equation into one or more algebraic equations that can be solved more easily. We will use this approach to calculate outflow hydrographs for a couple of interesting and informative flood routing problems.

The first step in obtaining a numerical solution to the flood routing problem is to re-write Equation 5.1 in terms of finite differences. To do this, we choose a computational time interval, Δt, separating two instants in time, t_n and t_{n+1}. Let the subscripts n and $n+1$ refer to values of each variable at these times. Equation 5.1 can then be approximated as:

$$\frac{V_{n+1} - V_n}{\Delta t} = \frac{I_n + I_{n+1}}{2} - \frac{O_n + O_{n+1}}{2}. \tag{5.2}$$

At each time step, the terms at time n are known; the terms at time $n+1$ are the unknowns we are solving for. At the initial time step ($n=1$), volume and outflow must be specified. We refer to the values of V_1 and O_1 as initial conditions. In addition to the initial conditions, we have to specify the inflow hydrograph during the interval of interest. That is, I_n and I_{n+1} are known at each time step. The unknown terms being solved for at each time step are V_{n+1} and O_{n+1}. Rearranging Equation 5.2 to put the unknowns on the left side of the equation gives:

$$\frac{2V_{n+1}}{\Delta t} + O_{n+1} = I_n + I_{n+1} + \frac{2V_n}{\Delta t} - O_n. \tag{5.3}$$

Given conditions at time t_n, the quantity on the left-hand side of Equation 5.3, $(2V_{n+1}/\Delta t) + O_{n+1}$, can be calculated. We need a second equation, as described below, to separate this into values of V_{n+1} and O_{n+1}. Once these are determined, all of the quantities on the right-hand side of Equation 5.3 are known for time t_{n+1}. The procedure is then repeated to solve for the values on the left side at the next time step.

Two flood routing problems are considered to illustrate the effect of storage on outflow hydrographs and the use of a numerical method to solve hydrological problems. The first problem is the routing of a flood through a reservoir. This problem has important applications to flood management and the maintenance of water supplies for public and agricultural use. The second is an empirical approach to the routing of flood waves through a river channel. River routing is important for flood prediction.

5.4.1 Reservoir routing

Reservoirs are an effective tool in flood control because of their large storage capacity. Hydrograph peaks downstream of a reservoir are smaller in magnitude and delayed in time compared to those on the same stream upstream of the reservoir. Understanding how a flood wave is modified as it passes through a reservoir makes it possible to design and manage reservoirs to protect downstream areas.

Reservoirs are formed when a river is dammed. The simplest dam is one designed so that when reservoir depth exceeds a certain value, water will flow over a spillway (Figure 5.7). The spillway is like a weir (see Section 4.3). There is a fixed relationship

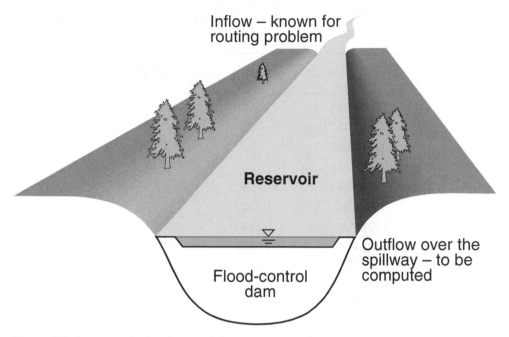

Figure 5.7 An example flood-control dam and reservoir.

between the depth of water above a weir crest (h_{weir}) and discharge over it. This relationship provides the second equation we need to complete the reservoir routing problem. Discharge over a weir is given by

$$Q = C_q \sqrt{2g}\, h_{weir}^{3/2} w_c,$$

where w_c is the width of the weir crest and C_q is an empirical coefficient. With $C_q = 0.5$, typical for a dam spillway, the weir equation gives us an expression for reservoir outflow

$$O = 2.2 h_{weir}^{3/2} w_c, \tag{5.4}$$

with O expressed in m^3 s^{-1} and both h_{weir} and w_c in m. Water depth in the reservoir h is given by the sum of h_{weir} and the height of the weir (spillway) crest, $h_{spillway}$. Therefore, Equation 5.4 allows us to relate water depth in the reservoir to the outflow over the spillway.

Water depth in a reservoir is also related to the volume of water in the reservoir V in a manner dictated by the geometry or topography of the reservoir basin. For example, a 2011 bathymetric survey was used to establish the relationship between depth and volume in Totten Reservoir, a small reservoir in southwest Colorado (Kohn 2012; Figure 5.8). This reservoir has a spillway that is roughly 20 m long at a height of 10.1 m above the reservoir bottom. Reservoir depth and volume above the spillway are provided in Table 5.1.

The application of these relationships to the reservoir routing problem is best illustrated by example. The reservoir in the example has the geometry indicated in Figure 5.8. The computational time step Δt is dictated by the resolution of the input hydrograph. For

Table 5.1. Depth and volume for Totten Reservoir in Montezuma County, CO (Kohn 2012), and related quantities needed for reservoir routing calculation

Depth, h (m)	Volume, V (m³)	h_{weir} (m)	O (m³/s)	$(2V/\Delta t)$ + O (m³/s)
10.1	3.69E+06	0.00	0.00	341
10.4	3.98E+06	0.30	7.4	376
10.7	4.30E+06	0.61	21	419
11.0	4.61E+06	0.91	38	466
11.3	4.93E+06	1.22	59	516
11.6	5.27E+06	1.52	83	570
11.9	5.60E+06	1.83	107	627
12.2	5.95E+06	2.13	137	687
12.5	6.29E+06	2.44	167	750
12.8	6.64E+06	2.74	200	814

Note: Assuming $\Delta t = 21600$ s (=0.25 days), outflow O is given by Equation 5.4, $h_{weir} = h - h_{spillway}$, and $h_{spillway} = 10.1$ m.

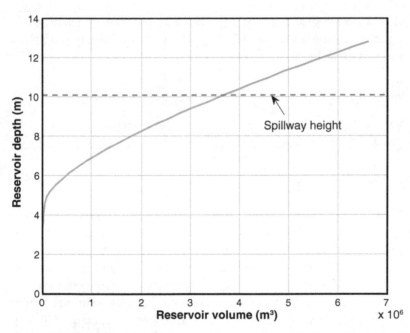

Figure 5.8 Relationship between reservoir volume and depth for Totten Reservoir in Montezuma County, CO. Such relationships are typically obtained from topographic and bathymetric surveys of the reservoir area.

Data from Kohn (2012).

Table 5.2. Inflow hydrograph for reservoir routing example

Time (days)	Inflow (m³ s⁻¹)	Time (days)	Inflow (m³ s⁻¹)
0.00	0.70	2.00	2.80
0.25	0.90	2.25	2.30
0.50	2.00	2.50	1.80
0.75	5.00	2.75	1.50
1.00	9.00	3.00	1.20
1.25	6.70	3.25	1.00
1.50	5.00	3.50	0.85
1.75	3.80	3.75	0.70

this example we will use the input hydrograph given in Table 5.2, which has a time step $\Delta t = 6$ hr or 2.16×10^4 seconds. We also must specify the initial values of O_1 and V_1 before we begin the calculation. The spillway height $h_{spillway} = 10.1$ m (Figure 5.8) and we will assume $h_{weir} = 0.0$ m initially, corresponding to a reservoir volume $V_1 = 3.7 \times 10^6$ m³ and initial outflow $O_1 = 0$ m³s⁻¹ (Table 5.1). In other words, the water level of the reservoir is just at the top of the spillway at the initial time t_1.

The solution to Equation 5.3 can be found by constructing Table 5.3. The quantities in each column are identified by the expressions at the tops of the columns. To begin, values for I_n and $I_n + I_{n+1}$ from the inflow hydrograph (Table 5.2) are entered into columns **A** and **B**, respectively, for each time t_n. Based on the initial values V_1 and O_1 given above, the first entry in column **C** of Table 5.3 is $[(2 * 3.7 \times 10^6 \text{ m}^3)/(2.16 \times 10^4 \text{ s})] - 0$ m³ s⁻¹ $= 342.6$ m³ s⁻¹. The entry in column **D** is computed using Equation 5.3:

$$\frac{2V_2}{\Delta t} + O_2 = I_1 + I_2 + \frac{2V_1}{\Delta t} - O_1 = 344.2 \text{ m}^3 \text{ s}^{-1}.$$

That is, **D = B + C**.

Now we need to figure out how to separate the term in **D1** (column **D**, row **1**: $2V_2/\Delta t + O_2$) into values of V_2 and O_2. From Table 5.1 (and Figure 5.8), we know reservoir volume V for a range of values of water depth, h. We can convert the values of h to corresponding values of O using Equation 5.4 with $h_{weir} = h - h_{spillway}$. This gives us values of V and O for each value of h. From these we can calculate $2V/\Delta t + O$ (column **D**) for a given Δt. The resulting values of h_{weir}, V, O, and $2V/\Delta t + O$ are listed in Table 5.1 for $\Delta t = 2.16 \times 10^4$ s and $h_{spillway} = 10.1$ m. A graph of $2V/\Delta t + O$ versus O (Figure 5.9) allows us to determine the value of O for any value of $2V/\Delta t + O$. In practice, it is often more convenient–particularly when doing the calculation in a spreadsheet—to fit a curve to the relationship shown in Figure 5.9. Since we don't expect outflows that are larger than inflows, we focus on getting a good fit to small values of outflow, O (Figure 5.9):

Table 5.3. Reservoir routing computation

Step n	Time t_n (days)	I_n	$I_n + I_{n+1}$	$\frac{2V_n}{\Delta t} - O_n$	$\frac{2V_{n+1}}{\Delta t} + O_{n+1}$	O_{n+1}	t_{n+1}
		A	B	C	D	E	
1	0.00	0.70	1.60	*342.6*	*344.2*	*0.06*	0.25
2	0.25	0.90	2.90	*344.1*	*347.0*	*0.60*	0.50
3	0.50	2.00	7.00	*345.8*	*352.8*	*1.79*	0.75
4	0.75	5.00	14.00	*349.2*	*363.2*	*4.15*	1.00
5	1.00	9.00	15.70	*354.9*	*370.6*	*5.98*	1.25
6	1.25	6.70	11.70	*358.6*	*370.3*	*5.91*	1.50
7	1.50	5.00	8.80	*358.5*	*367.3*	*5.15*	1.75
8	1.75	3.80	6.60	*357.0*	*363.6*	*4.25*	2.00
9	2.00	2.80	5.10	*355.1*	*360.2*	*3.45*	2.25
10	2.25	2.30	4.10	*353.3*	*357.4*	*2.81*	2.50
11	2.50	1.80	3.30	*351.8*	*355.1*	*2.29*	2.75
12	2.75	1.50	2.70	*350.5*	*353.2*	*1.89*	3.00
13	3.00	1.20	2.20	*349.4*	*351.6*	*1.55*	3.25
14	3.25	1.00	1.85	*348.5*	*350.4*	*1.29*	3.50
15	3.50	0.85	1.55	*347.8*	*349.4*	*1.08*	3.75

$$O = -2.64 \times 10^{-6}(2V/\Delta t + O)^3 + 0.00408(2V/\Delta t + O)^2$$
$$-1.68(2V/\Delta t + O) + 202.6. \tag{5.5}$$

Note that Equation 5.5 may not be a good fit for values of $O > 50$ m^3 s^{-1}.

Returning to Table 5.3, we find from Equation 5.5 (or Figure 5.9) that the value of O_2 (**E1**) is 0.06 m^3 s^{-1} for a **D1** value of 344.2 m^3 s^{-1}. We enter t_2 in the last column to keep track of the proper time for the outflow hydrograph.

At time step $n = 2$, we can calculate the term **C2** in Table 5.3 as the term **D1** $- 2 \times$ the term **E1**. That is,

$$\frac{2V_2}{\Delta t} - O_2 = \frac{2V_2}{\Delta t} + O_2 - 2O_2 = 344.1 \text{ m}^3 \text{ s}^{-1}.$$

This is added to $I_2 + I_3$ (**B2**) to obtain the value for column entry **D2** $= 347.0$ m^3 s^{-1}, the value of the left side of Equation 5.3 for $n + 1 = 3$. We repeat these steps for each n until the table is complete. The calculated outflow hydrograph is tabulated in the last two columns of Table 5.3. To summarize, the series of steps for completing Table 5.3 (using i to represent the row number and letters to represent the variables in each column) is:

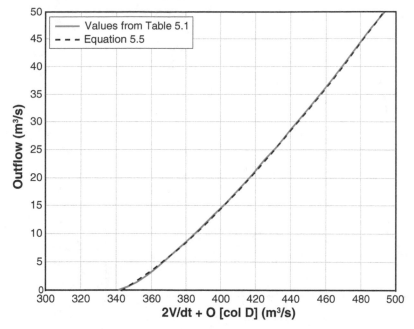

Figure 5.9 Outflow versus $(2V/dt) + O$ for the reservoir routing example based on low outflow values in Table 5.1. Equation 5.5 provides a fit to the observed relationship useful for completing the calculations in the reservoir routing example.

1. $B_i + C_i = D_i$.
2. $D_i +$ Equation 5.5 or Figure 5.9 produces E_i.
3. $D_i - 2 \times E_i = C(i+1)$.
4. return to step 1 for next i.

The predicted outflow hydrograph shows how the reservoir has worked to reduce the effect of the inflow flood (Figure 5.10). The reservoir stores a portion of the inflow during high inflow periods and releases the water when the inflow subsides. Again, note that the effect of this storage is to delay the peak discharge and to reduce its magnitude.

It is instructive to consider how the outflow hydrograph for the reservoir routing problem changes with reservoir characteristics. First, if the reservoir were larger, it would store more water. For the same inflow, the rise in water level in the reservoir would be smaller and, therefore, the increase in outflow would be smaller. The outflow hydrograph would have a lower peak value and would be spread over a longer period of time. Second, we might conceive of an outflow mechanism that released water more slowly than the spillway (or of a spillway with a lower weir coefficient). This also would have the effect of decreasing the peak discharge because the water would have to build up to a higher level to get the same outflow. If the water level in the reservoir were initially below the level of the spillway, then there would be no outflow over the spillway until the inflow

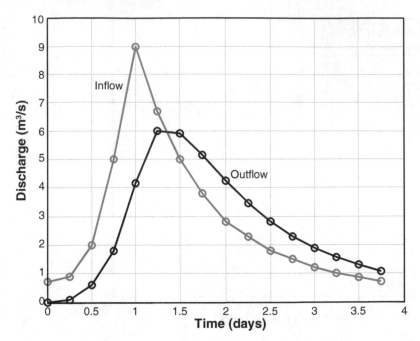

Figure 5.10 Hydrographs for the reservoir routing example.

could raise the water level in the reservoir to the level of the spillway. In the example shown in Figure 5.10, 9.6×10^5 m^3 of water enters the reservoir over the course of 3.5 days, an amount roughly equivalent to the volume of a 1.0 m-thick slab of water in the reservoir just below the height of the spillway. Therefore, if water depth in the reservoir were initially > 1.0 m below the spillway, the water surface would never reach the level of the spillway during this flood.

This example of a reservoir with a spillway is reasonable for reservoirs located on relatively small rivers or streams. However, large dams, such as the Grand Coulee Dam on the Columbia River, Glen Canyon Dam on the Colorado River, and Folsom Dam on the American River, have outlets that are more complex than simple spillways. Water is released through carefully controlled outlets and over spillways. This allows the volume of water in the reservoirs behind the dams to be varied seasonally and the discharge of water through the dams to be adjusted according to the operational policy for the dam. By gaging inflow and considering the trade-offs between volume of water in the reservoir and outflow rates, reservoirs can be managed for maximum benefit. For example, during periods of drought, extra water can be stored. However, if water levels in a reservoir are kept high in anticipation of drought, the reservoir capacity for storing floodwaters is decreased.

Some of the difficulties in optimally managing a reservoir are illustrated by an example from the American River. In 1986, before the flood on the American River, the operating procedures for Folsom Dam called for a flood-storage capacity of 4.93×10^8 m^3 beginning on November 17 and extending until February 8. Subsequent to February

8, the allowable storage was varied according to how much the accumulated precipitation for that season had been. The operating policy also called for a maximum controlled release from the dam of 3.26×10^3 m^3 s^{-1} up until the reservoir level reached full pool. Subsequently releases were dictated by emergency spillway procedures designed to protect the structure from failure.

When the flood began in February 1986, Folsom Reservoir's flood-storage capacity was about three-fourths of that called for in the operating policy; that is, the amount of water stored in the reservoir exceeded the permissible storage (Figure 5.11). Outflows reached 3.68×10^3 m^3 s^{-1}, some 13% above the maximum controlled release. Fortunately, disastrous flooding was avoided because the storm abated and because the levee system downstream of Folsom Reservoir stood up to the "excess" releases from the reservoir. The effect of the storage reservoir on the flood is clear—the peak inflow discharge, for example, was 5.44×10^3m^3 s^{-1}, compared to the outflow peak of 3.68×10^3 m^3 s^{-1} (Figure 5.11).

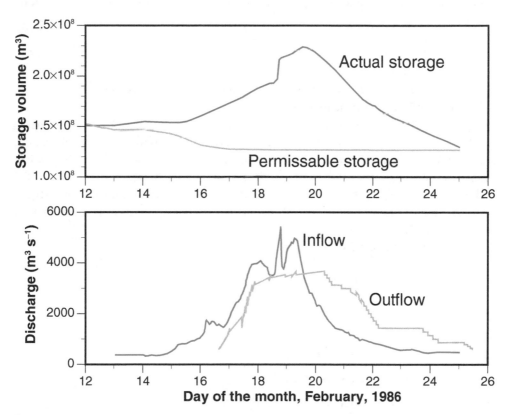

Figure 5.11 Flows and storage in Folsom Reservoir for the storm of February 1986. Data from NRC (1995a).

5.4.2 Flood routing in rivers

A comparison of Figure 5.10 with Figure 5.6b suggests that a flood wave is affected by passage through a segment of a river in much the same way as it is affected by passage through a reservoir. A detailed analysis of the changes to a flood wave as it moves through a river section is somewhat more involved than reservoir routing, but the principles used are the same in each case. The conservation of mass equation (Equation 5.1), or its finite difference equivalent (Equation 5.3), serves as the starting point for river routing. The difference is that rather than using a weir equation to relate storage and outflow, we need an equation relating the outflow from a river reach to the volume of water stored in the reach. A complete analysis of the relationship between storage volume and outflow in a river reach is beyond the scope of this book. Instead, we adopt a simpler approach in which observed inflow and outflow hydrographs are used to develop an empirical relationship between storage and outflow.

The **Muskingum method** of flood routing in rivers assumes that the volume of water stored in a river reach can be related to the inflow and outflow for the reach by the equation

$$V = K_t[xI + (1-x)O], \tag{5.6}$$

where K_t is a travel time constant and x is a constant that weights the contributions of inflow and outflow on storage in the reach. Although these two constants are not entirely independent, K_t primarily determines the rate at which the flood peak moves through a reach and x primarily determines the attenuation of the peak flow. The best values of K_t and x for a specific reach are obtained using measured inflow and outflow flood hydrographs. We can get reasonable estimates of the parameters, however, if we take $x = 0.2$ and set K_t equal to the length of a reach, L, divided by mean velocity through the reach, U. Note that a rapid flow through a long reach could have the same K_t as a slower flow through a shorter reach.

Substituting Equation 5.6 into Equation 5.3 gives

$$O_{n+1} = C_0 I_{n+1} + C_1 I_n + C_2 O_n, \tag{5.7}$$

where

$$C_0 = (-K_t x + 0.5\Delta t)/(K_t - K_t x + 0.5\Delta t).$$
$$C_1 = (K_t x + 0.5\Delta t)/(K_t - K_t x + 0.5\Delta t).$$
$$C_2 = (K_t - K_t x - 0.5\Delta t)/(K_t - K_t x + 0.5\Delta t).$$

The values of the coefficients C_0, C_1, and C_2 must be positive (for $x = 0.2$, $0.75\Delta t \leq K_t \leq 2.0\Delta t$) and the sum $C_0 + C_1 + C_2 = 1$. Like the reservoir problem, once the initial outflow O_1 and the inflow hydrograph are specified, the calculations can be stepped through to obtain the outflow hydrograph.

The flood routing procedure presented above allows one to compute an outflow hydrograph given a known inflow hydrograph. To illustrate the method, the inflow hydrograph used in the reservoir routing problem (Table 5.2) is routed through a river reach using the Muskingum method. The value of x is taken to be 0.2. Outflow hydrographs for $K_t = 6$ hr and 12 hr are shown with the inflow hydrograph in Figure 5.12. For a given flow rate, larger values of K_t correspond to increasingly long reach lengths. For example, if $U = 1$ m s^{-1}, the reach length for $K_t = 6$ hr is approximately 22 km; the reach length is 44 km for $K_t = 12$ hr. Alternatively, for a single value of reach length, say 40 km, the two values of K_t correspond to flow rates of approximately 1.9 and 0.9 m s^{-1}, respectively. Because we are disregarding lateral inflow to the reach, the choice of reach length for the river routing calculation would depend on the distance between major tributaries and the limitations on values of Δt and K_t necessary to keep the coefficients in Equation 5.7 positive.

The hydrographs pictured in Figure 5.12 show that the lag in the timing of the outflow peak relative to peak inflow is approximately equal to the travel time constant. As the travel time constant increases, the flood wave attenuation increases. For $K_t = 12$ hr, the peak outflow is about 70% of the peak inflow. It is interesting to compare the attenuation of the flood wave by friction and storage in a river channel (Figure 5.12) to the attenuation of the same flood wave by storage in a reservoir (Figure 5.10). A flood wave must move

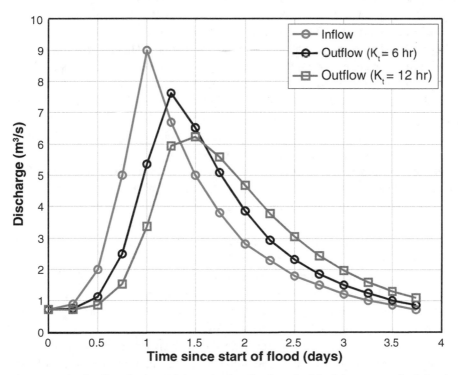

Figure 5.12 Outflow hydrographs calculated using the Muskingum method for the same inflow hydrograph used in the reservoir routing example (Figure 5.10). As the travel time constant K_t increases, the time lag and attenuation increase.

through a long stretch of a river before peak discharge is reduced as much as a moderate-sized reservoir can accomplish in a relatively short distance.

Because the peak in the outflow hydrograph is delayed relative to the inflow hydrograph, there is a direct operational use for river routing techniques such as the Muskingum method. For example, a gaging station on a river can be set up to transmit data automatically to the office of a person responsible for flood warning. These data can be augmented with forecasts of river inflow for the subsequent 10 days (based on weather forecasts) and then fed continuously to a routing program that gives the corresponding forecast for the outflow. Warnings about likely flood crests can be issued on the basis of such forecasts. During early parts of a flood event, forecasts of crest height downstream will be uncertain because the peak inflow value is not yet known. After the inflow flood has crested and the river has begun to subside, however, the forecasts of outflow crests will become more accurate because the inflow hydrograph up through the time of peak discharge is now available to the routing program.

5.5 Flood Frequency Analysis

Flood routing can be used to determine the effectiveness of a given flood-control structure. There will always be floods that exceed the design specifications of a structure. The question is, what is the probability of those floods occurring in any given year, and is that probability acceptably low? There are a variety of issues that demand the use of probabilities in analyzing floods. For example, suppose a person erects a building in a river valley and wishes to insure it against damage from floods. The insurance company is faced with the problem of setting the premiums so that it will have a reasonable hope of securing a profit. If the future sequence of storms over the life of the structure could be predicted with certainty and the resulting precipitation routed through the watershed with certainty, then the company could determine the exact time that the house would be damaged in a flood and calculate the premiums based on that prediction. Of course, this cannot be done, so an alternate solution must be found.

As with any insurance, the calculation of premiums involves the probability of an event occurring. That is, if a determination was made that the structure would be flooded once every 50 years (on the average), then the premiums would be set to cover the cost of damage this frequently. This is another way of saying that the probability of the house being flooded is 1/50 in any year. This situation does not guarantee that the company will make money on each individual policy. The structure may be flooded in the first year of the policy or it may be flooded three times in the first 10 years. On the other hand, it may not be flooded at all in 75 years. The probability statement does not specify that floods will occur exactly 50 years apart but only that one is *expected* to occur an average of once in 50 years (see also Section 2.2.4.)

The same notion of probability is needed in the analysis of flood-protection measures. At the beginning of this chapter, the problem of flooding on the American River system at Sacramento was introduced. Mention was made that Folsom Dam was originally thought to offer flood protection from a 1-in-500-year storm. Analysis of the level of protection in this case required: an estimate of the 1-in-500-year inflow hydrograph and a routing procedure for this inflow hydrograph through the reservoir.

The probability of a flood equaling or exceeding a given magnitude can be approximated using discharge records from a gaging station. For example, if there were 20 years of record for a certain station we might intuitively expect that the largest recorded flood is an approximation to the "20-year flood"—a flood that occurs, on average, once in 20 years. The formal procedure is summarized in the following steps:

1. The highest discharges recorded in each year for the *n* years of record are listed, similar to the case of rainfall extremes discussed in Chapter 2.2.4. These peak discharges are called floods and the series of the largest flood in each year is termed the *annual series*.

2. These floods are ranked according to magnitude. The largest flood is assigned rank 1, the second largest rank 2, and so on.

3. An initial determination of the flood statistics can be found by plotting the logarithm of discharge for each flood in the annual series against the fraction of floods greater than or equal to that flood; this fraction, termed the exceedance probability, is given by $r/(n+1)$, where r is the rank of the particular flood.

4. If the data conform to a lognormal distribution (i.e., the logarithm of the data are normally distributed), they will plot along a straight line if a normal probability scale is used on the y-axis as in Figure 5.13.

5. If a log-normal distribution does not adequately fit the data, another distribution, such as a log Pearson type III extreme value distribution, might provide a better fit to the data (see Haan, 2002.) By fitting a suitable distribution to the data, the exceedance probability and return period for a flood of any magnitude can be estimated.

6. The best fit to the data, whether lognormal or otherwise, defines the *exceedance probabilities* for floods of any given discharge. The *return period*, the average span of time between any flood and one equaling or exceeding it, is calculated as $T_{return} = 1/$ (exceedance probability).

As an example, we apply this procedure to the Powell River at Big Stone Gap, VA, for the period from 1945 through 1994. (There is a data gap between 1994 and 2002, so we confine our analysis to the 50-year period from 1945 to 1994.) The annual flood peaks range from 48 to 680 m^3 s^{-1} (Table 5.4). A line fit to the natural logarithm of annual peak discharge versus exceedance probability, plotted using a normal probability scale (Figure 5.13), gives a reasonably good fit to the data. We can estimate the 100-year flood by finding the value of discharge indicated by the straight line fit to the data for an exceedance probability of 0.01. For the example shown in Figure 5.13, we estimate a discharge of approximately 520 m^3 s^{-1}. (The log discharge from the graph is about 6.25; e raised to the 6.25 power is about 520.)

The extrapolated value for the 100-year flood probably is not realistic given that the highest few discharge values fall below the line fit to the data. Because of this, we have to be cautious about estimating return periods of floods much larger than those in the record. This is a typical problem that arises when fitting a lognormal distribution to extreme flood

Table 5.4. Annual flood peaks for Powell River, Big Stone Gap, VA, 1945–1994

Water year, Oct 1 to Sep 30	Q_{max}, maximum discharge for the year, m³ s⁻¹	ln(Q_{max})	r, flood rank	r/n+1, year[1]
1945	138.75	4.93	28	0.55
1946	260.51	5.56	7	0.14
1947	150.08	5.01	19	0.37
1948	240.98	5.48	9	0.18
1949	135.92	4.91	29	0.57
1950	140.45	4.94	25	0.49
1951	135.92	4.91	30	0.59
1952	148.95	5.00	20	0.39
1953	151.50	5.02	18	0.35
1954	81.55	4.40	45	0.88
1955	151.78	5.02	17	0.33
1956	185.19	5.22	12	0.24
1957	254.85	5.54	8	0.16
1958	106.75	4.67	37	0.73
1959	140.17	4.94	26	0.51
1960	115.53	4.75	35	0.69
1961	125.16	4.83	33	0.65
1962	146.11	4.98	22	0.43
1963	475.72	6.16	2	0.04
1964	105.91	4.66	38	0.75
1965	148.10	5.00	21	0.41
1966	223.42	5.41	10	0.20
1967	302.99	5.71	4	0.08
1968	90.61	4.51	43	0.84
1969	47.57	3.86	50	0.98
1970	286.00	5.66	6	0.12
1971	124.31	4.82	34	0.67
1972	107.89	4.68	36	0.71
1973	179.53	5.19	14	0.27
1974	181.79	5.20	13	0.25
1975	144.42	4.97	23	0.45
1976	105.91	4.66	39	0.76
1977	679.60	6.52	1	0.02
1978	84.95	4.44	44	0.86
1979	139.89	4.94	27	0.53

Table 5.4. (*continued*)

Water year, Oct 1 to Sep 30	Q_{max}, maximum discharge for the year, m³ s⁻¹	$\ln(Q_{max})$	r, flood rank	$r/n+1$, year[1]
1980	93.73	4.54	42	0.82
1981	57.77	4.06	49	0.96
1982	168.49	5.13	16	0.31
1983	62.86	4.14	48	0.94
1984	319.98	5.77	3	0.06
1985	63.43	4.15	47	0.92
1986	104.77	4.65	40	0.78
1987	99.96	4.60	41	0.80
1988	70.79	4.26	46	0.90
1989	172.17	5.15	15	0.29
1990	131.39	4.88	31	0.61
1991	143.57	4.97	24	0.47
1992	197.09	5.28	11	0.22
1993	129.12	4.86	32	0.63
1994	288.83	5.67	5	0.10

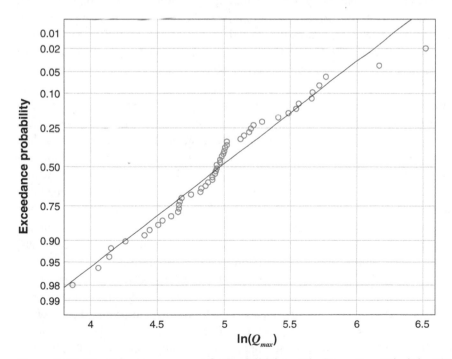

Figure 5.13 Flood frequency curve for Powell River, Big Stone Gap, Virginia, 1945–1994.

records. A better fit and, therefore, more reliable estimates of return period are often obtained using an extreme value distribution (e.g., Hahn, 2002). In addition, the length of the record strongly affects the accuracy of return period estimates. As records become longer, estimates of extreme events become better. Unfortunately, we are often faced with the exact problem of wanting to estimate a 100- or 200-year event from a relatively short record.

5.6 Concluding Remarks

Dams, such as Folsom Dam on the American River, offer important flood protection to people and property located on floodplains. Evaluation of protection measures requires the routing of hypothetical floods with specified exceedance probabilities. The U.S. Army Corps of Engineers (USACE) is responsible for the analysis of flood-protection alternatives for the American River system and many others. The USACE derives the "design hydrographs" for various exceedance probabilities by using a flood-frequency analysis similar to that described in this chapter. Technical evaluation of dams or alternative flood mitigation measures, such as natural wetland restoration, demands that we know how to "route" rainfall through the appropriate storage zone, be it a reservoir or a wetland. A comprehensive analysis of these routing problems requires that all hydrological processes that can affect streamflow be understood, including the movement of water in the subsurface. Chapters 6, 7, and 8 explore flow in the saturated and unsaturated portions of the subsurface. We return to consider the dynamics of water movement in catchments in Chapter 10.

Dams serve many valuable functions, including water storage and flood protection, but they are not without problems. A dam regulates the downstream flow of water in such a way that the discharge is more constant than it would be naturally. Damming of streams can alter water temperature and stream chemistry and fragment the stream itself. While regulation of flow in streams and rivers is beneficial in many respects for the human populations they affect, it may have far-reaching consequences for other parts of the river and riparian ecosystems. For example, fragmentation and reduced flow rates in dammed streams impede the upstream spawning migrations of salmon. They also affect the transport of sediment in the stream. One consequence of this is accumulation of fine sediment in reservoirs and increased water clarity downstream of dams. The latter effect is pronounced in the Colorado River downstream of the Glen Canyon Dam. Where the water was once warm and muddy, it is now clear and cold. This has transformed this part of the Colorado River into an excellent trout fishing stream, to the detriment of the native fish populations. Stream regulation also affects the riparian zones. For example, reduced overbank flooding allows time for less flood-tolerant plants to become established and for greater stabilization of the stream banks by vegetation.

Analysis of the world's largest river systems (292 rivers) revealed that over half (172) are strongly to moderately affected by fragmentation by dams and flow regulation by reservoirs (Nilsson et al., 2005). The large impacts of such fragmentation and regulation on river hydrology and ecosystems have prompted efforts to understand the scope of the changes and, in some cases, to mitigate the changes by introducing controlled floods (periods of high release intended to mimic naturally occurring floods) or even removing dams entirely. For example, the Elwha and Glines Canyon Dams on the Elwha River in the Olympic Peninsula, western Washington, were removed in 2011–2013, representing the largest dam removal project in U.S. history. A primary goal for removing the dams was habitat restoration, particularly spawning grounds for salmon and trout that used to be abundant in the Elwha River, but had become reduced in numbers or even locally extinct. Other goals included increasing the supply of sediment to the coastal area near the mouth of the Elwha River—sediment that had been trapped behind the dams—and renewing cultural traditions of the Lower Elwha Klallam Tribe, whose native home is along the Elwha River.

5.7 Key Points

- Hydrographs record the time history of flow in streams. Hydrographs typically are measured by continuously recording stream depth (stage) at a gaging station and combining it with a rating curve relating depth and discharge to obtain a continuous record of discharge. {Section 5.2}

- Peaks in discharge are termed floods, regardless of whether the water overtops the banks. Flood hydrographs are characterized by a steeply rising section followed by a less steeply falling segment as the flood passes. Background low flow conditions are termed baseflow. {Section 5.2}

- As a flood wave propagates downstream, its peak is attenuated owing to friction and storage of water in the channel. {Section 5.3}

- Storage of water in a reservoir increases as inflow increases. The conservation of mass equation, $dV/dt = I - O$, tells us that the greater the rate of change of storage, the greater the difference between inflow at the upstream end of a reservoir and outflow at the downstream end. {Section 5.4}

- Outflow from a reservoir or a river reach can be calculated from an inflow hydrograph using the conservation of mass equation and a second equation relating outflow to storage volume. The outflow hydrograph can be found by numerically solving the finite difference form of the conservation of mass equation. {Section 5.4}

- Reservoirs are effective for flood control because they can store a large volume of water, thereby producing a significant decrease in the magnitude of the outflow from the reservoir relative to the inflow at peak flood conditions. {Section 5.4.1}

- The Muskingum method of flood routing can be used to estimate the transformation of a flood wave as it moves through a river channel. The resulting outflow hydrograph depends on the inflow hydrograph and two constants: 1) a weighting coefficient that determines attenuation, $x \approx 0.2$; and 2) a travel time constant that determines the time lag between inflow and outflow peak discharge, $K_t \sim$ (reach length)/(mean velocity). {Section 5.4.2}

- The exceedance probability associated with a given peak annual flood discharge in an n-year-long record can be estimated graphically by plotting the logarithm of discharge against $r/(n+1)$ on a normal probability axis (see Figure 5.13), where r is the rank of each year's peak discharge from 1 (largest) to n (smallest). If the logarithms of the floods are normally distributed, then they will fall on a straight line. This line defines the exceedance probabilities for a flood of any magnitude. The return period is the inverse of the exceedance probability. {Section 5.5}

- Estimates of return periods for time spans longer than the existing record or probabilities of floods exceeding the largest recorded value can be obtained by fitting a suitable probability distribution to the data, which can then be used to estimate the return period of discharges larger than those found in the record. The farther the data are extrapolated (e.g., for return periods much longer than the record length), the larger the potential error will be in the resulting values of return period or flood magnitude. Errors also result if the data are not well fit by the chosen probability distribution. While a lognormal distribution is relatively easy to fit to an annual series of peak discharge values, an extreme value distribution is often needed to provide accurate estimates of floods with return periods much longer than the record. {Section 5.5}

5.8 Example Problems

Problem 1. If the flood used in the reservoir example delivered the same volume of water in a shorter amount of time (shorter duration with higher peak discharge), as given by the inflow hydrograph in the table below, how would the outflow hydrograph change? Complete the table below, assuming the initial conditions and other reservoir parameters remain the same.

Step N	Time t_n (days)	I_n	$I_n + I_{n+1}$	$\dfrac{2V_n}{\Delta t} - O_n$	$\dfrac{2V_{n+1}}{\Delta t} + O_{n+1}$	O_{n+1}	t_{n+1}
		A	B	C	D	E	
1	0.00	0.7	4.7	342.6			0.25
2	0.25	4.0	22.0				0.50
3	0.50	18.0	28.0				0.75
4	0.75	10.0	15.6				1.00
5	1.00	5.6	8.6				1.25
6	1.25	3.0	4.5				1.50
7	1.50	1.5	2.2				1.75
8	1.75	0.7	1.4				2.0

Problem 2. The Muskingum routing coefficients for a stream reach are determined to be: $C_0 = 0.26$, $C_1 = 0.55$, $C_2 = 0.19$. For the inflow hydrograph given in the table below, complete the calculation of the predicted outflow hydrograph.

Time (hr)	Inflow (m³ s⁻¹)	Outflow (m³ s⁻¹)
0000	10	10
0600	50	
1200	130	
1800	110	
2400	70	

Problem 3. In Figure 5.14, a 40-year annual series (1950–1989) of floods on the Eel River, California, is plotted against the fraction $[r/(n+1)]$ of floods with discharges greater than or equal to each value.

A. Fit a line through the data and determine the return period of an 8000 m³s⁻¹ flood.

B. Estimate the magnitude of the 100-year flood.

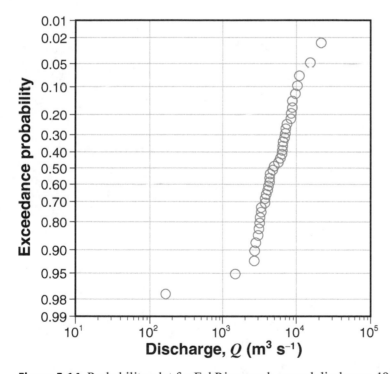

Figure 5.14 Probability plot for Eel River peak annual discharge, 1950–1989.

5.9 Suggested Readings

Dunne, T., and L.B. Leopold. 1978. *Water in environmental planning*. San Francisco: W. H. Freeman. Chapter 10, pp. 279–391.

Haan, C.T. 2002. *Statistical methods in hydrology*. New York: Wiley.

Mays, L.W. 2011. *Water resources engineering*, 2nd ed. Danvers, MA: John Wiley & Sons. Chapter 9, pp. 331–358.

Nilsson, C., C.A. Reidy, M. Dynesius, and C. Revenga. 2005. Fragmentation and flow regulation of the world's large river systems. *Science* 308:405–408.

6 Groundwater Hydraulics

6.1 Introduction

Of the total amount of freshwater on this planet, about 30% is contained beneath the surface of the Earth. This figure is even more impressive when one considers that all but a few tenths of a percent of the remaining freshwater is held in ice caps and glaciers (Table 1.1). Most subsurface water is found in rocks and soils that are saturated with water; that is, materials in which water occupies all pores, openings, and fractures. We refer to water in the saturated region of the subsurface as **groundwater.**

Groundwater is an important resource across the globe. Irrigated agriculture in many areas is based on withdrawal of groundwater. For example, Shah et al. (2000) estimate that 60% of irrigated grain production in India depends on the use of groundwater. Half of the world's mega cities (population of 10 million or more) are dependent on groundwater (Giordano 2009). Even in areas where surface water is available, groundwater often is a preferable source because of water temperature, quality, or accessibility. For example, in Dhaka, a city of 10 million in Bangladesh, groundwater is the predominant source of freshwater even though the city is located in the delta region of the Ganges and Brahmaputra rivers and is bordered by major rivers.

The study of groundwater is motivated partly by practical considerations of water supply. In addition, an understanding of the hydrological cycle for a catchment requires a quantitative description of how the groundwater reservoir functions. For example, recall that streamflow (baseflow) is maintained in perennial streams between precipitation events. We can hypothesize that this is in large part the result of the discharge of groundwater into stream channels. In this chapter and in Chapter 7 we discuss subsurface pathways through which precipitation eventually may reach a surface water body. We return to these ideas again in Chapter 10, when we examine mechanisms of runoff generation.

A number of environmental issues involve groundwater, especially the remediation of sites that have been contaminated by poorly controlled dumping practices and the identification of and planning for sites to safely dispose of hazardous wastes. An example is planning for the disposal of radioactive wastes.

The Waste Isolation Pilot Plant (WIPP) in New Mexico, United States, is an underground repository built for certain radioactive wastes that were generated during the construction of nuclear weapons. These wastes, including work gloves and laboratory glassware, have been stored in 55-gallon drums at facilities of the Department of Energy. The wastes are not *high-level*, but they contain isotopes that remain radioactive for very long periods of time (tens of thousands of years). These wastes must be isolated for millennia to ensure that they do not pose risks to human health. In the U.S. these wastes are being buried in a repository in a 600-m thick salt formation (the Salado Formation) some 700 m below the ground surface. Rock salt is thought to be a good host rock for radioactive wastes because it flows and over time will seal the wastes off from the environment. The only path that would lead to the release of radioactivity to parts of the environment where it might adversely affect people or ecosystems is through dissolution of the waste by water and transport of the dissolved constituents by groundwater. The Culebra dolomite, a regional aquifer, overlies the Salado Formation. Assessment of the suitability of the WIPP site for disposal of radioactive waste requires knowledge of how groundwater flows in the aquifer above the repository. Several questions might be raised. What causes or drives the movement of groundwater? What physical characteristics of subsurface fluids and porous media determine the rate of fluid movement?

Because we are again interested in the flow of water in relation to imposed forces, the equations of fluid mechanics provide the basis for the quantitative description of groundwater flow. At the scale of the pores in rocks and soils, however, the paths along which groundwater flows are complex, with many twists, turns, contractions, and expansions. So we recognize immediately that simplifications will have to be made to enable useful equations to be derived.

6.2 A Conceptual Model

Let us try to picture the flow of water in a **porous medium**, for example, like sand. The path a "parcel" of water might follow in moving through a material containing pores or void spaces is convoluted (Figure 6.1). Not only does the water follow tortuous paths, but the geometry of the channels of flow is extremely complex and cannot be specified completely (i.e., the position, size, and shape of all of the sand grains cannot be known). Finally, we recognize that the openings in which water flows are very small. Therefore, we might expect that frictionless flow is totally meaningless in this situation and that head losses will play the predominant role.

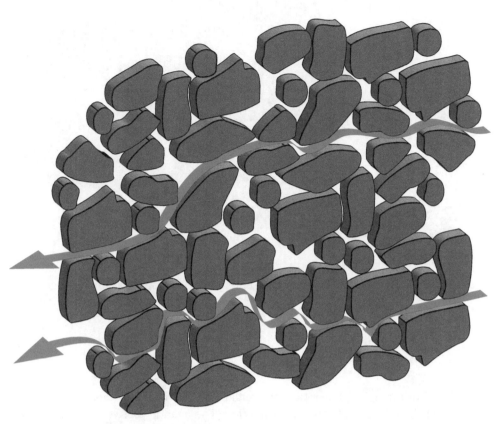

Figure 6.1 Schematic of a thin section of a porous medium and the tortuous flow path of two water "parcels." (Note that a real medium is three-dimensional with flow through the open spaces within the three-dimensional matrix.)

This picture of groundwater flow is exceedingly complex. To try to understand the fluid mechanics of the flow, we must resort to a conceptual model. We first consider the tortuous flow path (Figure 6.1) to be "straightened" by somehow stretching the path. The opening through which flow occurs then might be depicted as a pipe with a continually varying cross section (Figure 6.2). Of course, there are actually many flow paths and the entire system would consist of many different variable-radius tubes. Next we replace the variable-radius tubes with an *equivalent* set of constant-radius tubes. Thus, our conceptual model of flow through a porous medium is flow through a bundle of very small (capillary) tubes of different diameters. To be sure, the model is almost unacceptably oversimplified. Nevertheless, we can derive certain insights into flow in rocks and soils by examining this conceptual model.

Consider flow through a capillary tube. Because the tube is very narrow and the velocities are relatively small, it is reasonable to assume that the flow in the capillary tube is laminar. Moreover, because the velocities are very small, we can neglect the effect of changes in velocity and velocity head through the capillary tube, and assume that the diameter is constant. In Section 3.7.1, we found an expression for the average velocity,

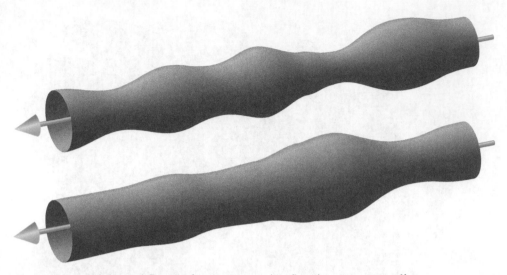

Figure 6.2 Straightened flow "tubes" representing flow in a porous medium.

$$U = -\frac{d}{dl}\left(\frac{p}{\rho g} + z\right)\frac{D^2 \rho g}{32\mu} \qquad (3.38)$$

of laminar flow through a pipe or tube of circular cross section in the direction l; D is the diameter of the tube and l is the flow direction. The quantity between parentheses is the sum of pressure head and gravitational or elevation head and is also known as the **hydraulic head** [L]

$$h = \frac{p}{\rho g} + z. \qquad (6.1)$$

The hydraulic head is a quantity, measurable at every point in a groundwater flow system. Because in these slow flows the velocity head is negligible, h represents the fluid mechanical energy per unit weight.

The negative sign in Equation 3.38 indicates that the flow is down the hydraulic head gradient from high to low hydraulic head. The discharge through the tube is given by the product of average velocity and the cross-sectional area of the tube:

$$Q = AU = \frac{\pi D^2}{4}\left(-\frac{dh}{dl}\frac{D^2 \rho g}{32\mu}\right) = -\frac{\pi D^4 \rho g}{128\mu}\frac{dh}{dl}. \qquad (6.2)$$

Equation 6.2 is a form of Poiseuille's law for the flow of a viscous fluid through a capillary tube. Discharge is directly proportional to the hydraulic head gradient (or hydraulic gradient), inversely proportional to the fluid viscosity, and directly proportional to the fourth power of the radius of the tube.

The equations derived from this conceptual model have several important implications. First, Equation 6.2 indicates that for a given fluid and given hydraulic gradient (dh/dl), the discharge varies as the fourth power of the radius of the tube. For example, the discharge from a tube of radius 10 mm will be 10^4 times greater than that from a tube of radius 1 mm, if all other conditions are the same. Because the size of the capillary tubes in our conceptual model is related (on an intuitive basis) to the texture or grain size of the porous medium, flow rates will depend on the texture. In addition, Equation 3.38 specifies that the average velocity, U, is proportional to the hydraulic gradient, dh/dl,

$$U = -\frac{D^2}{32} \frac{\rho g}{\mu} \frac{dh}{dl}. \tag{6.3}$$

Equation 6.3 is the form of Poiseuille's law that provides the most useful analogy for flow through a porous medium. As we will see, an equation almost identical to Equation 6.3 is the basis for studies of groundwater flow. Because Equation 6.3 was derived in Chapter 3 using well-known principles and stated assumptions, we can use the analogy between the conceptual model and the actual porous medium to great advantage in the sense that we can apply, in qualitative terms at least, knowledge about the factors that control laminar flow in tubes directly to the flow of groundwater.

6.3 Darcy's Law

In 1856, a French hydraulic engineer named Henry Darcy published an equation for flow through a porous medium that today bears his name. In designing a water treatment system for the city of Dijon, Darcy found that no formulas existed for determining the capacity of a sand filtration system. Consequently, Darcy performed a series of experiments on water flow through columns of sand.

Darcy packed sand into iron pipes and systematically measured parameters that he expected to influence the flow. Consider flow through a cylindrical volume of sand, with cross-sectional area = A and length = L (Figure 6.3). The sum of elevation head (z) and pressure head ($p/\rho g$) is represented as h and varies from h_1 to h_2 along the column. The hydraulic head at each end of the column is measured with an open tube, a simple manometer, as shown in the diagram. By varying L and the hydraulic head difference across the column ($h_1 - h_2 = \Delta h$), Darcy found that the total discharge Q varies in direct proportion to A and to Δh and inversely with L. That is,

$$Q = KA\frac{h_1 - h_2}{L}, \tag{6.4}$$

where K is a constant of proportionality called the **hydraulic conductivity** [L T^{-1}]. Equation 6.4 can be rewritten as:

$$\frac{Q}{A} = -K\frac{h_2 - h_1}{L} = -K\frac{h_2 - h_1}{l_2 - l_1}. \tag{6.5}$$

This can be written more generally as:

$$q = -K\frac{dh}{dl},$$
(6.6)

where $q = Q/A$ is the **specific discharge**. Equation 6.6 is the form of **Darcy's law** that we will use in our studies. Although q has dimensions of velocity [L T^{-1}], keep in mind that we obtained this term by dividing the discharge by the *total area* and that water flows only through a fraction of the area, the spaces between the solid grains of the medium.

In Equation 6.6, dh/dl is the **hydraulic gradient** and the negative sign indicates that positive specific discharge (indicating direction of flow) corresponds with a negative hydraulic gradient. Thus, Darcy's law states that specific discharge in a porous medium is in the direction of decreasing head and directly proportional to the hydraulic gradient (Figure 6.4).

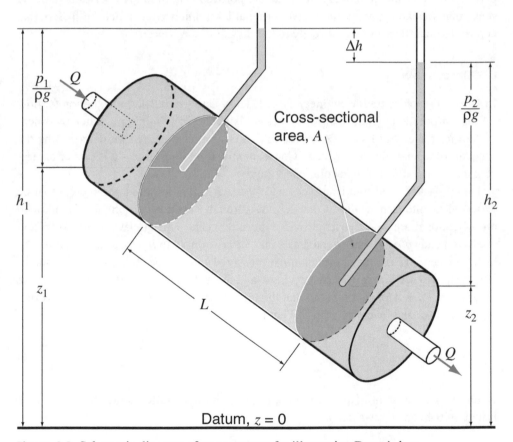

Figure 6.3 Schematic diagram of an apparatus for illustrating Darcy's law.

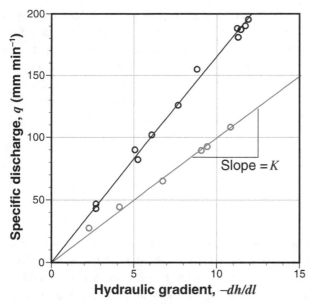

Figure 6.4 Darcy's (1856) original data showing a linear relationship between specific discharge and hydraulic gradient for two different sands.

6.3.1 Hydraulic conductivity, intrinsic permeability, and porosity

We can envision the hydraulic conductivity as the slope of a line relating specific discharge, q, with hydraulic gradient, dh/dl (Equation 6.6). Imagine a set of experiments using a given sample of material and a fluid of constant density and viscosity. By varying the hydraulic gradient and measuring the discharge, q can be plotted against dh/dl. According to Darcy's law a straight line should be the result, the slope of which will be K (Figure 6.4). Using a fluid with a constant density and viscosity, the slope of the relationship between q and dh/dl will depend only on the material and generally will increase with the coarseness of the material. Now imagine repeating the experiments with another fluid having different properties (e.g., a greater viscosity). We would expect the more viscous fluid to move more slowly if everything else remains constant.

The example above suggests that hydraulic conductivity depends on the nature of both the fluid and the porous material. The way in which K depends on these properties can be inferred by reference to the conceptual model discussed earlier. We already have remarked on the similarity between Equations 6.3 and 6.6, which are reproduced below (for one-directional flow in the l direction) to facilitate direct comparison.

$$U = -\frac{D^2}{32}\frac{\rho g}{\mu}\frac{dh}{dl}.$$ **(Poiseuille's law)**   (6.3)

$$q = -K\frac{dh}{dl}.$$ **(Darcy's law)**   (6.6)

This analogy suggests that $\rho g/\mu$ should describe the variation of K with fluid density and viscosity, and that $D^2/32$ should describe the variation of K with pore diameter. Of course, the "diameter" of the pores is not measurable (or even well defined), so the analogy is not perfect. However, qualitatively we might expect the grain size of the material to give an indication of the size of the openings (i.e., we would expect larger pores in a boulder field than in a silt deposit) and we can define and measure the average grain size of a granular material such as sand. Experimental evidence supports this analogy, at least for simple granular porous media. Based on experimental results, the following empirical relationship for hydraulic conductivity can be written:

$$K = (Nd^2)\left(\frac{\rho g}{\mu}\right), \tag{6.7}$$

where $N =$ a factor to account for shape of the passages [dimensionless]; $d =$ the mean grain diameter [L]; $\rho g =$ the unit weight of the fluid [M L^{-2} T^{-2}]; $\mu =$ the viscosity of the fluid [M L^{-1} T^{-1}]. From this we see that K is composed of two factors, one representing fluid properties and the other representing properties of the medium. Darcy's law can be written in a manner that clearly separates these two influences:

$$q = -k\left(\frac{\rho g}{\mu}\right)\left(\frac{dh}{dl}\right) \tag{6.8}$$

where k is referred to as the **intrinsic permeability** [L^2] of the porous medium. The factor Nd^2 in Equation 6.7 is equivalent to k. Once the intrinsic permeability of a certain rock formation is known, Equation 6.8 can be used to describe the flow of any fluid (oil, gas, or water) through that formation.

In addressing issues such as the movement of contaminants in the subsurface, it should be apparent that the hydraulic conductivity or intrinsic permeability of natural materials plays a major role, with higher values of K or k resulting in faster transport. Measuring or estimating these properties is a fundamental step in applying Darcy's law to a natural setting. There are a variety of techniques and methods for either directly or indirectly determining the permeability of a sample of porous material. A small sample can be placed in a device called a **permeameter**, not unlike Darcy's original column. The flow rate through the sample can be measured for a known hydraulic gradient and the permeability can be calculated directly using Darcy's law. In some situations, for example in looking at flow deep underground, samples may be difficult to obtain. There are a variety of methods, known as aquifer tests or "pump" tests, for determining permeability in such cases. By withdrawing or adding water to a well, and measuring the water level in that well or other wells nearby as a function of time, we can calculate the permeability. Finally, there are indirect methods that are based on measuring some other parameter, such as grain size, that is related to the permeability.

Literally thousands of permeability measurements have been made in different materials. The results show that the range of permeability of natural materials is quite large (Figure 6.5). Recalling the discussion of the WIPP site in the introduction, one reported range of hydraulic conductivities for salt deposits, such as the Salado Formation, is 10^{-12} to 10^{-10} m s^{-1} (Domenico and Schwartz, 1990). Relative to the other types of materials shown in Figure 6.5, this range is near the low end and is similar to shales and unfractured crystalline rocks. The resulting slow rate of groundwater flow is one reason that salt formations have been considered for waste repositories. For example, a typical hydraulic gradient of 1/100 in a salt formation with a hydraulic conductivity of 10^{-10} m s^{-1} will produce a specific discharge of 10^{-12} m s^{-1}, or less than 1 mm per 30 years!

Darcy's law indicates that for a given value of the hydraulic gradient (dh/dl), the specific discharge will be greater for a permeable material, such as a sand or gravel, than for a granite. The difference can be several orders of magnitude (Figure 6.5). As mentioned earlier, the specific discharge, despite having dimensions of velocity [L T^{-1}], is not a velocity. We could not use specific discharge to determine how long it will take a parcel of water to move from one point to another. The cross-sectional area available to the water is smaller than the actual cross-sectional area, such that the solid portion of the porous medium acts as a constriction. This constriction means that a tagged parcel of water, or, better, many tagged parcels that are averaged together, will appear to move through a porous medium at a speed that is faster than the specific discharge. The effect is similar to the constriction in a pipe discussed in Chapter 3; the constricted flow has a greater mean velocity for the same value of discharge.

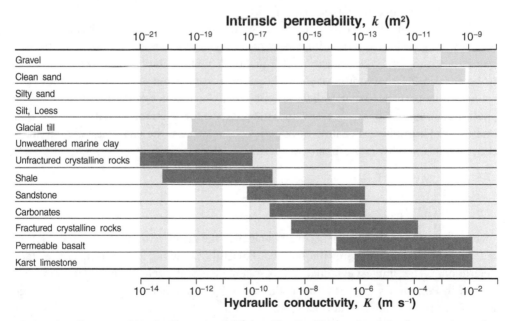

Figure 6.5 Ranges of intrinsic permeability and hydraulic conductivity for a variety of rocks (*gray bars*) and sediments (*blue bars*).
Data from Freeze and Cherry (1979).

We can determine the mean pore water velocity if we have some idea of the amount of "constriction." For porous materials, this property is the **porosity** [dimensionless], which is simply the fraction of a porous material that is void space:

$$\phi = \frac{V_v}{V_t},$$
(6.9)

where V_v is the volume of void space [L^3] and V_t is the total volume [L^3]. If we have measured the porosity, which by definition must be between 0 and 1, we can determine the mean pore water velocity, or **average linear velocity**, $\bar{v}$ [L T^{-1}]:

$$\bar{v} = \frac{q}{\phi}.$$
(6.10)

6.3.2 Restrictions on Darcy's law

Darcy's law is widely used for almost all situations involving motion of a fluid through soil or rock in the natural environment. Although the conceptual model we used to introduce our discussion of Darcy's law can apply (to a limited extent) only to granular materials such as sand, Equation 6.6 is applied in materials ranging from clay to limestone to fractured crystalline and metamorphic rocks. The openings through which fluid flows in most of these materials cannot be envisioned as capillary tubes. Nevertheless, Darcy's law usually can be applied with success.

There are some general limitations on the use of Equation 6.6, and because of the analogy we developed we can infer how these restrictions arise by examining the assumptions implicit in the derivation of Equation 6.3. First, Darcy's law has been found to be invalid for high values of Reynolds number. Experiments using very high flow rates in very permeable materials have found that when values of the Reynolds number exceed the range 1–10, Darcy's law does not describe the relationship between flow rate and hydraulic gradient. For flow through porous materials, the characteristic length in the Reynolds number is the mean grain diameter, d:

$$\mathbf{R} = \frac{\rho q d}{\mu},$$
(6.11)

where q, the specific discharge, is used in place of the mean velocity, U. This restriction can be explained by virtue of the fact that Poiseuille's law neglects inertial forces (accelerations). The simplest evidence of this statement is that a velocity head term does not appear in Poiseuille's law (or in the definition of hydraulic head) and that this term derives from consideration of acceleration.

Darcy's law also may fail to hold at very low values of hydraulic gradient in some very low-permeability materials, such as clays. Equation 6.6 implies that any hydraulic gradient, no matter how small, will cause some motion of water. In certain clay materials it is observed that below some threshold value, a small hydraulic gradient will not cause motion; that is, the force due to pressure and gravity must exceed some critical value before motion of water ensues. Poiseuille's law, which is our analogy for Darcy's law, is an expression of Newton's second law, and is based on the idea of balanced forces. However, the only forces considered are pressure and shear. Deviation from the law can be the result of the presence of other forces that we failed to consider (e.g., the inertial terms discussed in the previous paragraph). In the case of clays, there may be important electrostatic forces resulting from charge imbalances within the mineral structure and at low hydraulic gradients these may contribute to a force balance.

6.4 Water in Natural Formations

The subsurface is a complicated assortment of different materials, some of high permeability and some of low permeability. Hydrologists studying groundwater flow are interested in how the distribution of different materials in the subsurface influences patterns and rates of groundwater movement, and how the water interacts chemically with natural materials. The first topic will be addressed more fully in Chapter 7, but for now we introduce some terms that are applied to different soil or rock units, based on whether they are relatively permeable or not, and where they reside in the subsurface. An **aquifer** is a saturated geological formation that contains and transmits "significant" quantities of water under normal field conditions. "Significant" is a vague term but the implication is that aquifers are formations that can be used for water supply. Obviously, whether the supply is significant or not depends upon whether one is referring to a supply for a single rural dwelling or a large municipality.

Many aquifers are unconsolidated materials, mainly gravel and sand. Examples of this type of aquifer include those in coastal plain settings and intermontane valleys. Limestones, partially cemented sandstones and conglomerates, and permeable volcanic and igneous rocks are also important as aquifers. The limestone aquifers in the Paris basin in Western Europe and those underlying most of the Florida peninsula in the United States are critical resources as are the Deccan basalt groundwater system in India and the basalts of the Hawaiian Islands. Similarly, unconsolidated materials form important aquifers in the North China Plains in China and in the High Plains in the central United States.

Of course, not all formations are aquifers. An **aquiclude** is a formation that may contain water but does not transmit significant quantities. Clays and shales are examples of aquicludes. An **aquifuge** is a formation that neither contains nor transmits significant quantities of water. Unfractured crystalline rocks would fall into this category. The more general term **aquitard** is often used to denote formations that are of relatively low permeability and that may include both aquicludes and aquifuges.

Aquifers are classified according to hydraulic conditions and type of material. One type of aquifer is an **unconfined** or **water-table aquifer**. If an excavation is made in soil, then the near-surface material is usually not saturated (the unsaturated or vadose zone, Chapter 8). Deeper in the soil profile, saturated conditions prevail (**saturated zone**); groundwater by definition refers to water in the saturated zone of the subsurface.

The **water table** or phreatic surface is defined as a surface of zero gage pressure within the subsurface, and separates the saturated and unsaturated zones. Water will flow into an excavation or well up to this level; the water table is equivalent to a free surface. An aquifer with the water table as the bounding surface of the top of the aquifer is an unconfined aquifer.

The second type of aquifer is called **confined** or **artesian**. This type of aquifer is found when permeable material (the aquifer) is overlain by relatively impermeable material (an aquiclude; Figure 6.6). The water in a confined aquifer is under pressure and, in a well penetrating the aquifer, will rise above the top of the aquifer (see Figure 6.6). The height to which water rises in a well defines the **piezometric** or **potentiometric surface**. Note that a well penetrating a confined aquifer can be thought of as a **piezometer**, or single tube manometer. The water level in a piezometer is a measure of water pressure in the aquifer. Thus, the elevation of the potentiometric surface above an arbitrary (horizontal) datum is the sum of pressure head and elevation head, or the hydraulic head, h, of Darcy's law.

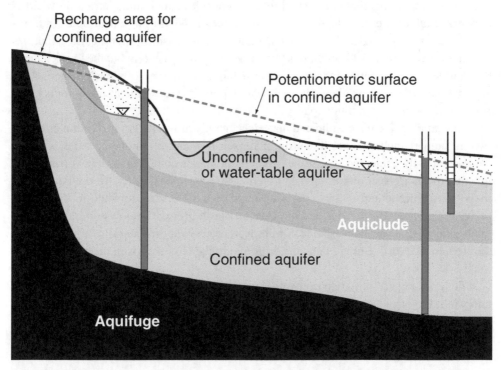

Figure 6.6 Hydrogeological units. Three piezometers are depicted, which are open in either the confined or unconfined aquifer, as indicated by the short horizontal lines. Note that in unconfined aquifers, the water level in the piezometer (*far right*) indicates the height of the water table; in confined aquifers, the water level in the piezometers (*left* and *center*) rises above the top of the aquifer and indicates the position of the potentiometric surface.

6.4.1 Construction of wells and piezometers

We derive information about the subsurface in large part by constructing wells and piezometers. A well is an opening, generally a cylindrical opening, from the ground surface down to a geological formation containing water. The formation is typically an aquifer, at least for wells constructed to supply water or to monitor a water source. In unconsolidated materials such as sand and gravel aquifers, a **well casing** is installed as part of the construction. A casing is made of solid material intended to support the opening and keep the well open. The well is screened over some interval to allow water to enter the otherwise impermeable casing. A **well screen** can be as simple as slots cut into a polyvinylchloride (PVC) pipe that is used as a casing. Wells typically are screened over a considerable depth of the aquifer to allow easy entry of water. A piezometer, by contrast, typically is screened over a narrow interval to reflect groundwater head at one location in an aquifer. If there is little vertical flow of groundwater, the head is sensibly constant over the depth of the aquifer and water levels recorded by wells screened over large vertical intervals will provide the same measurements as piezometers.

There are many methods for constructing wells including hand digging, augering, hammering, and jetting (using water under pressure to "erode" material and excavate a hole). The maximum depth that can be reached in the excavation of a well strongly depends on which of these techniques is used (only a few meters for hand dug wells, several tens of meters for jetted wells). One common method for constructing wells and piezometers in unconsolidated material involves the use of a hollow-stem auger. A **hollow-stem auger** is just what it sounds like (Figure 6.7a): auger blades surround an open core. There is a removable plug at the bottom and once the desired depth is reached, the plug is removed and a casing is inserted into the hollow stem. The auger is withdrawn, leaving the casing in place. The screened area of the well outside of the casing is backfilled with coarse material to allow ready flow of water to the well through the screen but to prevent fine material from clogging the screen. The upper part of the hole outside the casing is backfilled with a relatively impermeable material such as clay to prevent downward leakage of surface water into the well (or upward flow in the case of a confined aquifer). Near the surface a concrete cap is generally installed to complete the seal of the casing and prevent surface water and contaminants from entering the well (Figure 6.7b).

6.4.2 Water-level measurements

Key measurements that we derive from wells and piezometers are the elevations of the potentiometric surface or the water table. It is these measurements that allow us to establish head gradients that can be used with estimates of hydraulic conductivity to calculate groundwater flow rates. Once the elevation of the top of the well casing is established by surveying or the use of satellite positioning, we can determine the elevation of the water surface in a well by measuring the distance from the top of the well casing. One instrument that we routinely use to make these measurements is a **pressure transducer**. This is a device, as the name suggests, that measures water pressure. The transducer is lowered a known distance from the top of the well casing and below the water

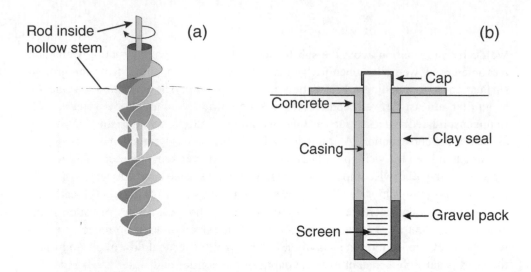

Figure 6.7 Schematic of a hollow-stem auger (*a*), one technique used to drill wells. Once wells are drilled (*b*), construction typically is completed as follows: coarse material is placed around the well screen to facilitate water inflow to the well, backfill is packed around the rest of the casing to seal off the screened region from possible surface contamination, and a grout and concrete cap is constructed to provide surface security.

level in the well and the pressure is measured. The hydrostatic Equation 3.5 can be used to compute the pressure at the depth of the transducer. Pressure transducers can be left in place for extended periods and connected to a data logger to record well hydrographs (Figure 6.8).

6.4.3 Geophysical techniques

There are other, indirect techniques that are used to estimate subsurface conditions that cannot be "seen" in the conventional sense of the word. Geophysics is a field that uses measurements of physical processes and properties to infer information about the structure of the Earth, and in the case of hydrogeology, the relatively shallow solid Earth in particular. That is, we use measurements involving the propagation of sound (seismic vibrations), electricity, or radar, for example, to infer the distribution of Earth properties that affect transmission. These include density for seismic vibrations, electrical resistivity for electricity, and the dielectric constant for radio waves (radar). These physical properties often can be correlated with hydraulic properties because the differences in rock type and structure that affect the physical properties also affect the hydraulic properties such as porosity and hydraulic conductivity. In a sense, use of geophysical techniques can be considered to be a way of "seeing into the Earth" (NRC, 2000). The array of quite sophisticated techniques used in geophysical investigations is extensive and a description of methods is beyond the scope of this book. Suffice it to say that the visualization of subsurface structure can provide significant information for understanding

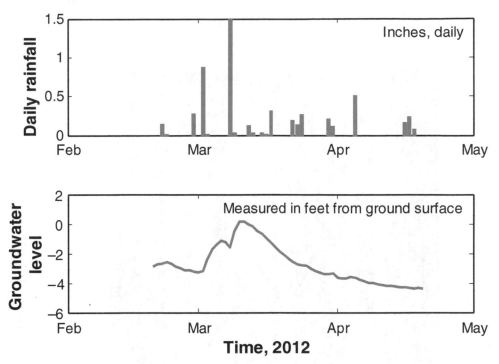

Figure 6.8 Well hydrograph for USGS Well 351428085003600, Hamilton County, Tennessee, USA.

Provisional data from USGS.

hydrogeology. Even a subtle underground feature may have important implications for the flow of water in soils and rocks (Figure 6.9).

6.5 Steady Groundwater Flow

The concepts developed in Section 6.3 can be applied readily to flow in natural formations. Consider first a simple example in which a confined aquifer is bounded by two channels (Figure 6.10). The height of the water in the channel on the left defines the hydraulic head at that boundary of the aquifer and the same is true on the right boundary. If the flow is steady and the confined aquifer is of constant thickness, b, then the specific discharge through the aquifer from $x = 0$ to $x = L$ is constant and we can write Darcy's law for flow in the x-direction as:

$$q = -K \frac{dh}{dx}. \tag{6.12}$$

We can rearrange this expression and integrate over the length of the confined aquifer to obtain an expression for hydraulic head, h:

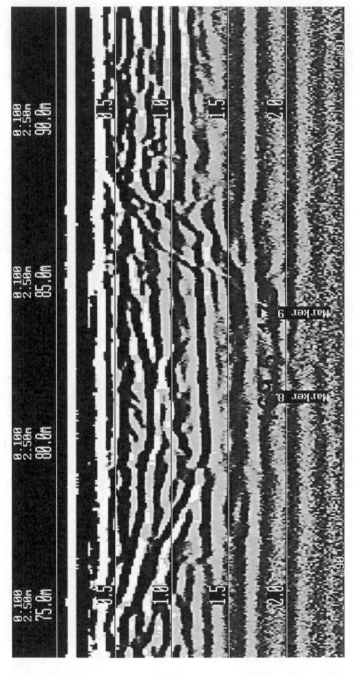

Figure 6.9 Trace from a ground-penetrating radar transect showing a buried channel. Such features often may be important for guiding groundwater flow.

Image courtesy of Alan D. Howard.

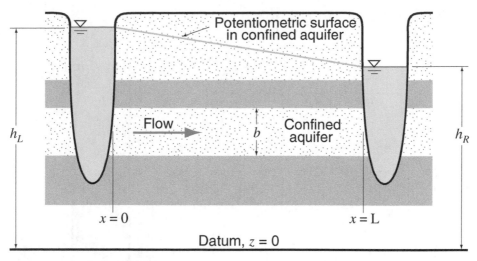

Figure 6.10 Horizontal flow in a confined aquifer.

$$dh = -\frac{q}{K}dx$$

$$\int_{h_L}^{h} dh = -\frac{q}{K}\int_{0}^{x} dx$$

$$h - h_L = -\frac{q}{K}x$$

$$h = h_L - \frac{q}{K}x. \tag{6.13}$$

Equation 6.13 specifies that head decreases linearly with distance from the left boundary; that is, the potentiometric surface is a plane sloping from left to right. This is shown as a solid gray line in Figure 6.10. Water flows in the downhill direction of this surface.

On a plan view of the channel-aquifer system considered above, the contour lines of the potentiometric surface are parallel to the channels (Figure 6.11). The spacing of the contours indicates the slope of this surface, just as do the contour lines of a standard topographic map. Once these lines of equal hydraulic head, or **equipotentials**, have been established, lines that indicate the direction of flow can be sketched in by constructing *perpendiculars* to the equipotentials because this represents the downhill direction of the potentiometric surface, the direction of flow specified by Darcy's law. These lines are called **streamlines**. Together, the equipotentials and the streamlines constitute a **flow net**. Flow nets can be applied to great advantage in actual field problems where groundwater flow patterns are to be established based on the measured water levels in a series of wells (see Fetter, 2000, for examples).

We will discuss the use of flow nets in greater detail in a later section of this chapter, but at this point we need to consider the physical reasoning behind the idea that streamlines

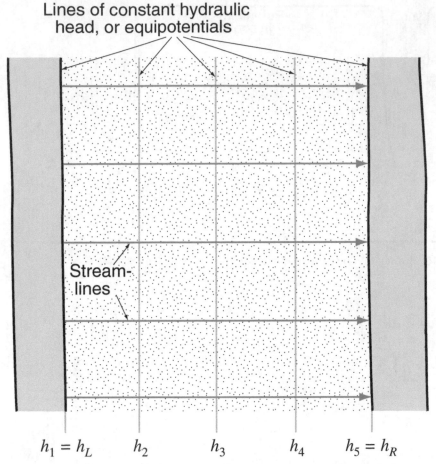

Figure 6.11 Plan view of the aquifer shown in Figure 6.10: a simple flow net. Note that the streamlines are perpendicular to the equipotentials.

must be perpendicular to equipotentials. Darcy's law indicates that flow should always be from high values of the hydraulic head to low values. In fact, flow must follow the *path of steepest descent*. Picture the potentiometric surface or the water table as a topographic surface, with hills (high h) and valleys (low h). Now imagine a drop of water moving along this surface in response to the pull of gravity. We have to imagine this water drop moving rather slowly, without the momentum we might expect a ball bearing or other object to have on a steep surface. The water drop moves down the slope and perpendicular to the "topographic" contours rather than following some arbitrary path down the slope; that is, it follows the path of steepest descent. The rate of descent is proportional to the slope, just as the specific discharge is proportional to the hydraulic gradient.

An example of the use of flow nets to obtain an understanding of hydrological phenomena is presented by Hubbert (1940). Consider an idealized valley in which the groundwater flow from the hillslopes is perpendicular to the axis of the valley (Figure 6.12). A flow net for a cross section through the hillslope shows flow along "U-shaped" paths from the ridge to the valley bottom. Although the homogeneity of the material and two-dimensional nature of the flow would never exist in nature, the general picture of flow is instructive. Note that water flow near the stream is *not* horizontal (as one might expect intuitively) but has a significant upward component. Groundwater flow is not concentrated near the water table with a large volume of stagnant water at depth—the flow patterns are actually such that flow occurs throughout the saturated zone. The fact that water flows along "U-shaped" paths is of practical and theoretical importance. The U.S. Geological Survey conducted a study on the presence and the movement of agricultural chemicals in shallow groundwater on the Delmarva (Delaware-Maryland-Virginia) Peninsula (Hamilton and Shedlock, 1992). They found nitrate (partly from applied fertilizers) at almost all depths sampled in the groundwater and attributed the patterns of contamination to land-use practices and to groundwater flow paths.

The construction of accurate flow nets by hand is a task requiring considerable practice. More commonly today, flow nets are constructed using numerical solutions to the equations governing groundwater flow. Regardless of how a flow net is constructed, it

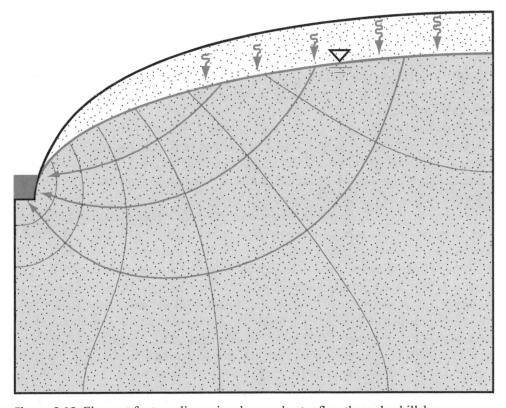

Figure 6.12 Flow net for two-dimensional groundwater flow through a hillslope.
Redrawn from Hubbert, 1940, fig. 45.

provides not only a visualization of the groundwater flow paths, but also information on the rate of groundwater flow in a particular region.

6.5.1 Quantifying groundwater flow using flow nets

Returning to the example of horizontal flow through an aquifer bounded by two channels separated by distance L (Figure 6.10), we find that

$$q = K \frac{h_L - h_R}{L}. \tag{6.14}$$

Because the hydraulic gradient, hydraulic conductivity, and aquifer thickness are constant, the value calculated using Equation 6.14 represents the specific discharge at any point in the aquifer. Quite often the total discharge in an aquifer, rather than specific discharge at a point, is of primary concern. Discharge per unit length of stream can be calculated from Equation 6.14 by multiplying by the thickness of the aquifer, b:

$$\textbf{Discharge per length} = Kb \frac{h_L - h_R}{L} = T \frac{h_L - h_R}{L}, \tag{6.15}$$

where $T = Kb$ [L^2 T^{-1}] is called the **transmissivity** of the aquifer. This is an important parameter when considering the development of a water supply from an aquifer. For example, a highly permeable formation 10 mm thick may not provide a usable supply of water but a formation 100 m thick with only a moderate value of hydraulic conductivity very likely will be usable.

We can use the geometry of a flow net like that in Figure 6.11 to calculate the discharge through the aquifer as well. This may not seem terribly important for this simple example, but when we consider more complicated flow patterns (e.g., Figure 6.12) we see that some simple calculations can be performed that allow us to quantify flow even in these settings.

In Figure 6.11, the simple flow net was constructed such that a series of squares was created, each bounded by a pair of equipotentials and a pair of streamlines. Within the flow net, water moves from high to low hydraulic head and cannot cross a streamline. The area between a pair of streamlines is referred to as a **streamtube**. In more complicated flow nets, these squares might become "curvilinear squares," as can be seen in Figure 6.12. If we isolate one of these squares (Figure 6.13) and make use of Darcy's law, then we can calculate the discharge through the square and extend this to determine the discharge through the aquifer. The square has sides of length ds (in the direction of flow) by dm (perpendicular to flow). Knowing that the aquifer thickness is b, we can apply Darcy's law to determine the total discharge through this box:

$$Q_S = qA = K(dm\, b)\frac{dh}{ds}, \tag{6.16}$$

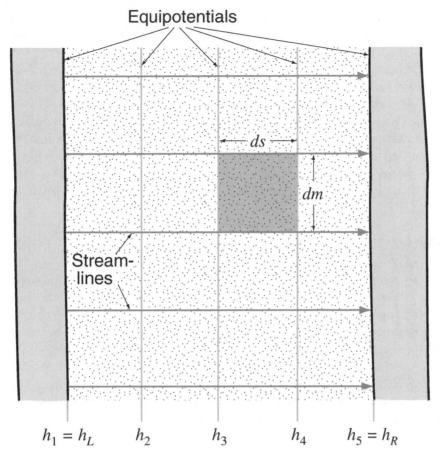

Figure 6.13 A simple flow net showing terms used to quantify flow in the aquifer. The gray square of dimension *ds* by *dm* is analyzed in the text.

where Q_s refers to the total discharge [$L^3\,T^{-1}$] through a streamtube and *dh* refers to the head difference across the box, or in this case $h_4 - h_3$. Because the domain is square, $dm = ds$, and equation (6.16) becomes:

$$Q_S = Kbdh. \tag{6.17}$$

In other words, if we know the hydraulic conductivity (or the transmissivity, $T = Kb$), we can simply look at our flow net to see what contour interval is used for hydraulic head (*dh*) and multiply that value by the transmissivity to determine the amount of water moving through each streamtube. If the aquifer is bounded at upper and lower ends, then we can count the number of streamtubes and multiply by Q_s to determine the total amount of water flowing through the aquifer.

We now look at a somewhat more complicated example. One effect of placing a dam in a stream or river is that a hydraulic gradient is created beneath the dam. At the upstream end, the hydraulic head (relative to the base of the dam) at the bottom of the reservoir is equal to the depth of water in the reservoir (Figure 6.14). On the downstream side, the hydraulic head is equal to the height of water in the river. A flow net can be constructed, as shown in Figure 6.14, to determine the pattern and rate of steady groundwater flow beneath the dam. Note that a low-permeability layer exists at depth, which represents the bottom of the aquifer beneath the dam. We can envision this boundary, as well as the base of the dam itself, as streamlines, because there will be relatively little flow across them. For this example, we assume that the dam is 100 m wide (in the direction into the page), and that the hydraulic conductivity of the material beneath the dam is 10^{-10} m s^{-1}. We can use Equation 6.17 to calculate the total discharge beneath the dam. We use the length of the dam (100 m) in place of the aquifer thickness (b). The contour interval for hydraulic head, dh, is 2 m. Then,

$$Q_S = Kbdh = 10^{-10} \text{ m s}^{-1} \times 100 \text{ m} \times 2 \text{ m} = 2 \times 10^{-8} \text{ m}^3 \text{ s}^{-1}. \tag{6.18}$$

We can multiply this value by the number of streamtubes (3) to obtain the total discharge, 6×10^{-8} m^3 s^{-1}, which is approximately equal to 1.9 m^3 yr^{-1}. In this case, the flow beneath the dam is fairly small, which is what we would hope for; the dam would be fairly ineffi-cient if water was constantly leaking around it! A larger value would result if the hydraulic conductivity was higher, or the difference in head across the dam was greater.

Before we conclude this chapter with a discussion of some of the complexities that enter into our evaluation of groundwater flow, we mention several other points regarding the use of flow nets as we have described it. First, we have considered only *steady ground-water flow*. It is not possible to construct a flow net for unsteady or transient groundwater

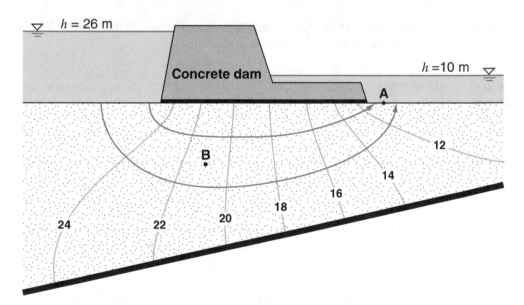

Figure 6.14 Flow net for groundwater movement beneath a concrete dam. The equipo-tentials (*gray lines*) are labeled with values of hydraulic head.

flow. Also, we can use a flow net to describe the variation in specific discharge, and, therefore, flow velocity, in an aquifer. You can imagine each streamtube as a "pipe," because water cannot cross a streamline. Therefore, from the principle of conservation of mass described in Chapters 3 and 4, we can state that within a streamtube, the specific discharge will be greatest where the streamtube is narrowest. The total discharge through the streamtube must be the same at any cross section. Referring to Figure 6.14, the flow velocity will be greater at point A than at point B. Finally, we have considered only homogeneous porous media, that is, materials with a spatially constant k. In the next section, we discuss the implications for flow through materials that are not homogeneous.

6.5.2 Heterogeneity and anisotropy

Virtually all natural materials through which groundwater flows display variations in intrinsic permeability from point to point. This is referred to as **heterogeneity**. The natural processes that create and modify rocks, sediments, and soils give rise to heterogeneity at all scales—from minor variations in grain size and the small holes called vugs in carbonate rocks (mm scale) to sedimentological and soil features (m scale) to variations in fracture spacing and lithological layering as depicted in Figure 6.6 (m to km scale). These variations in permeability from point to point complicate flow net construction. Permeable zones tend to focus groundwater flow, while, conversely, flow tends to avoid less permeable zones. It is possible to construct flow nets for some simple cases, such as layered aquifers and aquitards. However, when the permeability distribution is more complicated, hydrogeologists often rely on numerical simulation to depict the pattern and rate of groundwater flow.

Another complication often arises when measuring the permeability of natural materials. Rocks, sediments, and soils often have textural features that cause the permeability at a point to *vary with the direction of measurement*. Materials that display this trait are referred to as **anisotropic**, whereas **isotropic** refers to the condition in which the permeability does not depend on the direction of measurement. Consider a fractured rock aquifer, in which the fractures are predominantly horizontal. In this case the permeability will be higher when measured in the horizontal direction than in the vertical direction. You might picture this situation by referring to Figure 6.2 and imagining that the capillary tubes are aligned in the horizontal direction with relatively little hydraulic communication between them. Other features that produce anisotropy include (but are not limited to) the orientation of: platy minerals and small-scale layering in sedimentary rocks, such as clays; cooling cracks and lava flow tubes in basalts; large pores due to animal burrowing and plant roots in soils; and schistosity and fractures in metamorphic and igneous rocks.

Anisotropy, like heterogeneity, makes the job of constructing flow nets difficult, so that once again we must rely on numerical models that are capable of including this aspect of natural porous media. In the case of anisotropy, because water will tend to flow in a "preferred" direction, that is, in the direction of maximum permeability, Darcy's law must be modified to include this preference. As a result, in an anisotropic medium, the streamlines and equipotentials may not be perpendicular to one another at all points.

In introducing the concept of a flow net in Section 6.5, we described streamlines as paths of steepest descent in the "topographic landscape" of hydraulic head. We can extend this analogy to anisotropic media if we now imagine a series of ridges running at some angle with respect to the equipotentials. These ridges cause the water drop to move at some angle other than straight "downhill," since the water wants to follow the ruts between the ridges at least part of the time. (The flow will be straight downhill if the ridges are either straight downhill or follow the equipotentials.) These ridges in the hydraulic "topography" have the same effect as anisotropy in porous media. The flow direction will be altered from the normal direction parallel to the hydraulic gradient toward the direction of maximum intrinsic permeability.

6.6 Concluding Remarks

Darcy's law (and the law of conservation of mass implied in the construction and use of flow nets) provides the basis for computing steady groundwater flow patterns and rates. In some cases a simple "back-of-the-envelope" calculation is sufficient to gain a rough estimate of flow rates. In other cases, the aquifer may have a complex geometry or be heterogeneous and anisotropic and simple calculations will not suffice. Today hydrologists routinely use groundwater models (by which the groundwater flow equations are solved with the assistance of computers) to address environmental issues (e.g., Konikow et al., 2006). The computational techniques used in groundwater models are much more sophisticated than those we have used in this chapter. However, the ideas behind the computation are essentially the same. The geometry, conditions at the boundaries, and hydraulic parameters (intrinsic permeability, porosity) of the aquifer must be specified before predictions can be made with the models.

One use of groundwater models is in making assessments of compliance with environmental regulations. For example, before the WIPP site was licensed to begin operation, a set of regulations issued by the U.S. Environmental Protection Agency (EPA) had to be met. For WIPP, the U.S. Department of Energy had to show that the probability of significant releases to the accessible environment over 10,000 years into the future will be very small. The "accessible environment" for WIPP means groundwater in the Culebra formation "down gradient" of WIPP, that is, in the direction of decreasing hydraulic head. The demonstration of compliance with the EPA standard is done using a performance assessment analysis (Helton et al., 1997). In short, performance assessment uses a set of scenarios—sequences of hypothetical events that might occur in the future—and models to predict the impact of these scenarios. For WIPP, one scenario that is of concern involves someone in the distant future drilling a well into the repository and allowing some of the radioactive wastes to flow up the well and into the Culebra dolomite. A groundwater model must be used to "route" the hypothetical contaminant through the aquifer to decide whether significant quantities of contaminant might reach a human population in this scenario. These and many other environmental problems require the kind of knowledge of groundwater hydraulics that we have introduced in this chapter.

6.7 Key Points

- The term "groundwater" refers to subsurface water found in completely saturated porous media. {Section 6.1}

- Flow through porous media may be approached using the analogy of laminar frictional flow through a bundle of small capillary tubes, where the flow through an individual tube is described using Poiseuille's law. {Section 6.2}

- Hydraulic head is a measurable quantity that can be used to describe flow in porous media. Groundwater always flows from regions of high hydraulic head to regions of low hydraulic head: $h = (p/\rho g) + z$. {Section 6.2}

- Henry Darcy published a report in 1856 describing a set of experiments in which he measured flow rates through sand-filled columns. His experiments resulted in an empirical equation relating the specific discharge in porous media to the hydraulic gradient, referred to as Darcy's law: $q = -K(dh/dl)$. {Section 6.3}

- The proportionality constant K in Darcy's law is called the hydraulic conductivity, and is a measure of the ability of the medium to transmit fluid. Hydraulic conductivity depends on both material and fluid properties, and is often separated into two groups of terms, $K = k\,(\rho g/\mu)$, where k is the intrinsic permeability. {Section 6.3.1}

- The properties of porous materials need to be known or estimated to solve groundwater flow problems: the porosity (φ), which is the fraction of total medium volume occupied by pore space; the hydraulic conductivity (K), which varies by at least 13 orders of magnitude for natural materials; and the intrinsic permeability (k), which varies similarly to hydraulic conductivity. {Section 6.3.1}

- Darcy's law may be used for many situations of flow through porous media, except when values of the Reynolds number are high or other forces (accelerations, electrostatic forces) are significant. {Section 6.3.2}

- Geological formations may be classified as: aquifers, saturated geological formations that contain and transmit "significant" quantities of water under normal field conditions; aquicludes, formations that may contain water but do not transmit significant quantities; aquifuges, formations that neither contain nor transmit significant quantities of water; or aquitards, a general term for formations that are of relatively low permeability, which may include both aquicludes and aquifuges. {Section 6.4}

- Aquifers may be either unconfined (bound at the top by the water table) or confined (bound at the top by an overlying aquitard). {Section 6.4}

- The water table is a surface separating saturated and unsaturated zones in the subsurface. The fluid pressure at the water table is zero (gage). {Section 6.4}

- Wells and piezometers are constructed to measure water levels that represent the elevation of the potentiometric surface at a location. {Section 6.4.1 and 6.4.2}

- Geophysical techniques are useful for "seeing into the Earth" and learning about possible controls on water flow through soils and rocks. {Section 6.4.3}

- Flow nets are used to depict patterns of steady two-dimensional groundwater flow, and are made up of two families of orthogonal lines: equipotentials (lines of constant hydraulic head) and streamlines (representing the path of flowing groundwater). Flow nets may be used to calculate the rates of groundwater movement. {Sections 6.5 and 6.5.1}

- Most natural porous media are heterogeneous (permeability varies from point to point) and anisotropic (the permeability measured at a point depends on the direction of measurement). {Section 6.5.2}

6.8 Example Problems

Problem 1. You are charged with designing a very simple filtration system for a community water supply, using cylindrical sand columns ($K = 5.0$ m day^{-1}). The filter needs to be 3.0 m long to adequately trap particulates in the water, and since the system will be driven by gravity, the pressure heads at the top and bottom of the (vertically-oriented) filter will be zero.

A. What diameter filter is required to treat 4.0×10^3 gallons of water per day? Is this value feasible (anything larger than about 1 m is not feasible)?

B. Consider each of the alternatives and how you might modify your design:
   i. Lengthen the sand filter (how long?)
   ii. Raise the hydraulic head at the inflow (how high?)
   iii. Use several filters (how many? what size?)

Problem 2. A permeameter is used to measure the hydraulic conductivity of a porous medium (see Figure 6.3). The permeameter is perfectly round in cross section, with a diameter of 50.0 mm. The following parameters are measured: $z_1 = 220.0$ mm, $z_2 = 150.0$ mm, $p_1/\rho g = 230.0$ mm, $p_2/\rho g = 280.0$ mm, with $L = 200.0$ mm, and a discharge at the lower end of the column $Q = 500.0$ mm^3 min^{-1}.

A. Which way is water flowing in the column? Is water flowing from high to low hydraulic head? From high to low pressure?

B. Calculate the specific discharge, q (mm min^{-1}).

C. Calculate the hydraulic conductivity, K, of the material (m s^{-1}).

D. Calculate the intrinsic permeability, k (m^2).

Problem 3. Consider two piezometers placed side by side but open in different aquifers at depth (i.e., the two piezometers on the right side of Figure 6.6). The following measurements are made:

	Piezometer #1	Piezometer #2
Elevation of piezometer (m above mean sea level)	200	200
Depth of piezometer (m)	60	20
Depth to water in piezometer (m)	20	18

A. Calculate the elevation and hydraulic heads (relative to mean sea level), pressure head, and fluid pressure in the two piezometers, and fill in the table below.

	Piezometer #1	Piezometer #2
Elevation head, z (m)		
Pressure head, $p/(\rho g)$ (m)		
Hydraulic head, h (m)		
Pressure, p (Pa)		

B. Calculate the *vertical* hydraulic gradient between the two piezometers. Is the flow of water upward or downward?

Problem 4. Consider the flow net for a drainage problem shown in Figure 6.15. Drains such as pipes and culverts placed in a wet field may be used to remove groundwater by creating a "sink" or area of low hydraulic head. In the figure, a cross section through such a field is shown. The hydraulic conductivity, K, of the surficial material is 1.0×10^{-5} m s^{-1}. The thick black lines represent impermeable boundaries; a constant

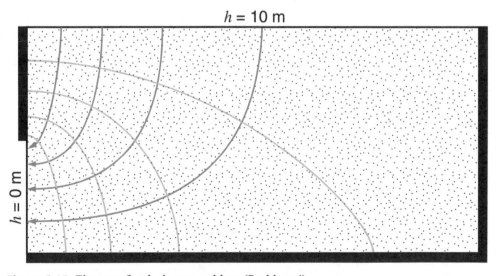

Figure 6.15 Flow net for drainage problem (Problem 4).

head is assigned to the top and lower left side. The cross section is 20 m long by 10 m deep. The gray lines are equipotentials and the blue lines arc streamlines.

A. Place labels on the equipotentials, indicating the value of the hydraulic head along the line.

B. Calculate the discharge through each streamtube, and the total discharge or rate at which the field is drained (m³ day⁻¹ per width of material).

6.9 Suggested Reading

Fetter C.W. 2000. *Applied hydrogeology*, 4th ed. Upper Saddle River, NJ: Prentice Hall, Chapters 3 and 4.

7 Groundwater Hydrology

7.1 Introduction

Concerns about groundwater arise in many settings, including, for example, developing plans for sustainable water uses in urban areas with limited surface water. The question of how to deal with the resource problems associated with urbanization—including water resources—is one that taxes responsible governments in almost every part of the world. In the United States, for example, more than 40% of the water used by people in Orange County south of Los Angeles comes from groundwater. Because precipitation and surface runoff vary with the seasons, Orange County uses spreading basins (gravel-lined areas) over which they spread water so it can infiltrate to recharge the groundwater that is in high demand. During most years there is not enough natural flow in the rivers to accommodate the demand for recharge. In those years, reclaimed wastewater may be used for recharge. For example, in the city of Anaheim the Santa Ana River is diverted into spreading basins to allow the water to infiltrate and recharge the underlying aquifer. In the summer over 90% of the flow in the river is treated wastewater that is returned to the river upstream of the spreading basins (NRC, 2012c). Trade-offs between amount and quality of water are inevitable and careful plans must be made to sustain the resource.

One example of a difficult problem in the use and protection of groundwater resources can be found in Mexico City, which is now one of the world's "megacities." Water must be supplied—and wastewater treated—for about 20 million people. Usable surface water is scarce in the Mexico City Basin, so groundwater from the Mexico City Aquifer is a primary source of freshwater. Near Mexico City, the aquifer is a sequence of alluvial fill sediments interstratified with basalt deposits and overlain by clays. The principal aquifer is from 100 to 500 m thick and the overlying clay aquitard is approximately 100 m thick. The water table below Mexico City has been declining at a rate of 1 to 1.5 m per year; the rate at which water is being withdrawn by pumping wells is much greater than the natural rate of replenishment. The practice of withdrawing groundwater faster than it can be replenished, referred to as "overdraft," has been common since about 1900. Furthermore, poor waste-disposal practices have adversely affected water quality in the aquifer. Clearly, if the water resource for Mexico City is to be made sustainable, changes in the pattern of use based on an understanding of the basin hydrology will be required.

In the United States, the High Plains Aquifer (often referred to as the Ogallala Aquifer) has experienced similar problems (McGuire, 2011). The High Plains Aquifer is an important source of water for much of the central United States, including parts of Colorado, Kansas, Nebraska, New Mexico, Oklahoma, South Dakota, Texas, and Wyoming. About 20 percent of the irrigated land in the United States is in the High Plains, and about 30 percent of the groundwater used for irrigation comes from the High Plains Aquifer. Overdrafts of groundwater have resulted in a decline in water levels throughout much of the aquifer. Between 1950 and 2009, the decline was more than 4 m (on average) and exceeded 45 m in some places. Declining water levels (which are expected to continue into the future) increase the cost of water, because pumps require more energy to lift the water a greater distance.

In places such as the High Plains, Mexico City, northern India, the North China Plain, and the southeast of Spain, groundwater is a limited resource and is being depleted (Wada et al., 2010). In other words, the rate of groundwater withdrawal exceeds the rate at which it is naturally replenished. As a result there is a net reduction in groundwater storage. In some cases, modern societies are using groundwater that accumulated in aquifers over geological time scales and under different climate conditions. This unsustainable use of non-renewable groundwater resources is sometimes called **groundwater mining**. The water removed from the ground is ultimately released to the ocean, and contributes—in addition to the effects of climate change—to sea-level rise (Wada et al., 2010; Konikow, 2011). Global estimates of groundwater depletion vary between 27 km^3 yr^{-1} (Margat et al., 2006) and 283 km^3 yr^{-1} (Wada et al., 2010). Table 7.1 shows an intermediate estimate (145 km^3 yr^{-1}) along with the major hotspots of groundwater depletion in the world. As a term of comparison, globally, the irrigation water used by crops is about 545 km^3 yr^{-1} (Siebert et al., 2010).

To understand limitations on the sustainable use of groundwater, we must examine the water balance for the subsurface. How and at what rate is groundwater replenished or recharged? Hydrological basins (the rocks, sediments, and soil underlying catchments) exhibit natural rates of groundwater flow as a result of the infiltration of precipi-

Table 7.1. Total net groundwater depletion and average groundwater depletion in 2001–2008

Aquifer	Net depletion in 2001–2008 (km³)	Average annual rate (km³ yr⁻¹)
In the USA		
Atlantic Coastal Plain	2.8	0.3
Gulf Coastal Plain	67.4	8.4
High Plains (Ogallala) Aquifer	94.7	11.8
Central Valley, California	31.4	3.9
Western Alluvial Basins	2.1	0.3
Western Volcanic Systems	2.9	0.4
Deep Confined Bedrock Aquifers	2.6	0.3
Agricultural and Land Drainage	0	0
Total (all USA systems)	203.9	25.5
Non-USA Aquifer Systems		
Nubian Aquifer System	18.9	2.4
North Western Sahara Aquifer System	17.6	2.2
Saudi Arabia Aquifers	109.1	13.6
North China Plain	40.0	5.0
Northern India and Adjacent Areas	423.5	52.9
Indirect Estimates for Other Areas	350.0	43.7
Total Global	1163.0	145.4

Source: According to Konikow (2011).

tation, exfiltration or discharge of groundwater, and evapotranspiration. Flow nets provide a tool for evaluating the natural rate of groundwater flow. How is that flow altered by perturbations such as pumping? Groundwater pumping in the Mexico City Basin has changed the pattern of flow as well as the water balance. Because the flow pattern determines how pollutants move within the subsurface, flow nets are also important in evaluating the risk of contamination.

Our best information about the water balance of an aquifer system, information that is essential for water resources planning, usually comes from records of water levels and groundwater pumping. This is one of the reasons why groundwater hydrology, as interpreted from flow nets and water-level records, is an important area of study.

7.2 Flow Nets and Natural Basin Yield

Flow nets can be constructed for any setting where the approximations of steady and two-dimensional flow are valid, as long as the conditions at the boundaries and the distribution

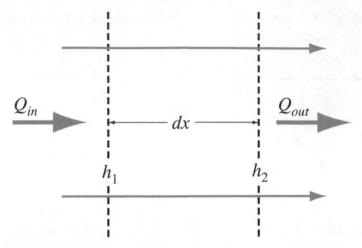

Figure 7.1 Definition sketch for the derivation of the steady groundwater flow equation.

of hydraulic conductivity are known. Later in this chapter we explore cases where flow is not steady. For the present discussion, we consider steady flow to be an approximation to the real system, such that temporal variations in the height of the water table are small relative to the thickness of the flow system.

Flow net construction is based on a mathematical model of groundwater flow. The form of this model is a differential equation that describes the physical process of groundwater motion. We can develop such an equation using conservation of mass and Darcy's law. Consider a single curvilinear square within a flow net as a control volume (Figure 7.1). This volume is bounded by two equipotentials and two streamlines. The streamlines can be thought of as impermeable boundaries, because no water crosses them. For steady flow, the inflow (Q_{in}) must equal the outflow (Q_{out}), or the change in storage must be zero. For the streamtube (the area bounded by a pair of streamlines), this may be written:

$$\frac{Q_{out} - Q_{in}}{dx} = \frac{dQ}{dx} = 0,$$

where Q [L^3 T^{-1}] refers to the total discharge at any section through the streamtube. For uniform flow in the x-direction (both streamlines are horizontal), the specific discharge, q_x, also must be a constant:

$$\frac{dq_x}{dx} = 0. \tag{7.1}$$

Darcy's law may be written for the specific discharge as:

$$q_x = -K\frac{dh}{dx}, \tag{7.2}$$

where dh is equal to $h_2 - h_1$. Substituting Equation 7.2 into Equation 7.1 and assuming that K does not vary in space (i.e., the material is homogeneous):

$$\frac{dq_x}{dx} = -K \frac{d}{dx}\left(\frac{dh}{dx}\right) = 0$$

or

$$\frac{d^2h}{dx^2} = 0. \tag{7.3}$$

In two spatial dimensions, this would become:

$$\frac{\partial^2 h}{\partial x^2} + \frac{\partial^2 h}{\partial y^2} = 0. \tag{7.4}$$

Equation 7.4 is referred to as the **Laplace equation** and may be solved to determine the distribution of hydraulic head (the equipotentials) in two dimensions, provided that the region is homogeneous (constant K). Of course in the two-dimensional case, the equipotentials (and streamlines) are not straight parallel lines as in the one-dimensional case in Figure 7.1. The conditions at the boundaries of a domain are required to solve this equation. A flow net may be thought of as a graphical solution to the Laplace equation. The Laplace equation is simply a mathematical statement of the law of conservation of mass (Equation 7.1) combined with Darcy's law (Equation 7.2). It is now common for the Laplace equation to be solved using computers. Computer-based groundwater models have become important tools for solving problems of groundwater hydrology and contaminant transport.

For steady flow in two dimensions through a hillslope, conservation of mass requires the inflow of water to be balanced by the outflow, because there can be no change in storage. In other words, the same number of streamtubes must leave the hillslope as enter it. This idea gives rise to naturally occurring **recharge** and **discharge areas**. A recharge area occurs where water is crossing the water table downward, hence, recharging the groundwater system. A discharge area occurs where groundwater is moving upward across the water table, thereby discharging into the unsaturated zone above, or to the land surface or a surface-water body such as a lake or stream. Consider a hillslope bounded by an impermeable bottom (for example, the bedrock beneath an unconfined aquifer) and two impermeable sides, called **groundwater divides**. The upper boundary is the water table. Often, the water table is a subdued replica of the land surface topography, such that the water table is higher beneath hilltops than beneath a valley. A flow net shows a balance between recharge in the upland portion and discharge in the valley (Figure 7.2a).

Flow nets must obey certain rules at the boundaries. For example, it is probably apparent that streamlines cannot cross impermeable, or "no-flow," boundaries. Equipotentials must be perpendicular to impermeable boundaries, because otherwise some flow across the boundary would be implied. A constant-head boundary is represented by an

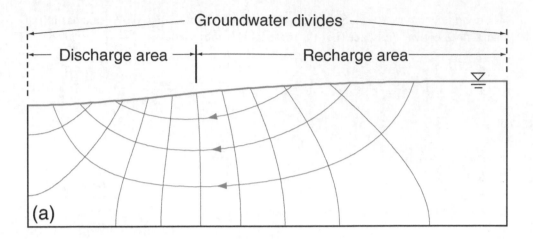

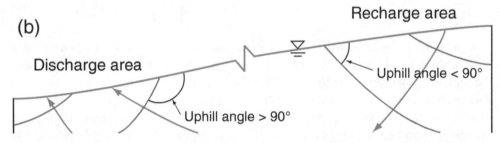

Figure 7.2 The natural pattern of groundwater flow in a simple basin (*a*). Equipotentials are gray and streamlines are blue. Recharge and discharge areas may be distinguished by looking at the angle that an equipotential makes with the water table (*b*).

equipotential, and streamlines must be perpendicular to a constant-head boundary. At the water table, the hydraulic head is everywhere equal to the elevation, because the gage pressure is zero. Equipotentials and streamlines may both intersect a non-horizontal water table, and the orientation of these lines provides a means of delineating recharge and discharge areas (Figure 7.2b).

The hydraulic head along any equipotential is a constant by definition. One way of labeling an equipotential is by noting where it intersects the water table (Figure 7.3). Water in a piezometer that is open at a particular depth will rise to an elevation equal to the hydraulic head at that point. For example, in Figure 7.3 the piezometer is open at the depth depicted. . The elevation head (z) at that point is 20 m. Because the open interval intersects the 40-m equipotential, the hydraulic head (h) at that point is 40 m. The pressure head ($p/\rho g$), therefore, is equal to $h - z$, or 20 m.

Flow nets have a "dimensionless" quality. They are a picture of the *pattern* of steady groundwater flow and depend only on the physical features of the basin. For example, the pattern shown in Figure 7.2a does not depend on the actual size of the basin or the (homogeneous) permeability of the basin rocks or sediments. However, quantities determined from the flow net *do* have particular dimensions (and units). Recalling the discus-

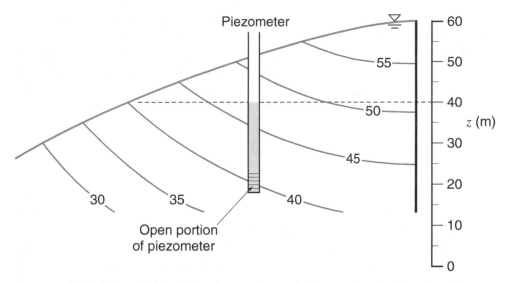

Figure 7.3 The hydraulic head along any equipotential is equal to the elevation of its intersection with the water table.

sion in Chapter 6 on quantifying flow using flow nets, the discharge through each streamtube may be calculated knowing only the hydraulic conductivity and the difference between adjacent equipotentials (dh):

$$Q_S = Kbdh. \tag{6.17}$$

Equation 6.17 is valid as long as the flow net has been constructed using curvilinear squares. For a horizontal flow net, b is the aquifer thickness (see Chapter 6). The total volume of flow through a vertical cross section depends on the cross section width, however, and so b refers to that width. If we let $b = 1$ unit [L], the discharge calculated from Equation 6.17 is simply the discharge *per unit basin width*.

Consider how we might use the flow net shown in Figure 7.2a. We would have to know the length of the basin, for example $L = 100$ m, in which case the hill is 30 m ($0.3L$) high on the right side. The difference in hydraulic head between adjacent equipotentials is equal to $0.005L$, or 0.5 m. If the hydraulic conductivity is 0.5 m day^{-1} (approximately 5.8×10^{-6} m s^{-1}), the discharge through each streamtube is:

$$Q_S = Kbdh = (0.5 \text{ m day}^{-1})(1 \text{ m})(0.5 \text{ m}) = 0.25 \text{ m}^3 \text{ day}^{-1}. \tag{7.5}$$

There are four streamtubes in the flow net. Therefore, the total discharge through the section is 4×0.25 or 1.0 m^3 day^{-1}. This is the discharge per meter basin width ($b = 1$ m). If the basin is 500 m wide, then we could simply multiply 1.0 m^3 day^{-1} by the 500-m width to find the total discharge from the basin (500 m^3 day^{-1}).

The above calculation shows that the basin discharge is proportional to the hydraulic conductivity of the basin material. Knowledge of the permeability of basin materials is essential to determining the water balance for a groundwater system. Freeze and Witherspoon (1968) referred to a calculation such as that above as the **natural basin yield**. Under natural or undisturbed conditions (i.e., in the absence of anthropogenic groundwater withdrawals or changes in climate or vegetation), this is the average rate of discharge from a hillslope or basin.

7.3 Regional Groundwater Flow

The simple flow net above (Figure 7.2a) provides a template for understanding regional groundwater flow. However, flow patterns are conditioned by variation in the shape of the basin and the water table, and spatial patterns of hydraulic conductivity. This section explores the primary controls on the pattern and rate (natural basin yield) of steady groundwater flow in a basin bounded by divides at the sides, the water table at the top, and a low-permeability unit at the base. The terms introduced in the previous section (recharge and discharge areas, natural basin yield) provide some keys to exploring the importance of each of these controls.

7.3.1 The effect of basin aspect ratio

Hydrological basins occur in a variety of shapes, and we would expect the basin shape to exert some influence over the groundwater flow pattern. Tóth (1962, 1963) examined the influence of **basin aspect ratio** (length to depth) on the pattern of groundwater flow in a homogeneous two-dimensional basin with a gradual water-table slope. Imagine a basin of constant length (L), but with a depth that might vary, depending on the depth to a low-permeability unit, such as crystalline bedrock. The pattern of flow will be similar to that shown in Figure 7.2a, with single recharge and discharge areas. Our experience with Darcy's law should tell us that the natural basin yield will be large for deeper basins (smaller aspect ratio), because the volume through which groundwater is flowing is larger. This assumes that we are comparing basins with identical water-table profiles. An analogy might be constructed using columns of different cross-sectional area. For a given hydraulic gradient (analogous to the slope of the water table in a basin), the discharge will be greater through a larger column. Recall that Darcy found that, for the simple case of one-dimensional flow, the total discharge was proportional to the cross-sectional area.

Compare Figure 7.4a and Figure 7.4b. The natural basin yield in each case is directly proportional to the number of streamtubes and to the hydraulic conductivity of the basin materials. For the shallow basin (large aspect ratio, Figure 7.4a), there are only two complete streamtubes and a small fraction of a third. For the deep basin (small aspect ratio, Figure 7.4b), almost eight complete streamtubes occur. This observation matches our expectation, although the total discharge is not perfectly proportional to the basin depth. In a catchment setting, we might expect the thickness of permeable material near the surface to have a big influence on the rate of groundwater flow. Where thin soils are found, the rate of groundwater flow might be relatively small.

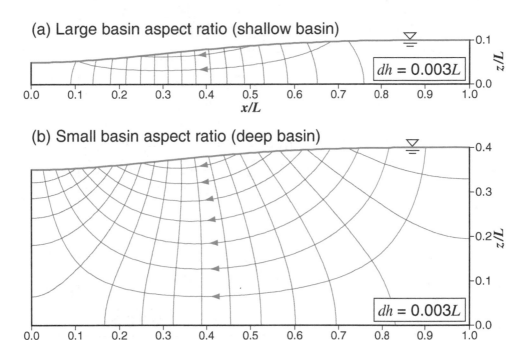

Figure 7.4 The effect of basin aspect ratio (length to depth) on natural patterns of groundwater flow. The water-table profile is the same for the shallow (*a*) and deep (*b*) basins.

There is some difference between the patterns of flow in the two basins in Figure 7.4. For the deep basin (Figure 7.4b), vertical hydraulic gradients (and, therefore, vertical flow) exist over a large portion of the basin. In the shallow basin (Figure 7.4a), the flow is essentially horizontal over most of the basin.

7.3.2 The effect of water-table topography

Variation in hydraulic head resulting from topographic relief of the water table is in most instances the driving force for groundwater flow. Complex land-surface topography should produce similarly complex water-table topography, and so we need to consider the flow systems produced by such topography. Although real topography can be quite complex, we might consider a general picture in which "local" topography (small-scale undulations or hills and valleys) is superimposed on a "regional" slope. The end-member cases are shown in Figures 7.4 and 7.5, for regional and local topography, respectively. These patterns are similar in some respects, because flow occurs from highs to adjacent lows. A single hill-valley flow system (dashed boxes in Figure 7.5) resembles the larger flow systems produced by regional water-table topography (Figure 7.4). These flow systems appear to have a lower depth limit, below which flow is very slow or non-existent, as may be seen particularly well in Figure 7.5b. However, the flow doesn't actually become zero, but only slows as streamtubes become wider and the hydraulic gradient along the

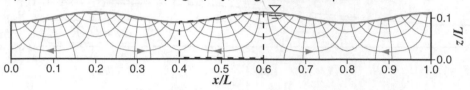

(a) Local watertable topography, large basin aspect ratio

(b) Local watertable topography, small basin aspect ratio

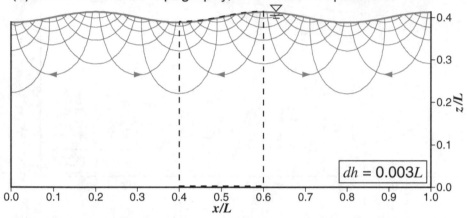

Figure 7.5 Regional groundwater flow patterns for the case of local water-table topography. The dashed boxes indicate an individual local flow system. The same equipotential spacing (dh) of $0.003L$ is used for both the large (a) and small (b) basin aspect ratios.

streamlines decreases. It could also be shown that the apparent "depth" of these local flow systems depends on the local hydraulic gradient.

Tóth, in the same studies cited earlier, referred to these flow cells (dashed boxes in Figure 7.5) in which water flows from topographic high to adjacent low as **local flow systems**. Larger flow systems, from regional high to low, he referred to as **regional flow systems**. The similarity in appearance between these types of flow systems suggests that the use of different terms to describe them depends only on one's definition of "regional." However, the distinction becomes clearer when we superimpose local hill-and-valley topography on top of a regional slope (Figure 7.6). For the case of a deep basin (Figure 7.6b), both local and regional systems develop (streamtubes labeled "L" and "R" in the figure), as well as what Tóth referred to as an **intermediate flow system** (streamtube labeled "I" in the figure). If the basin is relatively shallow (Figure 7.6a), a regional system may still exist, but may be attenuated as a result of the dominant influence of the local flow systems.

The conclusions that we can draw from these flow nets are as follows:

1. If local relief is negligible, but a regional water-table slope exists, only a regional flow system will develop.

2. Conversely, if local hill-and-valley topography exists, but no regional slope, only local flow systems will develop.

(a) Regional and local watertable topography, large basin aspect ratio

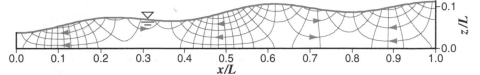

(b) Regional and local watertable topography, small basin aspect ratio

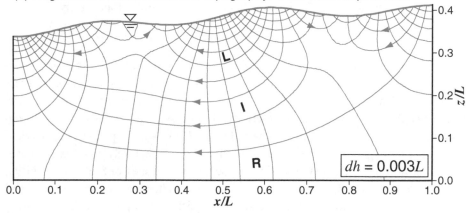

$dh = 0.003L$

Figure 7.6 Groundwater flow patterns for the case of combined regional and local water-table topography. Streamtubes labeled "L," "I," and "R" indicate the local, intermediate, and regional flow systems, respectively. The same equipotential spacing (*dh*) of 0.003*L* is used for both the large (*a*) and small (*b*) basin aspect ratios.

3. If both local and regional topography exists in a basin, all three types of flow systems (local, intermediate, and regional) will develop. As a result, precipitation infiltrating on a hilltop may eventually discharge at an adjacent low or follow a longer flow path toward the regional low point.

Complex water-table topography will necessarily produce complex patterns of groundwater flow, but the general observations discussed above can be applied to other settings. Referring to Figure 7.7, can you identify local, intermediate, and regional flow systems?

7.3.3 The effect of geological heterogeneity

Geological materials are always heterogeneous, a fact that complicates analysis. Aquifers and other permeable zones in the subsurface are capable of capturing and focusing groundwater flow. In some cases, recharge and discharge areas may develop in locations that are not predicted from the configuration of the water table. Hillside springs are one good example. Freeze and Witherspoon (1966, 1967) performed a series of numerical "experiments" in which they calculated flow nets for a variety of configurations of the water table and subsurface geology. These calculations demonstrated the ability of subsurface aquifers to alter the flow system (Figure 7.8).

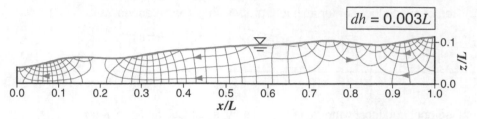

Figure 7.7 A flow net for a shallow basin with complex water-table topography.

A highly permeable unit such as an aquifer causes flow to become more nearly vertical within the overlying unit, and may narrow the discharge area (Figure 7.8b). Aquifers act as conduits that are capable of rapidly transmitting water to the principal discharge area. Because the mean hydraulic conductivity of the basin is greater, the natural basin yield is increased (compare Figure 7.8a with Figure 7.8b). A large portion of this yield is transmitted through the aquifer. An aquitard at depth effectively reduces the basin depth and natural basin yield (Figure 7.8c). Discontinuous zones of high permeability, such as the truncated aquifer shown in Figure 7.8d, may alter the areas of recharge and discharge. The discharge produced at the end of the aquifer may appear as a surface spring or seep.

Hsieh (2001) produced a computer model that is useful for demonstrating the influence of the topography of the water table and heterogeneity on flow nets. Once the model is downloaded (http://water.usgs.gov/nrp/gwsoftware/tdpf/tdpf.html), it can easily be used to explore a variety of different flow conditions to gain insights about recharge and discharge areas of groundwater flow systems.

Mexico City lies within the discharge area for the Basin of Mexico, which is a "closed basin." This means that there are no natural surface water outflows for the basin, and all outflow occurs through evapotranspiration. The Basin is ringed by volcanic mountains that form the recharge areas for the Basin. Because of the heterogeneity of the materials filling the basin, an aquifer-aquitard system exists, and the pattern of flow from the mountains to the center of the basin looks something like Figure 7.8b. Prior to heavy pumping during the past 100 years, groundwater flowed from the recharge areas (mountains) into discharge areas in the Basin center through the aquifers, and discharged upward across the overlying aquitards. Surface springs and lakes resulted from this natural pattern of groundwater flow.

Potable groundwater under artesian pressures (i.e., heads that caused water flow to the surface in wells without pumping) was discovered in the Basin of Mexico in 1846, promoting the rapid development of this resource (NRC, 1995b). Since that time, the large overdrafts of groundwater have reduced the pressures within the confined aquifers and altered the natural pattern of groundwater flow. Now most of the water within the Basin is moving downward toward the heavily pumped aquifers; surface springs and lakes have dried up. In 1983, systematic monitoring of the water levels in wells began. These records of water levels in wells have provided important information on the water balance in the basin.

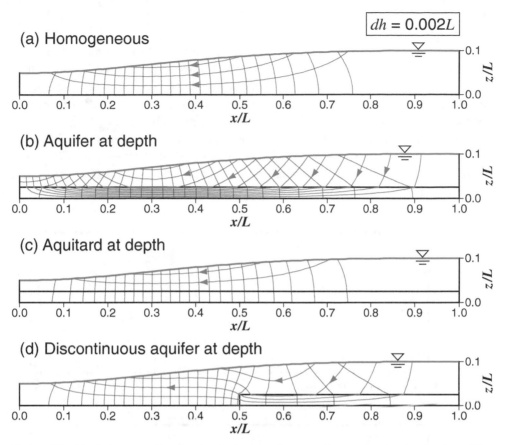

Figure 7.8 The influence of geological heterogeneity on patterns of groundwater flow. The homogeneous case (*a*). An aquifer at depth (*b*; hydraulic conductivity of the aquifer is an order of magnitude greater than the overlying unit). An aquitard at depth (*c*; hydraulic conductivity an order of magnitude less than the overlying unit). A discontinuous aquifer at depth (*d*). The equipotential spacing (*dh*) is the same for all, $0.002L$.

7.4 Well Hydrographs

The three major controls on patterns of groundwater flow described in the previous section concern *natural* groundwater flow. The natural system is assumed to be at steady state, such that recharge and discharge balance one another and the water table is approximately constant. Groundwater systems, however, are both dynamic (although sometimes sluggish) and subject to human alteration, as we have seen for the Mexico City region. The task of depicting patterns of flow in these circumstances is a difficult one. We often have observations, in the form of well hydrographs, that allow us to gain some understanding of these effects. As we will see in the next sections, confined and unconfined aquifers behave somewhat differently, and we consider each separately.

7.4.1 Unconfined aquifers

A **well hydrograph** shows the variation in water level in a well through time. In an unconfined aquifer, the water level in a well generally indicates the position of the water table.
Well hydrographs may show variations over different time scales. For example, daily
fluctuations may be observed in carefully monitored wells, and very gradual changes over
several years of observation may also be evident. The most obvious cause of water-level
variations is fluctuation in the inputs and outputs. In Coffee County, Tennessee, the water-
table elevation is observed to vary with the season (Figure 7.9). Precipitation in Tennessee
is not very seasonal (monthly precipitation averages are relatively constant). However,
recharge to groundwater does vary seasonally because it is influenced by factors other than

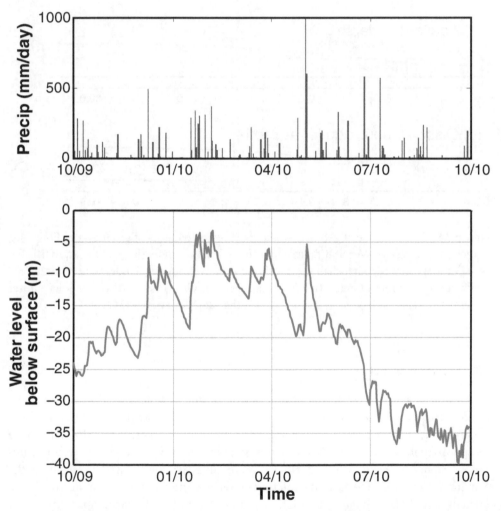

Figure 7.9 Daily precipitation (*a*) for a weather station in Coffee County, Tennessee, USA and water levels (*b*) for water year 2010 in a nearby well in an unconfined
aquifer.

the amount of rainfall, the most important of these being evapotranspiration. The variation in the water table in Coffee County can be explained by noting that evapotranspiration is at a peak in the summer months and greatly reduced in the winter months. Consequently, because the ground remains unfrozen during the winter in southern Tennessee, more water is available for recharge in the winter and groundwater levels rise. The converse is true in the summer—little recharge occurs and levels decline as the groundwater discharges to provide the baseflow in streams.

In areas where the water table is close to the ground surface, groundwater levels are influenced directly by the transpirational demands of plants. During the day when transpiration is high, water movement is *upward* from the water table and the level declines. At night, transpiration is reduced, groundwater flows laterally from locations upslope that are relatively unaffected by direct transpiration effects, and the water table recovers (Figure 7.10).

Recharge to aquifers does not occur solely from direct infiltration of rainfall. For example, water can seep from surface-water bodies, such as rivers, ponds and lakes, into the ground. Artificial recharge (recharge induced by activities of people as opposed to that which occurs naturally) can be implemented by introducing water into wells (recharge

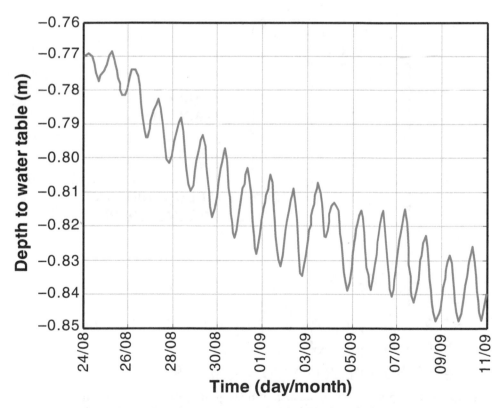

Figure 7.10 Water levels for a late summer period in 2010 in a shallow well adjacent to White Clay Creek at the Stroud Water Center near Avondale, Pennsylvania, USA.

wells) or by routing water into infiltration basins in permeable material (the idea mentioned in regard to Orange County in the introduction).

Unconfined aquifers store water; a change in the elevation of the water table indicates a change in the amount of water stored in the aquifer. Consider a portion of an unconfined aquifer of (in plan view) area 1 m² (Figure 7.11). A 1-m drop in the water table would result from the removal of a certain volume, V, of water. The volume of water removed depends on the porosity of the aquifer and how much water would remain behind after gravity has drained the upper cubic meter of aquifer material. (The water that remains is held by forces that will be described in the next chapter.) Hydrologists refer to this volume, V, per square meter of aquifer per meter drop in the groundwater table as the **specific yield** (S_y) of an unconfined aquifer. Because it is the volume of water produced per unit aquifer area per unit decline in hydraulic head, it is dimensionless. The specific yield is characteristic of a given aquifer and allows determinations to be made of the change in the volume of water stored over time. Values of specific yield typically are less than the porosity, and most range from 0.01 to 0.30. Not all of the water in an aquifer volume will drain under the influence of gravity. Some will be retained by forces causing adhesion of water to particle surfaces. This is why the specific yield is less than the porosity.

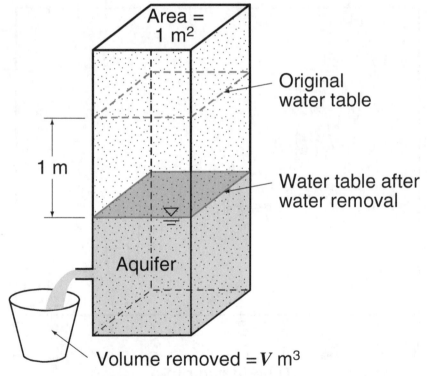

Figure 7.11 The specific yield of an unconfined aquifer. For a 1-m decline in the water table, the volume of water produced per unit aquifer area is the specific yield, S_y.

The specific yield is a hydrological parameter that determines the response of the water table to changes in inputs and outputs. In the case of an increase in evapotranspiration (Figures 7.9 and 7.10), the change in water-table level may be fairly uniform over a given area, although variations will occur due to the lateral movement of groundwater and spatial variations in evaporation rate and vegetation. Pumping a well has a different effect. Pumping produces a decrease in hydraulic head at a point, which increases the hydraulic gradient toward the well. The change in water level in the pumping well, or in observation wells nearby, is referred to as a **drawdown**. The amount of this drawdown will decrease as one moves away from the pumping well, and the pattern that is produced is called a **cone of depression** because of its characteristic shape (see Figure 7.14). The shape and extent of the cone of depression within an unconfined aquifer depend on the pumping rate, the transmissivity and specific yield of the aquifer, and time. From the definition of the specific yield, however, we know that the *volume* of the cone of depression is equal to the volume of water removed (by pumping) divided by the specific yield:

$$V_{cone} = \frac{V_{pumped}}{S_y}.$$ \hfill (7.6)

7.4.2 Confined aquifers

In an unconfined aquifer, the response to pumping a well is a change in the water table. Water is drained out of the aquifer as the water table declines, and the specific yield provides a measure of the volume of water released from storage. However, the upper boundary of a *confined* aquifer, the overlying aquitard, does not move substantially in response to withdrawing water from a well within the aquifer. Instead, a cone of depression is created *within the potentiometric surface*. The aquifer material is not being drained, and the aquifer remains saturated. However, water is withdrawn from storage within the aquifer (otherwise, the well would become a dry hole almost immediately). Analogous to the specific yield of an unconfined aquifer, we can define the **storativity** (S) of a confined aquifer as the volume of water produced per unit aquifer area per unit decline in the potentiometric surface (Figure 7.12). The question becomes, how is water removed from a confined aquifer without de-watering (draining) the aquifer material?

Consider the forces acting on a horizontal plane within a confined aquifer (Figure 7.13). A downward force due to the weight of the overlying material exists, which when divided by the area of the plane (A) is called the **total stress** (σ_T). Because the plane is not undergoing an acceleration, Newton's second law tells us that this downward force must be balanced by an equivalent opposing force or forces. In other words, the weight of the overlying material must be supported or held up by something. In this case the opposing forces per unit area (stresses) are the pressure (p) of the water and the upward stress exerted by the aquifer solids, called the **effective stress** (σ_e):

$$\sigma_T = p + \sigma_e.$$ \hfill (7.7)

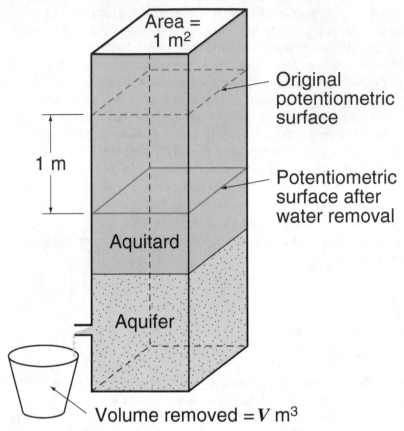

Figure 7.12 The storativity of a confined aquifer. For a 1-m decline in the potentiometric surface, the volume of water produced per unit aquifer area is the storativity, S. The aquifer material is not drained and remains saturated.

This idea may be stated in another way: the weight of the overlying material on a horizontal plane is supported in part by the fluid and in part by the solid. The total stress won't vary over time:

$$d\sigma_T = dp + d\sigma_e = 0.$$

Therefore, any change in pressure must be offset by a change in effective stress:

$$dp = -d\sigma_e. \tag{7.8}$$

If water is being withdrawn from a well within a confined aquifer, the hydraulic head, and, hence, the fluid pressure, is being reduced at that point (dp is negative). As a result, the fluid will expand slightly. This is because water is (slightly) compressible. This is one mechanism by which water is released from storage in a confined aquifer. The additional *volume* of water produced may flow to the well and be withdrawn by the pump.

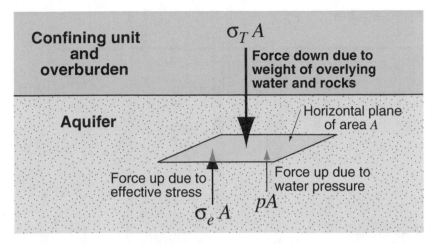

Figure 7.13 Force due to water pressure and effective stress balance the downward force due to the weight of the overburden.

Equation 7.8 indicates that the decrease in fluid pressure must be accompanied by an *increase* in the effective stress (positive $d\sigma_e$). A part of the weight of the overlying material is being transferred from the fluid to the solid. This results in the compression of the aquifer material, just as decreasing the fluid pressure resulted in expansion of the fluid. Compressing the aquifer material is similar to squeezing a sponge, and produces water that may be pumped from the well. This is the second mechanism by which water is removed from storage in a confined aquifer.

Water and most aquifer materials are not very compressible. As a result, storativity values tend to be lower than values of specific yield. Storativity generally ranges from 0.005 to 0.00005, as compared with 0.01 to 0.30 for specific yield. In this discussion of storage in confined aquifers, we have assumed that the aquifer is perfectly confined, or that all the flow occurs within the aquifer and none within the confining layers. If an aquifer is only semi-confined, with some water coming from surrounding aquitards, then it is referred to as a **leaky aquifer**. Because of this additional source of water, the storativity will often be higher in leaky aquifers.

The extent of the cone of depression depends, among other things, on the value of the storage parameter (either specific yield or storativity). Consider two aquifers, one unconfined and one confined, that are being pumped at the same rate. Because the specific yield of the unconfined aquifer will be greater than the storativity of the confined aquifer, the drawdown in the unconfined aquifer will be less than that in the confined aquifer (Figure 7.14). That is, the volume of the cone of depression in a confined aquifer is equal to the volume of water pumped divided by the storativity [change S_y to S in Equation 7.6], a very small number. Thus removal of water from confined aquifers produces substantial drawdown of the potentiometric surface. In confined aquifers that are heavily pumped over long time periods, the cone of depression can be quite extensive, approaching tens of kilometers in lateral extent, with tens of meters of drawdown in the vicinity

(a) Unconfined aquifer

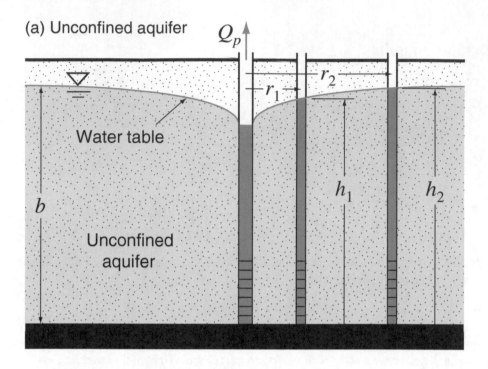

(b) Confined aquifer

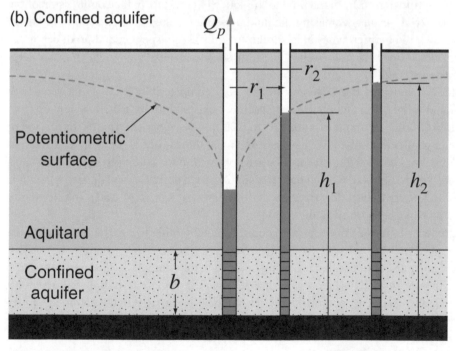

Figure 7.14 The cone of depression in an unconfined (*a*) and confined (*b*) aquifer. The wells are being pumped at a constant rate (Q_p), and hydraulic heads are being monitored at two piezometers at radial distances r_1 and r_2 from the pumping well.

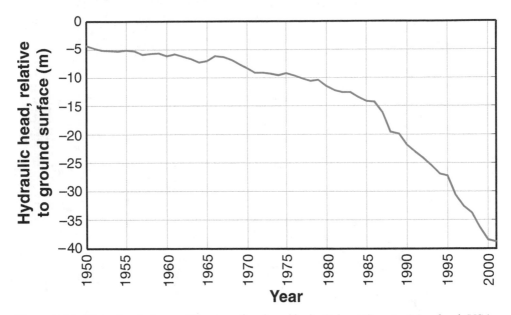

Figure 7.15 Water levels in a well in a confined aquifer in Calvert County, Maryland, USA. Data from DePaul et al. (2008).

of the well (Figure 7.15) and a cone of depression that extends a long distance from the major pumping center (Figure 7.16).

Other things can affect the balance between water pressure and effective stress in aquifers and so affect the water levels in wells. Atmospheric pressure varies as low-pressure and high-pressure systems move across the landscape. An increase in atmospheric pressure creates an additional force on the top of a confined aquifer that is distributed between water pressure and effective stress. That is, only a portion of the force from the increased atmospheric pressure will be transmitted to the water in the aquifer. The increased atmospheric pressure also acts on the water standing in the well. In the well, the full increase in the atmospheric pressure is transmitted to the water. Thus, for a confined aquifer, an increase in atmospheric pressure causes a decrease in the water level in a well and, conversely, a decrease in atmospheric pressure results in an increase in water level. For an unconfined aquifer, changes in atmospheric pressure are transmitted equally to the water table and the water in a well, and so do not produce changes in water levels.

7.5 Well Tests to Estimate Aquifer Properties

The rate at which water levels change in wells depends in part on the aquifer properties. For example, if a quantity of recharge water is suddenly added to an unconfined aquifer, the water levels will decline as the water drains out to streams. The rate of decline will depend on the hydraulic conductivity and the storativity of the aquifer. Thus, if the water table or potentiometric surface in an aquifer is purposefully disturbed, by removing or adding water for example, measurements of the time variation of water levels in wells

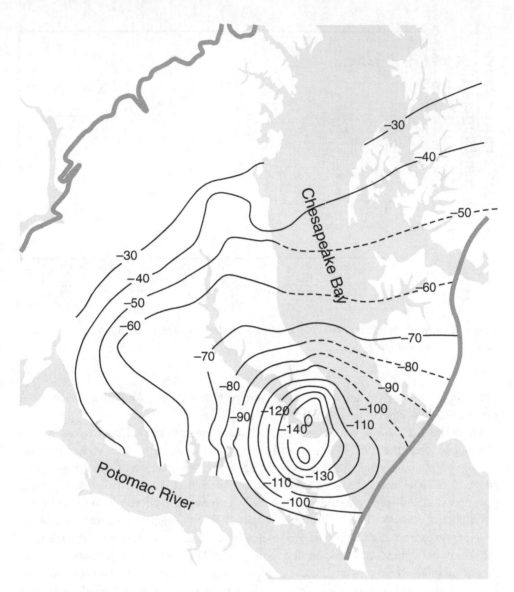

Figure 7.16 Map of the potentiometric surface around a groundwater pumping center in Maryland, United States. Contours are feet relative to mean sea level. (Original level was approximately 10 feet above mean sea level.)

Figure modified from Soeder et al. (2007)

can be used to back-calculate the aquifer properties. Below we briefly describe two well tests that are used in the field.

7.5.1 Slug tests

Slug tests use measurements made on a single well. The water level in the well is suddenly perturbed by removing a "slug" of water. This lowers the water level in the well, creating a head gradient toward the well. According to Darcy's law, water will then flow radially into the well until the water level in the well is restored. The level of the water in the well relative to the steady-state water table, H, is measured as a function of time.

The initial perturbation inside the well is a depression of the water level by H_0 and subsequently the level varies with time, $H(t)$. The well casing has radius r_c whereas the radius of the borehole is R. The length of the well screen (the non-impervious part of well casing through which water flows) is L (Figure 7.17).

The rate of water flow into the well, Q, can be calculated using Darcy's law.

$$Q = 2\pi r L K \frac{dh}{dr},$$

where r is the distance from the center of the well and h the saturated thickness of the unconfined aquifer at distance r (Figure 7.17).

This equation can be rearranged as:

$$\frac{Q}{2\pi KL}\frac{dr}{r} = dh. \tag{7.9}$$

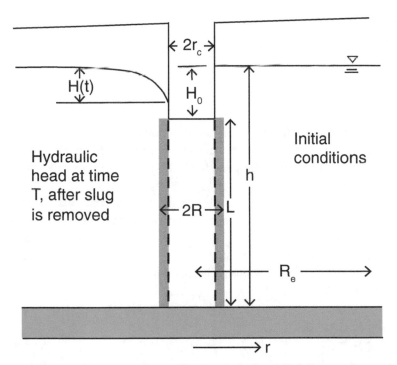

Figure 7.17. Schematic for a slug test. R_e is the radial distance beyond which the perturbation by the slug has no effect. Note that the right side of the figure depicts conditions initially, before the slug is removed, and the left side depicts conditions that evolve through time after the slug test is initiated.

Equation 7.9 can be integrated across the step change in H caused by the removal of the slug, letting r vary from R to R_e, the effective radius beyond which there is no effect.

$$\frac{Q}{2\pi KL}\ln\frac{R_e}{R} = (h_{R_e} - h_R) = H(t). \tag{7.10}$$

Because after the initial abrupt withdrawal no other water is extracted from the well, a mass balance in the well itself gives the following relationship.

$$\frac{dH}{dt} = -\frac{Q}{\pi r_c^2}. \tag{7.11}$$

Equations 7.10 and 7.11 can be combined to get

$$\frac{dH}{H} = -\frac{2KL}{r_c^2 \ln(R_e/R)} dt. \tag{7.12}$$

which can be integrated from time zero (when the slug is removed) to a later time, t, to give:

$$\ln\frac{H_t}{H_0} = -\frac{2KL}{r_c^2 \ln(R_e/R)}t, \text{ or}$$

$$\ln H(t) = -\frac{2KL}{r_c^2 \ln(R_e/R)}t + \ln H_0. \tag{7.13}$$

Equation 7.13 provides a relationship that permits estimation of hydraulic conductivity given measurements of time, t, and $H(t)$. The natural logarithm of H can be plotted versus time. Equation 7.13 indicates that the plot should be a straight line with slope $-2KL/[r_c^2 \ln(R_e/R)]$ and intercept $\ln H_0$.

Values for L, r_c, and R can be gleaned from records kept during well construction. The radial extent of the slug perturbation, R_e, is not known. One assumption is to set it equal to L. Because of the logarithmic dependence of K on R_e errors associated with this assumption have only a small impact on the estimation of K. With this assumption and an estimate of the slope of the straight line relationship between $\ln H$ versus t, the hydraulic conductivity can be calculated.

$$K = \frac{-(slope)\times r_c^2 \ln(L/R)}{2L}. \tag{7.14}$$

For example, data collected for a slug test on well near Oyster, Virginia (Figure 7.18) are fit with a straight line with slope $-0.19\,\text{s}^{-1}$. The well casing has an inside diameter of 7.6 cm and the borehole radius is estimated to be 12.7 cm. The length of the well screen

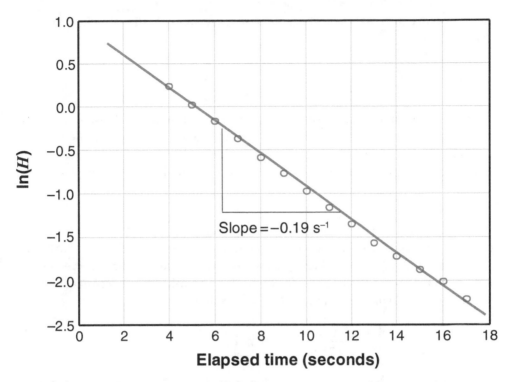

Figure 7.18 Slug test results for a well near Oyster, Virginia, USA.

is 5 m. From Equation 7.14, we estimate the hydraulic conductivity for this shallow (sand) aquifer to be

$$K = (0.19 \text{ s}^{-1})(0.076 \text{ m})^2 \ln(5 \text{ m}/0.127 \text{ m})/(10 \text{ m}) = 4.1 \times 10^{-4} \text{ ms}^{-1}.$$

7.5.2 The Cooper-Jacob method

Slug tests to determine hydraulic conductivity in the field are quite useful in many ways. They are easy to carry out, the duration of a test is short, and only one well is involved. The measurements reflect a relatively small volume of aquifer material around the well, however. In cases where an average response over a larger area is the aim, tests that reflect the aquifer properties between two wells may be preferred. The basic idea is that one well is pumped and the hydraulic head is measured as a function of time in observation wells, say at distances r_1 and r_2 from the pumping well (Figure 7.14). The heads h_1 and h_2 will change with time as the pumping proceeds. The equation that describes the variation of head with respect to distance from the pumping well and time is an extension of the Laplace equation (Equation 7.4) to account for time-varying conditions. (Details can be found in hydrogeology texts such as Fetter, 2000.)

$$\frac{S}{T}\frac{\partial h}{\partial t} = \frac{\partial^2 h}{\partial x^2} + \frac{\partial^2 h}{\partial y^2}. \tag{7.15}$$

The Cooper-Jacob (1946) method can be used to estimate S and T from a solution to Equation 7.15.

$$s = h_0 - h = \frac{Q_P}{4\pi T}\ln t + \frac{Q_P}{4\pi T}\ln\frac{2.25T}{r^2 S}. \tag{7.16}$$

In a field test, pumping is started and elapsed time is measured. The drawdown, s, is the difference between the original height of the potentiometric surface in the observation well and the measured hydraulic head as a function of elapsed time. Equation 7.16 indicates that a plot of s versus the $\ln t$ should be a straight line with slope $Q_P/4\pi T$ and intercept $\frac{Q_P}{4\pi T}\ln\frac{2.25T}{r^2 S}$.

Miah and Rushton (1997) report data for pumping tests conducted in the Madhupur aquifer in Kapasia, Bangladesh (Figure 7.19). A 10-cm diameter well was pumped at a rate of 5140 m^3 day^{-1} for several days. Drawdown was measured in an observation well 100 m from the pumped well. The slope of the drawdown-$\ln t$ plot is 0.36 m and the intercept is 1.82 m (Figure 7.19). The Cooper-Jacob solutions give

$$T = \frac{Q_P}{4\pi(\text{slope})} = \frac{5140 \text{ m}^3 \text{ day}^{-1}}{(4\times\pi\times 0.36 \text{ m})} = 1130 \text{ m}^2 \text{ day}^{-1}.$$

Once transmissivity has been estimated, the storativity can be obtained using the intercept of the straight line fit to the data.

$$S = \frac{2.25T}{r^2 \exp\left((\text{intercept})\dfrac{4\pi T}{Q_P}\right)} = \frac{(2.25)(1130 \text{ m}^2 \text{ day}^{-1})}{(100 \text{ m})^2 \exp\left[\dfrac{(1.82 \text{ m})(4\pi)(1130 \text{ m}^2 \text{ day}^{-1})}{5140 \text{ m}^3 \text{ day}^{-1}}\right]}$$

$$= 0.0016.$$

Notice that, unlike the Cooper-Jacob method, the slug test provides an estimate only of the hydraulic conductivity and not of the storativity. This is due to the fact that the slug test is based on a water balance equation for the well (Equation 7.11), in which the flow from the aquifer is calculated without accounting for the unsteady character of groundwater flow. In fact, Equation 7.9 is integrated assuming that Q is constant with respect to r. This assumption (i.e., $\partial Q/\partial r = K\partial^2 h/\partial r^2 = 0$) corresponds to setting to zero the left-hand side of Equation 7.15. Because the transient character of the slug test's response is expressed by the water balance of the well and not by the unsteady groundwater flow, this test does not provide any information on the "inertia" of groundwater flow (i.e., of storativity or specific yield).

7.6 Contaminant Hydrogeology

Among the very important environmental issues today are how to prevent contamination of groundwater and, should contamination occur, how to clean up aquifers. Groundwater

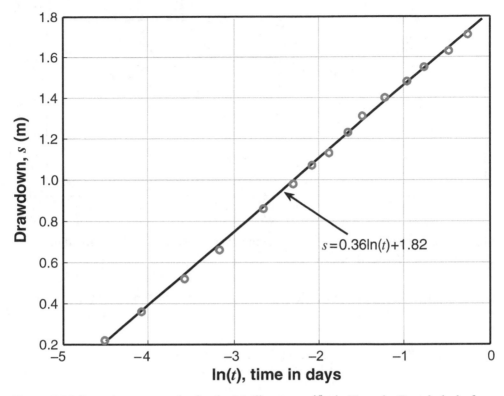

Figure 7.19 Pumping test results for the Madhupur aquifer in Kapasia, Bangladesh, for the Cooper-Jacob method.

flow is relatively slow in human terms, so once an aquifer is contaminated, clean up can be extremely difficult or even essentially impossible using reasonable resources. Knowledge needed to determine effective protective measures and to explore the feasibility of clean-up options includes insights from physical hydrologists regarding rates and directions of groundwater flow.

Advection is the process of transport of contaminants by the movement of the water in which they are dissolved or suspended. The contaminants "go with the flow" and travel along groundwater flow paths. For example, a contaminant injected into the groundwater at the water table will move by advection along a path determined by the flow net if the flow is steady (Figure 7.20). The speed at which the contaminant moves can be estimated using Darcy's law provided the hydraulic conductivity, K, and the porosity, ϕ, of the aquifer is known. Although the specific discharge, q, given by Darcy's law has dimensions of velocity, it is not a true velocity. The specific discharge is discharge per total cross-sectional area of aquifer, but the flow occurs only through the openings in the rock. Recalling that porosity represents the fraction of the total volume of rock that is available to water, the **average linear velocity** of the groundwater flow can be estimated as $v = q/\phi$. If we take the schematic in Figure 7.20 to represent a local flow system in a sandy aquifer with $L = 10\,\text{km}$, $\phi = 0.25$, $K = 10^{-6}\,\text{m s}^{-1}$, and

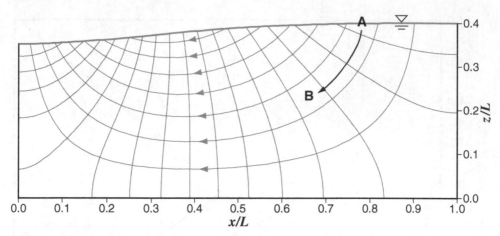

Figure 7.20　A contaminant entering the groundwater at point A will be carried by advection toward point B.

a head gradient along the flow path A → B of 0.033 m/m, the time for a contaminant to travel 100 m would be

$$t_{travel} = \frac{path\ length}{v} = \frac{100\ \text{m}}{(10^{-6}\ \text{m s}^{-1} \times 0.033\ \text{m/m})/0.25} = 7.5 \times 10^6\ \text{s} \approx 87\ \text{days}.$$

Even with this relatively high head gradient and hydraulic conductivity, we would estimate that the time for a contaminant to travel the length of the domain would approach 25 years. For lower gradients and longer paths, the times would be longer. Flushing a contaminated aquifer by natural recharge of fresh water can require very long time periods.

Advection is not the only process that occurs as contaminants move in groundwater. The flow paths through pores, cracks, and fractures in the aquifer are convoluted and water moving along various paths mixes as the paths converge, diverge, join, and split. These mixing processes result in spreading of the contaminant by **dispersion**. A pulse of contaminant will spread out to form a plume as it moves along the flow path (Figure 7.21).

Groundwater contamination can originate from many sources, for example from landfills that were constructed before we understood how contaminants can leach from them and move to groundwater. Over time, extensive plumes can form, threatening natural ecosystems and the quality of streams into which the groundwater drains. Hydrologists need to understand how transport occurs to address potentially serious environmental problems. The U.S. Geological Survey has been studying hydrological and biogeochemical process at a landfill near Norman, Oklahoma (Christenson and Cozzarelli, 2003). The leachate plume has spread several hundred meters from the landfill and has affected a nearby wetland. Determination of flow paths and rates of flow require estimates of hydraulic conductivity. At the Norman landfill site, slug tests were used to characterize the heterogeneity of the sediments and gage the impact on groundwater contaminant transport (Scholl and Christenson, 1998).

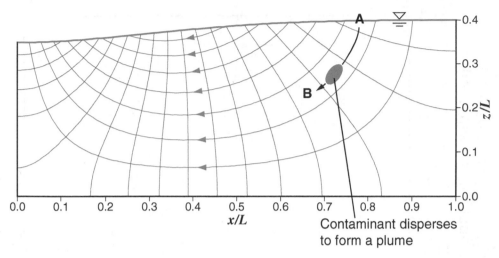

Contaminant disperses
to form a plume

Figure 7.21 Dispersion spreads a contaminant out as it is advected along a groundwater flow path.

7.7 Land Subsidence

A common misconception concerning confined aquifers has been that confining layers are perfectly impervious. However, no geological materials are truly impervious and confining units such as aquitards (e.g., clays, shales) may contain large amounts of water that can move given enough time. Figure 7.8b illustrates this possibility. In instances where overlying units (or interlayers within the aquifer system) are clays, pumping the confined aquifer has two effects. The first is to increase the effective stress on the aquifer and the second is to de-water the clay units. What happens when clays are de-watered? You may have observed the cracks that appear on the bottoms of mud puddles that have dried up. These cracks reflect the shrinkage of clays upon drying. The same thing can (and does) happen when clay units above or within aquifers have part of the water in them removed: they shrink. The result is called **land subsidence**.

As discussed, the cone of depression in confined aquifers can be quite large (e.g., Figure 7.16), and land subsidence due to pumping can be significant over considerable areas. In the San Joaquin Basin, California, subsidence has affected an area of at least 4200 square miles; maximum subsidence approaches 9 m in some areas (Helm, 1982).

In Mexico City the lowering of the potentiometric surface in the Mexico City Aquifer has resulted in the removal of water from the overlying clays. The land surface has subsided by some 7.5 m in the central part of Mexico City (Figure 7.22; NRC, 1995b). Old well casings in the city extend several meters above the ground surface because the land has subsided around the casings. The result of this subsidence has been extensive damage to the city's infrastructure, including building foundations and the sewer system. Another serious problem in Mexico City relates to flooding. The city is bordered to the east by Texcoco Lake—the natural low point of the southern portion of the Basin of Mexico. In 1900, the lake bottom was 3 meters below the median level of the city center.

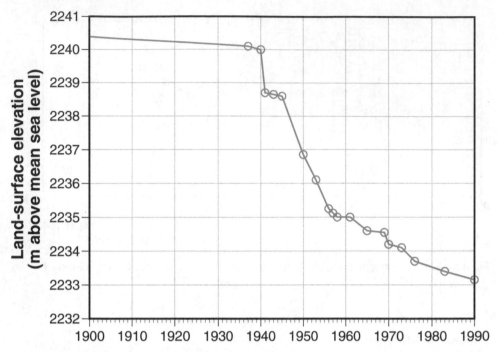

Figure 7.22 Land subsidence in Mexico City.

Data from Ortega-Guerrero et al. (1993).

By 1974, the lake bottom was 2 meters *higher* than the city! This change is the result of greater land subsidence within the city and has aggravated flooding problems. A complex drainage system (including excavations to lower Texcoco Lake) has evolved to control flooding.

7.8 Groundwater Recession

Groundwater levels in unconfined aquifers tend to rise after rainy periods when recharge occurs and to decline gradually as the water discharges into streams and other bodies of surface water following recharge events. The declining portion of a stream hydrograph following a storm, snowmelt, or wet season reflects the decline in groundwater levels through time because surface-water runoff to the stream ceases in a relatively short time after the precipitation stops. This decline in groundwater input to a stream is known as **groundwater recession**.

Recession curves often have the shape of a negative exponential. In regions with a very pronounced seasonal rainfall distribution this can be seen very clearly. In Pescadero Creek, south of San Francisco, for example, a semi-logarithmic plot of streamflow recession shows that a straight-line segment (an exponential relationship) fits the streamflow data reasonably well from early May through early July (Figure 7.23). A straight line also fits the data from July through August, but with a different slope than

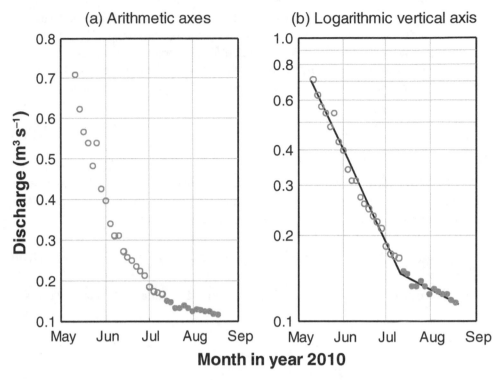

Figure 7.23 Groundwater recession for Pescadero Creek, California, showing both arithmetic (*a*) and logarithmic (*b*) axes.

the earlier data. This result suggests that, for Pescadero Creek, two groundwater reservoirs with different time constants control the base-flow recession. The form of the curve for the recessions is:

$$Q = Q_0 e^{-ct}, \tag{7.17}$$

where Q [$L^3\, T^{-1}$] is the discharge at time t after recession begins, Q_0 is the discharge at the beginning of the recession period, t is time measured from the beginning of the period, and c [T^{-1}] is a *recession constant*.

The appropriateness of this equation can be investigated by considering the conservation of mass. During recession, the inflow to the entire groundwater reservoir can be assumed to be zero and we would expect that the outflow would be a function of some measure of the groundwater elevation, $\bar{h}$ [L]. In other words, the hydraulic head (and hydraulic gradient) should decrease with time, and, thereby, reduce the driving force for groundwater discharge. We could write this as:

$$q_r = f(\bar{h}), \tag{7.18}$$

where q_r is the discharge per meter width of stream [$L^2\,T^{-1}$]. Conservation of mass requires the outflow to be balanced by the change in groundwater storage. The time rate of change of groundwater stored should depend on $\bar{h}$, the storativity S or specific yield S_y (depending on whether the aquifer is confined or unconfined), and the length of the aquifer L (Figure 7.24). The appropriate conservation equation would then be:

$$SL\frac{d\bar{h}}{dt} = -q_r = -f(\bar{h}). \tag{7.19}$$

The left side of this equation is the time rate of change of water stored in the aquifer per meter stream width. Now to integrate Equation 7.19 we need a specific form of $f(\bar{h})$. The simplest relationship that might be used is a direct proportionality:

$$q_r = cSL\bar{h}, \tag{7.20}$$

where c is a proportionality constant [T^{-1}]. Substituting Equation 7.20 into 7.19 gives:

$$\frac{d\bar{h}}{dt} = -ch, \tag{7.21}$$

which can be integrated to give:

$$\bar{h} = \bar{h}_0 e^{-ct}, \tag{7.22}$$

where $\bar{h}_0$ is the average hydraulic head at the initial time. Then, from Equation 7.20 we have:

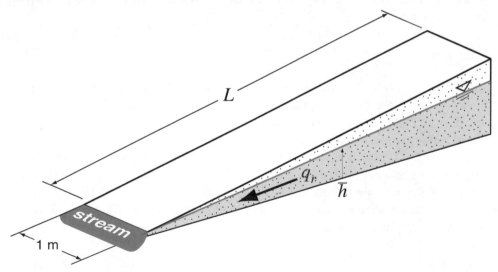

Figure 7.24 A simple model of groundwater recession.

$$q_r = (cSL\bar{h}_0)e^{-ct}.$$

or

$$q_r = q_{r0}e^{-ct} \tag{7.23}$$

Equation 7.23 is directly comparable with Equation 7.17. This comparison indicates that the use of 7.17 implies that the groundwater reservoir behaves as a *linear reservoir*, that is, that groundwater outflow is directly proportional to the amount of water stored. Analysis of baseflow recession curves can be useful for estimating flows throughout a dry season and thus can be the basis for planning water use in areas with seasonal precipitation regimes.

7.9 Concluding Remarks

Natural groundwater flow patterns and rates are controlled by basin geometry, water-table configuration, and how hydraulic conductivity is distributed in the subsurface. Flow nets, or the modern equivalent, numerical groundwater models, are useful tools for describing these patterns and rates. Groundwater hydrology also relies on the analysis of well hydrographs. The changes in water levels and the consequences of these changes can be of profound importance. An excellent example is the case of Mexico City. Mexico City is dependent on groundwater; only 2% of its water supply is from surface water within the basin. Several aqueducts bring surface water from other basins as far away as 125 km, but these supply only about 26% of the demand. Thus, about 72% of Mexico City's water supply comes from groundwater (NRC, 1995b). Planning for sustainable use of the aquifer is obviously critical.

Despite the realization in the early part of this century that groundwater resources in the Basin of Mexico were being rapidly depleted, over-exploitation of the aquifer continues to cause problems of land subsidence and increased vulnerability to contamination. To predict a useful life for the aquifer under current pumping rates (or to plan future uses to keep from depleting the resource), the drawdown rates of the aquifer must be measured accurately and the hydraulic parameters of the aquifer must be estimated. The National Research Council (1995b) recommended implementation of a program to develop the hydrological data necessary for supporting sustainable management of the resource. Analysis of well hydrographs will be an important part of this program.

In many places, conserving groundwater resources is an important concern. Numerical models can be used to construct flow nets describing past or present groundwater flow patterns. Several kinds of information are required to model groundwater flow accurately, some of which have been described in this chapter and Chapter 6, namely: porosity, intrinsic permeability, and storativity or specific yield of basin materials; basin geometry and water-table configuration; and a record of water levels that can be compared with model predictions. Once the present conditions have been established, these models can be used to predict the consequences of future or continued changes or management practices. The High Plains Aquifer in the central United States was mentioned in the introduction as a heavily exploited aquifer system. There is obvious concern for this resource because of the dramatic changes in water levels that have been observed over the past 50 years. A numerical modeling study of the impact of future (1998–2020) management strategies for this aquifer system predicted water-level declines of over 30 meters for some portion of the region (Luckey et al., 1999). The extent of the predicted region of declining water levels depends on the particular pumping rates assumed. Despite the vastness of the region underlain by this aquifer system, the groundwater resource does have a limit, and significant depletion remains a concern for the millions of people who depend on it.

7.10 Key Points

- Groundwater flow patterns are characterized by recharge and discharge areas; water moves downward across the water table in recharge areas, and upward in discharge areas. {Section 7.2}

- The water table is often a subdued replica of the land-surface topography, especially in humid areas. {Section 7.2}

- Flow nets may be used to determine the natural basin yield-the quantity of water, on average, that discharges from a basin. If the flow is steady, this quantity will be balanced by the recharge rate. {Section 7.2}

- The pattern and rate of groundwater flow are controlled by the basin aspect ratio, the water-table topography, and the distribution of permeability. {Section 7.3}

- The basin aspect ratio is the ratio of basin length to basin depth or thickness. Shallow basins are characterized by mostly horizontal flow, and deeper basins by significant vertical flow components and relatively higher natural basin yield. {Section 7.3.1}

- Variations in the height of the water table drive groundwater flow. The topography of the water table may produce patterns of local, intermediate, and regional flow. {Section 7.3.2}

- Hydrological basins are generally heterogeneous; the distribution of aquifers and aquitards influences the pattern of groundwater flow. Aquifers tend to focus flow and provide relatively rapid pathways for water movement toward discharge areas. {Section 7.3.3}

- A well hydrograph is a record of water level in a well through time. Well hydrographs show variations due to daily and seasonal variations in evapotranspiration rate, pumping of wells, and any other process affecting the subsurface water balance. {Section 7.4}

- Changes in groundwater storage are quantified using the specific yield (unconfined aquifers) and storativity (confined aquifers). Both may be defined as the quantity of water produced per unit aquifer area per unit decline in the hydraulic head. Specific yield values tend to be much larger than values of the storativity, because declines in head in unconfined aquifers are accompanied by drainage of a portion of the aquifer; in confined aquifers, the material is not drained and water is only released from storage due to expansion of the water and compression of the aquifer. {Sections 7.4.1 and 7.4.2}

- Water removed from aquifer storage through pumping produces a cone of depression in the water table or potentiometric surface. The size of the cone of depression is related to the pumping rate, the aquifer transmissivity, and the storage coefficient (specific yield or storativity). {Section 7.4.2}

- Withdrawing water from a well reduces the hydraulic head and fluid pressure at that point. As a result, the effective stress (the total stress minus the fluid pressure, or $\sigma_e = \sigma_T - p$) is increased; the effective stress is that portion of the weight of the overlying material that is borne by the solid material of the aquifer. Water is then "released from storage" in two ways: through fluid expansion as the pressure decreases and by expulsion from the compressed aquifer material. {Section 7.4.2}

- Well tests are used to determine aquifer hydraulic properties in the field. Slug tests performed in a single well give information about the hydraulic conductivity of the aquifer material surrounding the well. {Section 7.5.1}

- Tests involving more than one well use a pumping well to remove water from an aquifer and observation wells to measure changes in hydraulic head that occur as a result of the pumping. The Cooper-Jacob (1946) method can be used with the observation well measurements to estimate the transmissivity and storativity of the aquifer in the vicinity of the area of the wells used in the test. {Section 7.5.2}

- Contaminants in an aquifer are advected by the water flow at an average linear velocity of q/ϕ. Dispersion of the contaminant also occurs during flow, causing spreading of the contaminant into a plume. {Section 7.6}

- Land subsidence is one result of groundwater withdrawal from pumping wells. This subsidence is produced by compression of the aquifer material (an increase in effective stress) and by de-watering of clayey aquitards, causing shrinkage. {Section 7.7}

- Streamflow in times between precipitation or snowmelt events is generally produced by groundwater discharge. Because the groundwater reservoir must decrease as a result (assuming no new water is being supplied), the groundwater discharge decreases in time, often producing an exponential groundwater recession curve ($Q = Q_0 e^{-ct}$). {Section 7.8}

7.11 Example Problems

Problem 1. Determine the natural basin yield ($m^3 yr^{-1}$ per meter basin width) for the following cases.

A. For the basin in Figure 7.4a with $L = 5000\,m$ and $K = 30\,m\ yr^{-1}$

B. For the basin in Figure 7 4b with $L = 5000\,m$ and $K = 30\,m\ yr^{-1}$

C. For the basin in Figure 7.5a with $L = 200\,m$ and $K = 100\,m\ yr^{-1}$. You might want to begin by calculating the discharge through each local flow system (dashed box).

Problem 2. Answer the following questions for the basin in Figure 7.25.

A. Is the lower unit an aquifer or an aquitard?

B. What fraction of the total basin yield passes through the lower unit?

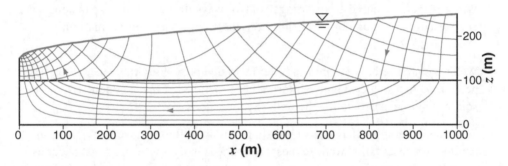

Figure 7.25 Flow net for Problem 2.

Problem 3. Consider the flow net shown in Figure 7.26. The sides of the region are groundwater divides, the top boundary is the water table and the bottom is an impermeable boundary. Streamlines are blue and equipotentials are gray.

A. Label the equipotentials with the appropriate value of hydraulic head (m).

B. Draw arrows on the streamlines indicating the direction of groundwater flow.

C. Label all recharge and discharge areas.

D. Indicate at least one area within the flow net where flow is relatively fast, and one area where flow is relatively slow.

E. Determine the water level (hydraulic head, m) in wells A and B.

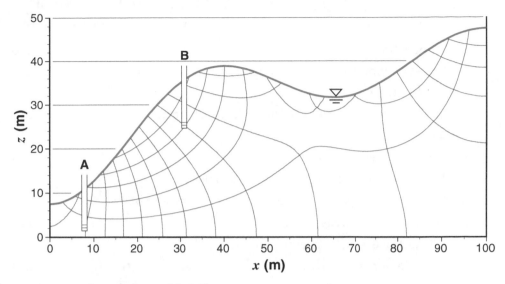

Figure 7.26 Flow net for Problem 3.

Problem 4. Well hydrographs for unconfined aquifers can be used to determine the change in water stored within the aquifer. Consider the change in water-table level observed in Coffee County, Tennessee, between mid-May and August 2010 (Figure 7.9).

A. What is the change in water-table level during this time period (m)? Does this indicate an increase or a decrease in water stored within the aquifer?

B. If the specific yield of the aquifer is 0.25, and the aquifer has an area of 600 km², what is the change in water stored over the same time interval (m³)?

Problem 5. Determine the recession constant, c (day^{-1}), for the two groundwater "reservoirs" contributing to Pescadero Creek during the spring and summer of 2010 (Figure 7.22).

7.12 Suggested Readings

Fetter, C.W. 2000. *Applied hydrogeology*, 4th ed. Upper Saddle River, NJ: Prentice Hall, Chapters 5 and 7.

National Research Council (NRC). 1995. *Mexico City's water supply: Improving the outlook for sustainability*. Washington: The National Academies Press.

8 Water in the Unsaturated Zone

8.1 Introduction

In most areas, with the exception of bogs and swamps, the water table is some distance below the ground surface. Between the ground surface and the water table is a region in which the pore spaces of the rock or soil may be partly filled with air and partly with water. This region is referred to as the **unsaturated zone** or **vadose zone** and water in this zone is referred to as **soil moisture**. Hydrologists want to be able to describe the flow of water in the unsaturated zone to deal with a number of important issues. For example, in the last chapter we discussed the concept of recharge to subsurface aquifers. Recharge takes place most often through the unsaturated zone, either overlying an unconfined aquifer or in the recharge zone of a confined aquifer. Changing climate and land use practices have strong effects on recharge through the vadose zone, and understanding these effects is particularly important in arid and semi-arid areas (Scanlon et al., 2006).

One important aspect of water flow in the vadose zone is the water balance of plants. Most terrestrial plants extract water from the vadose zone. Plants wilt when soils become too dry because the forces holding the water in the soil are too great to allow the plants access to the water (see also Chapter 9). Related to the water balance of plants is the practice of irrigation in agriculture. Agriculture accounted for 90% of freshwater use over the past century and water used for irrigation accounts for 90% of consumptive use (return of water to the atmosphere by evapotranspiration) of water that is withdrawn from streams, rivers, and aquifers (Scanlon et al., 2007). Given that water resources are already stressed in many regions and that food demand is expected to increase by 50% as the global population increases to 9 billion by 2050, there is a real need to improve the efficiency with which we use water to grow crops (Rockström et al., 2007). Understanding the movement of soil water, and its uptake by plants and "loss" through evapotranspiration and recharge to the groundwater system, is essential in this regard.

Another example of an important problem in vadose-zone hydrology is the use of semiarid locations with deep unsaturated zones for the disposal of wastes. The Low-Level Radioactive Waste Policy Act was passed by the United States Congress in 1980. This act provides for the formation of regional compacts by states to supply sites for the safe disposal of low-level radioactive waste (LLRW). LLRW includes test tubes, rags, rubber gloves, tools, and so forth used in medical research and treatment, in other research (for example, in research in environmental sciences to study the biodegradation of organic wastes), and in nuclear power plants. The major concern about finding a suitable site is groundwater contamination, because the pathway through groundwater is the one that is most likely to be the one that places human populations at risk of exposure to contaminants. Given the concern for transport of contaminants to groundwater, it has been recognized for some time that disposal of wastes in the unsaturated zone in a desert environment should be a safe alternative for such disposal (Winograd, 1981).

In response to the LLRW Policy Act, Arizona, California, North Dakota, and South Dakota formed the Southwest Compact and selected a site in Ward Valley in the Mojave Desert for the first disposal site for the Compact. Before any site can be used for the disposal of wastes, a license must be obtained. The licensing procedure requires careful study of the site and estimates of rates at which radionuclides might leach into the groundwater. One of the key parts of the analysis is the determination of flow rates in the unsaturated zone. Even with careful study, siting waste-disposal facilities is a contentious issue and opposition can be expected to any site. Opponents often challenge the scientific assumptions regarding hydrological processes. The relationships between the forces on and the flow of water are ingredients of arguments presented by both proponents and opponents of waste-disposal facilities. We will return to the case of Ward Valley at the end of this chapter after we have developed the ideas useful in describing the flow of soil moisture.

Finally, as we will see in our discussion of catchment dynamics in Chapter 10, the storage within and release of water from the vadose zone are quite important in determining the stormflow dynamics of a catchment. The infiltration and movement of water in the unsaturated zone represents one potential pathway for precipitation entering a stream. Variation in the ability of water to infiltrate a soil or rock is, therefore, an important aspect of catchment water dynamics.

The early literature recognized three divisions within the unsaturated zone: the capillary fringe, the intermediate belt, and the belt of soil water (e.g., see Meinzer, 1923). According to Meinzer, the **capillary fringe** is "a zone in which the pressure is less than atmospheric, overlying the zone of saturation and containing capillary interstices some or all of which are filled with water that is continuous with the water in the zone of saturation but is held above that zone by capillarity acting against gravity." That is, the capillary fringe is a saturated zone *above* the water table where water is affected by capillary

forces. The uppermost belt, or belt of soil water, is "that part of the lithosphere immediately below the surface, from which water is discharged into the atmosphere in per-
ceptible quantities by the action of plants or by soil evaporation." This definition recognizes that plants, for the most part, extract water from a portion of the soil (the "root
zone") near the surface. The intermediate belt is "that part of a zone of aeration [i.e., the
unsaturated zone] that lies between the belt of soil water and the capillary fringe." The
intermediate belt is distinguished mainly by the fact that something must be between
the root zone and the capillary fringe. The distribution of moisture above the water table

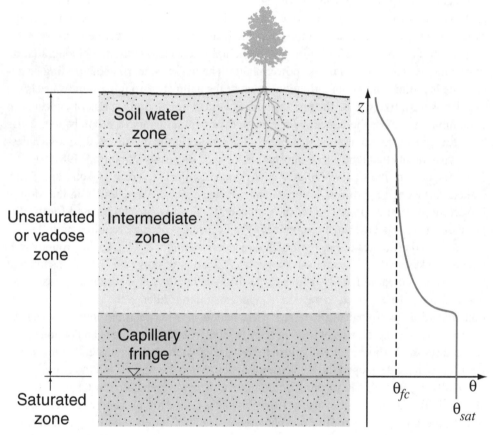

Figure 8.1 The distribution of moisture in the vadose zone and the classification of
waters according to Meinzer (1923). Water near the surface of the soil is available for
uptake by plant roots. After several days of fair weather, the moisture content in this belt
of soil water (or root zone) decreases substantially due to evapotranspiration. Directly
beneath the root zone, the moisture content tends to be fairly constant over a depth of up
to a meter or more. The relatively constant value of moisture content in this region is
referred to as the field capacity of the soil. Near the water table, the pores of the soil act
as "capillary tubes" and remain saturated even though the pressure head in the water is
negative. This saturated zone above the water table is the capillary fringe.

is what motivates the definition suggested above. The volumetric moisture content (or simply moisture content) in the capillary fringe is the **saturation value** (Figure 8.1); that is, the pores are completely filled with water. **Volumetric moisture content**, or more precisely volume wetness, is defined as the volume of water per bulk volume of soil sample. We will use the symbol θ to represent volumetric moisture content. After a rather rapid decrease from saturation, the moisture content in the intermediate belt may remain fairly constant. **Field capacity** is a term used to represent this "constant" moisture content. The moisture content in the soil water belt decreases rapidly from the field capacity due to the extraction of water by plant roots and to direct evaporation at the soil surface.

The divisions of the vadose zone often are useful in describing general observations of soil moisture. Of course, there are no sharp dividing lines marked off in the field. The physical principles that we use to quantify flows in the vadose zone do not change in moving, say, from the root zone to the intermediate zone. The terminology introduced above is used widely, however, and we will encounter these descriptive terms later. Our introduction to the physics of soil moisture will hold for all of the zones.

8.2 Forces on Water in the Unsaturated Zone

The unsaturated zone is a three-phase system consisting of soil, water, and air. The physical description of the system and of the flow of water in the system is thus more complex than for the two-phase system of the saturated zone. A full treatment on a microscale of the diverse forces acting on water in an unsaturated soil and the resultant motion of this water is not feasible. Fortunately, empirical work shows that, as in saturated soils, water flow in unsaturated soils is down a gradient of hydraulic head of soil water. For flow of groundwater in the saturated zone, the hydraulic head is composed of two terms, pressure head ($p/\rho g$) and head due to gravity, or elevation head (z). Darcy's law states that the flux of groundwater is proportional to the gradient in hydraulic head. Gradients in elevation head and in pressure head also drive flows in the vadose zone. All terrestrial water, including soil moisture, is within the Earth's gravitational field. Therefore, head due to gravity in the vadose zone is once again the potential energy per unit weight, z, that is, the elevation head. The main way that the physics for the vadose zone differs from that for the saturated zone is in the pressure head contribution to hydraulic head.

8.2.1 Pressure head

If water is withdrawn from a rock or soil matrix that does not shrink upon drying, air enters the pore space, and air-water interfaces (menisci) are present in the pore space. Such curved interfaces are maintained by **capillary forces**. Surface tension acting in the interfaces provides a mechanism of soil-water retention against externally applied suction. This phenomenon is seen in the rise of water in capillary tubes, for example, and is explained by the attractive forces between the glass walls of the tube and the water. The glass attracts the adjacent water molecules more strongly than do other water molecules themselves. The water is therefore "pulled" up the inside of the tube

(Figure 8.2). This "pull" is a tension which, in the terms that we are using, is a negative (gage) pressure. That is, the gage pressure in unsaturated soils is *negative*. Negative pressure heads are developed in unsaturated rock and soil matrices. The height of rise in a capillary tube (a measure of the negative pressure head) is inversely related to the diameter of the tube. Water will rise higher in a tube with a small diameter than it will in a tube with a large diameter. This observation translates to soil physics in that smaller diameter pores retain water against higher suctions than do larger pores (cf. the large capillary tube and the small capillary tube sketched in Figure 8.2). Thus, when water drains from a soil or rock, large pores empty first because it takes relatively less applied suction to pull water out of larger pores. The negative pressure produced by capillary forces, when divided by ρg, is referred to as the **capillary-pressure head**.

At first glance, the idea of negative pressure head does not seem intuitive to most people. Yet there are commonplace examples with which most of us are familiar. If asked about the pressure head at the surface of a pan of water in your kitchen, you should be comfortable with the answer "zero." But what happens if you bring a dry sponge just barely into contact with the water surface? Water *rises* into the pores of the sponge. The sponge pulls water up from a state of zero pressure head. The "pull" is capillary tension, a negative gage pressure. Conversely, if the sponge subsequently is withdrawn from the water surface, all of the water does not drain out. The sponge retains water against the downward

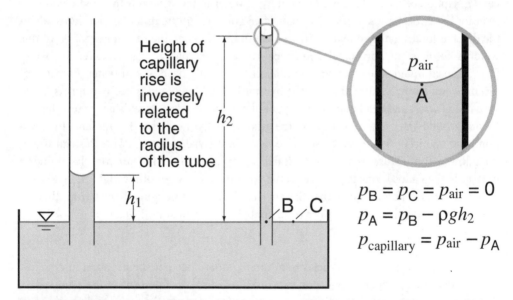

$$p_B = p_C = p_{air} = 0$$
$$p_A = p_B - \rho g h_2$$
$$p_{capillary} = p_{air} - p_A$$

Figure 8.2 Surface tension "pulls" water up into capillary tubes. Water pressure within the tubes is less than atmospheric pressure, or is negative in gage units. The height of water above the free surface in the tube is equal to the negative of the capillary-pressure head. The amount of negative pressure head with which a capillary tube can "hold" water is inversely related to the diameter of the tube. That is, small-diameter tubes (and by analogy, soil pores) hold water at a more negative pressure head than do large-diameter tubes (or soil pores).

action of gravity by counteracting the downward gradient in elevation head with an upward gradient in (negative) pressure head.

8.3 Capillary-Pressure Head and the Moisture Characteristic

For areas where there are moderate fluxes of water through the vadose zone, the two major driving forces on soil water are the gradients in the negative capillary-pressure head and the gradient in elevation head. This situation is exactly analogous to that for flow in the saturated zone, with the only change being that the positive pressure heads encountered in groundwater are replaced by negative capillary-pressure heads. Thus, the hydraulic head for the vadose zone is defined to be the sum of the head due to gravity and the (negative) capillary-pressure head. (In general, several forces act to create the negative pressure heads in the unsaturated zone. The treatment given here remains valid, but an "equivalent" negative pressure head that incorporates all important forces, rather than just capillarity, is used. See Childs (1969) or Guymon (1994) for further explanation. The material that we present is strictly valid for relatively moist soils and rocks and is valid for almost all circumstances with the extended definition of negative pressure head.)

A suction (or negative pressure relative to atmospheric pressure) must be applied to withdraw water from the unsaturated zone above the water table. The greater the applied suction, the more water is withdrawn, and the lower is the soil-moisture content when the soil has reached equilibrium with the applied suction. Our example of a sponge may provide insight. In the kitchen, we don't have devices to exert suction to remove water from a sponge. We can make an analogy between exerting suction and "squeezing" the sponge, however. If we exert small effort in squeezing a sponge, a relatively small amount of the water will be forced from the pores. To paraphrase the sentence referring to exerting suction on a rock, the greater the applied "squeezing" to the sponge, the more water is withdrawn, and the lower is the amount of water remaining in the sponge when "squeezing" has ceased. The relationship between the external suction applied to a rock and the amount of water per bulk volume (the moisture content) that the rock retains against that de-watering suction is called the **moisture characteristic**. The applied negative pressure is a measure of the water-retaining forces of the soil and represents the capillary-pressure head, ψ. The moisture characteristic generally is presented as a plot of ψ versus θ.

The moisture characteristic for a porous material can be determined using a *pressure plate* apparatus (Figure 8.3). The rock (soil) sample sits on a porous plate made of a fine-grained material (e.g., a ceramic) that remains saturated even at high negative pressure heads. The sample is allowed to equilibrate at a given negative pressure head and the moisture content associated with this capillary-pressure head is determined. The experiment is repeated at different negative heads to obtain other points on the moisture characteristic. The locus of all such points then defines the moisture characteristic (Figure 8.4). The moisture characteristic is one of the important curves that define the relationships among hydraulic variables in the soil-water system. Another is the relationship between moisture content and hydraulic conductivity.

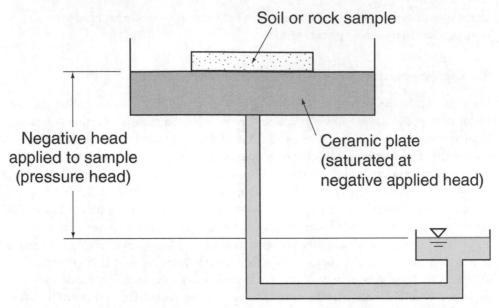

Figure 8.3 A pressure plate for measuring capillary-pressure head. The rock sample is placed in contact with the ceramic plate which is saturated with water under a negative pressure head that is set by the distance of the free water surface to the right of the diagram below the ceramic plate. The moisture content of the sample is recorded after the sample has come to equilibrium with the selected capillary-pressure head. This measurement gives one point on the moisture characteristic curve.

8.4 Darcy's Law

As water is drained from a saturated soil, the large pores fill with air but the smaller spaces remain filled with water. The water in the smaller capillaries is held by surface tension more tightly than is the water in the large pores. In a moist but unsaturated soil, water flows through the water-filled pores while avoiding the larger, air-filled spaces. For a given moisture content, we might conceptualize this flow as if the air-filled pores were filled instead with a solid (e.g., wax). We would thus have flow in an equivalent, saturated porous medium and would expect Darcy's law to be valid, but with the hydraulic conductivity appropriate for the equivalent medium. This is, in fact, found to be the case; Darcy's law is valid for unsaturated soil but each moisture content corresponds to a different equivalent saturated medium, and hence a different value of hydraulic conductivity. We therefore write the hydraulic conductivity as $K(\theta)$ indicating that it is a function of moisture content.

The hydraulic conductivity decreases very rapidly as the medium becomes unsaturated (Figure 8.5). This can be explained by recalling that Poiseuille's law (Equation 3.36) indicates that mean velocity is proportional to the second power of the diameter of a cylindrical tube. Hence, as we have seen, the intrinsic permeability of a porous medium is proportional to the square of the pore size (Equation 6.7). Because the discharge

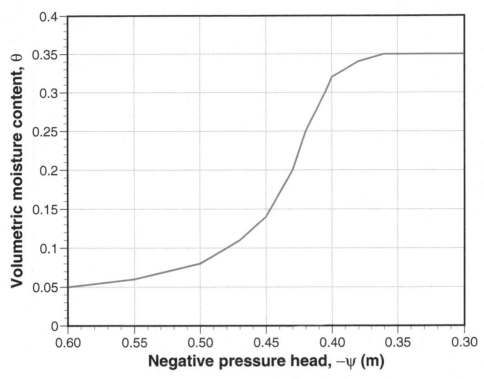

Figure 8.4 Moisture characteristic for a fine sand determined by starting at saturation and draining the sample. Note that for this sand, saturation is maintained for capillary-pressure heads between 0 and −0.36 m. The capillary fringe in such a material would be 0.36 m high. As the capillary-pressure head is reduced from about −0.40 m to −0.45 m, the moisture content drops sharply from the saturation value of 0.35 to about 0.15. This steep drop is typical for sandy soils. Much of the pore space is in large pores that drain once a critical suction is exceeded. The moisture content therefore drops abruptly. On the other hand, moisture content drops by only about 0.08 as capillary-pressure head drops from −0.45 m to −0.60 m.

is equal to the mean velocity multiplied by the cross-sectional area of flow (πD^2) in the case of a cylindrical tube), the discharge will be proportional to the fourth power of the tube diameter. The large pores of a soil become air-filled first as a suction is applied to a soil, thereby relegating flow to the smaller pores which (according to Poiseuille's law) can conduct water at *much lower flow rates* than could be handled by the emptied (larger) pores. This reasoning also can explain why the hydraulic conductivity of an unsaturated clayey soil may be appreciably higher than that of an unsaturated sandy soil (see Section 8.10).

Note that the hydraulic gradient driving flow in unsaturated materials is the hydraulic head appropriate for unsaturated materials, $\psi + z$:

$$h = \psi + z, \tag{8.1}$$

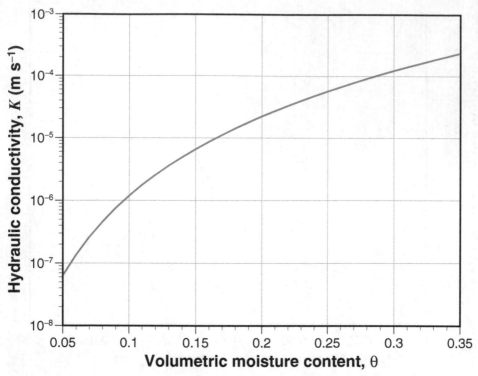

Figure 8.5 Variation of K with θ for a fine sand. Note that the scale for K is logarithmic. The hydraulic conductivity decreases by orders of magnitude as capillary-pressure head drops from −0.35 m to −0.55 m (moisture content decreases from saturation to about 0.06 as seen in Figure 8.4).

where ψ = capillary-pressure head and z = elevation head. Note that capillary-pressure head is a function of moisture content in this expression. Darcy's law for unsaturated conditions can thus be written:

$$q = -K(\theta)\frac{dh}{dl}.$$
$$(8.2)$$

8.5 Vertical Water Movement

An appreciation of many important aspects of the movement of water in the unsaturated zone can be gained by considering one-dimensional flow in the vertical direction. In this case, Equation 8.2 becomes:

$$q_z = -K(\theta)\frac{dh}{dz}$$
$$(8.3)$$

or, making use of Equation 8.1:

$$q_z = -K(\theta)\frac{d(\psi + z)}{dz}$$

or,

$$q_z = -K(\theta)\left(\frac{d\psi}{dz} + 1\right).$$
(8.4)

Equation 8.4 alone is sufficient to describe the steady flow of soil moisture. As with groundwater flow, description of transient or unsteady processes requires the addition of a continuity equation to Darcy's law (e.g., see Guymon, 1994). For the case of vertical flow of water the appropriate continuity equation is:

$$\frac{\partial \theta}{\partial t} = -\frac{\partial q_z}{\partial z}.$$
(8.5)

The left side of this equation represents the rate of change of mass in a small control volume and the right side is the difference between the inflow rate and the outflow rate, each expressed on a per unit volume basis. That is, the equation has the same conceptual basis as the continuity equation that we used in earlier chapters. Combining Equations 8.4 and 8.5 gives an equation that, along with information on the relationships among θ, ψ, and K, describes the flow of water in unsaturated rocks. The resulting equation, referred to as the **Richards' equation**, is:

$$\frac{\partial \theta}{\partial t} = \frac{\partial}{\partial z}\left[K(\theta)\left(\frac{\partial \psi}{\partial z} + 1\right)\right].$$
(8.6)

For steady flow, the time derivative of moisture content is zero and a single integration of the right hand side of Equation 8.6 recaptures Equation 8.4. That is, for steady flow the specific discharge, q_z, is constant and can be calculated from Equation 8.4. Of course, Equation 8.4 can be used to calculate the flux in unsteady flow as well. Under unsteady conditions, q_z changes with time so the calculation can be taken to give a "snapshot" of the flow at a given time. Solutions to the Richards' equation, which accounts for time variation explicitly, yield a complete time history of heads and fluxes for specified conditions.

A simple example illustrates the kind of calculation that can be done using Equation 8.4. Suppose that, at noon on a day in September, moisture content in a fine sand is measured to be 0.25 at an elevation of 3 m above the local water table and to be 0.15 at an elevation of 3.5 m above the water table. Assuming that the hydraulic relationships in Figures 8.4 and 8.5 hold for the sand in question, we can estimate the direction of flow of water and the magnitude of the flux. Consider first the direction of flow. From Equation 8.3 we see that, if $\partial h/\partial z > 0$, then the flow will be downward (because the calculated q_z will be negative). Conversely, if $\partial h/\partial z > 0$, then the flow will be upward. We recognize that this conclusion merely reiterates the main point of Darcy's law, that water flows *down* a gradient in hydraulic head. So, for our case, we approximate the derivative by a finite difference:

$$\frac{\partial h}{\partial z} \approx \frac{h_{3.5} - h_3}{0.5 \text{ m}} = \frac{(3.5 \text{ m} + \psi_{3.5}) - (3.0 \text{ m} - \psi_3)}{0.5 \text{ m}},$$

where the subscripts refer to conditions at the different elevations above the water table. Evaluation of the expression requires values for ψ. From Figure 8.4, we determine that for a moisture content of 0.25, the capillary-pressure head is about −0.42 m and for a moisture content of 0.15 it is about −0.45 m. The calculated hydraulic gradient is 0.94 (positive) so the water flow is downward. To estimate the magnitude of the flux, we multiply the gradient by the hydraulic conductivity. We know that hydraulic conductivity varies over the 3-m to 3.5-m interval because moisture content varies. We might determine an "average" value of hydraulic conductivity in different ways. For example, we could find $K(\theta = 0.25)$ and $K(\theta = 0.15)$ and average these. Or we might just use $K(\theta = 0.20)$. The first method gives a value for K of about 0.000023 m s^{-1} and the second gives a value of about 0.00002 m s^{-1}. Thus, the specific discharge is estimated to be approximately 0.00002 m s^{-1}.

8.6 The Equilibrium Profile above a Water Table

Now consider how we would expect the moisture content profile to look in a homogenous material above a static water table under conditions of zero vertical flux. This would be the situation we might expect from the capillary fringe through a good part of the intermediate zone after a prolonged period without rain. In this case, Equation 8.4 becomes:

$$q_z = -K(\theta)\left(\frac{d\psi}{dz} + 1\right) = 0. \tag{8.7}$$

Dividing through by K (we can assume that K is not zero) and rearranging, we have:

$$\frac{d\psi}{dz} = -1$$

or

$$d\psi = -dz. \tag{8.8}$$

This equation can be integrated from the water table ($z = 0$, $\psi = 0$) to some arbitrary point in the unsaturated zone (z, ψ) so that under zero-flux conditions:

$$\psi = -z. \tag{8.9}$$

That is, the capillary-pressure head balances the head due to gravity so there is no hydraulic gradient and thus (by Darcy's law) no moisture flux.

Equation 8.9 can be used to infer the moisture distribution if the moisture characteristic (ψ as a function of θ) is known. Clearly, the equilibrium moisture profile above a water table has exactly the same shape as the moisture characteristic because of Equation 8.9. Thus, for a sandy soil with a very steep moisture characteristic at the dry end of the curve, the field capacity represents the "nearly constant" moisture content associated with the steep portion of the curve (Figure 8.6). The moisture characteristic for a soil having a higher clay content, however, may not exhibit such a steep characteristic and in such cases the unambiguous definition of a field capacity is problematic.

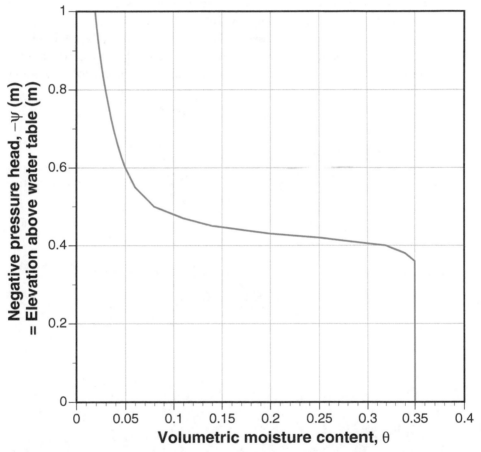

Figure 8.6 The profile of moisture content above a water table under conditions of zero vertical flux of water. The equilibrium profile under these conditions has the same shape as the moisture characteristic. For fine sand, the moisture content drops rapidly above the capillary fringe and remains relatively constant in the range of 0.04 to 0.08 over a sizable range of elevation. This relatively constant moisture content is referred to as field capacity.

8.7 The Profile of Capillary-Pressure Head as an Indicator of Flow

Measurements of capillary-pressure head can be used to infer how water is moving in an unsaturated soil or rock. In humid to subhumid climates, a **tensiometer** can be used to measure the capillary-pressure head. A tensiometer consists of a cylindrical tube, typically PVC, with a porous cup mounted on the end (Figure 8.7). The cup, typically made of ceramic or Teflon, is porous but with fine pores that remain saturated under the water tensions (i.e., capillary-pressure heads) to be measured. The tube is inserted into the soil, ensuring that a close contact is established between the porous cup and the soil. The tube is filled with water and tightly capped. A pressure gage is used to measure the pressure in the water. Suppose the tensiometer is placed in a soil in which the capillary-pressure head is -0.3 m. Immediately after the tube is filled with water and capped, the pressure head inside the porous cup can be assumed to be zero. Because the porous cup is in contact with the soil, water is drawn out of the (otherwise sealed) tensiometer through the porous cup, lowering the pressure in the water inside the tube. Only a small amount of water needs to be withdrawn from the tensiometer to lower the pressure due to the low compressibility of water; as the water pressure inside the tensiometer is low-

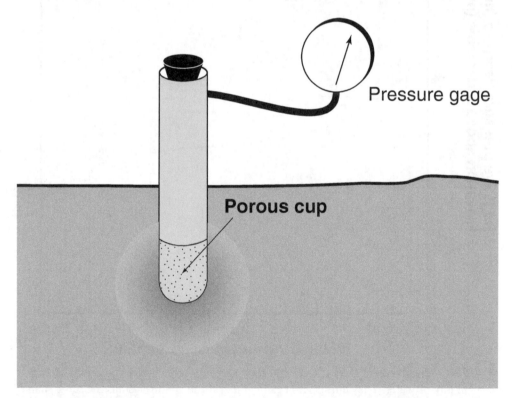

Figure 8.7 A tensiometer, consisting of a closed tube with a porous cup at the end. Negative pressure, or tension, in the unsaturated zone is recorded by the pressure gage as water in the tube is drawn into the partially saturated soil.

ered, the water expands (slightly) because the porous cup remains saturated and air cannot enter. Water will cease to flow out of the tensiometer when the (negative) pressure inside the tensiometer reaches −0.3 m, the tension in the surrounding soil water. Thus, by allowing a tensiometer to equilibrate with the surrounding soil and then reading the pressure gage, we get a measure of the capillary-pressure head in the soil.

If a pair of tensiometers is installed in a soil, it is easy to see how flow direction can be determined. The example at the end of Section 8.5 illustrates the procedure. Recall that in that example, the capillary-pressure head at an elevation of 3 m above a water table was −0.42 m and the pressure head at 3.5 m above the water table was −0.45 m. These heads might have been measured with tensiometers. The calculation then illustrates how measured tensions can be used with Darcy's law to infer flow direction.

In some instances, we might want to infer direction of flow using measurements from a single tensiometer. Is this possible? The calculation in Section 8.6 shows that, when there is no flow of moisture in the vertical direction, the capillary-pressure head decreases directly with depth, $\psi = -z$. Thus, the profile of pressure head with height above the water table is a straight line (Figure 8.8). In the field, if there is no vertical flow, a measured capillary-pressure head should plot on the straight line representing equilibrium. Suppose a measurement shows that capillary pressure at a spot in a soil is smaller than it would be under no-flow conditions (region A in Figure 8.8). In this case we can conclude that water flow will be upward if flow is steady. If measured capillary pressure is lower than the equilibrium value, the gradient in hydraulic head relative to the water table will be upward and so we expect that water flow will be upward because water is being drawn up from the water table. Similarly, if measured pressures are greater than equilibrium values (region B in Figure 8.8), we can reason that water is flowing downward away from the point of measurement (assuming steady flow).

8.8 The Infiltration Process

At depth in a soil profile (roughly in the "intermediate belt" of Figure 8.1), the analyses based on the equilibrium profile as presented above are often very useful because the steady-flow approximation is reasonable. Recharge to an aquifer through a thick unsaturated zone, for example, can be approximated reasonably as a steady flow process. Many important hydrological processes involve water fluxes right at the soil surface, however, and in this area (the "soil-water zone" of Figure 8.1) changes in soil moisture can be rapid and large. One of the most important of the surface hydrological processes is **infiltration**, the movement of rain and melting snow into the soil. The process of infiltration at the soil surface is inherently unsteady and a rigorous treatment of the problem usually is approached through the use of Richards' equation, Equation 8.6. A simplified treatment of the problem, first presented by Green and Ampt (1911), illustrates how soil physical principles can be used to understand the infiltration process. The Green-Ampt equation will be outlined after a physical description of the infiltration process is sketched.

The maximum infiltration rate (commonly called the **infiltration capacity**) of a soil is the rate at which water will move vertically downward when a supply of water at zero

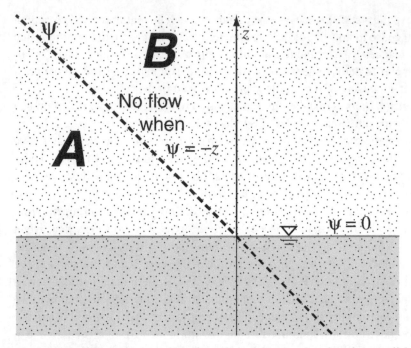

Figure 8.8 The vertical profile of capillary-pressure head under conditions of no flow is a straight line ($\psi = -z$). If a measurement of capillary-pressure head falls on this line, it is reasonable to infer that water flow in the vertical at that point is negligible. If measured capillary pressure is smaller (*region A*) than it would be at equilibrium (for a fixed value of z), it is reasonable to infer upward movement of water at that point. Conversely, if measured capillary pressure is greater (*region B*) than it would be at equilibrium, downward flow is indicated. The inferences are exactly correct under steady flow conditions.

(gage) pressure is maintained at the surface. The infiltration rate in a dry soil is initially very large. Even though the hydraulic conductivity is low, the capillary pressure gradient is extremely large because the soil only a few millimeters below the surface has a very low (and negative) pressure head. For example, if the capillary-pressure head 10 mm below the surface is initially −2.0 m, the finite-difference approximation of the gradient at the surface (assuming that the surface is at the saturation moisture content) would be (0.01 m + 2.0 m)/0.01 m = 201. Even with a hydraulic conductivity of $10-6$ m s^{-1}, the infiltration rate would be about 0.0002 m s^{-1} = 720 mm hr^{-1}. A rainfall rate of 720 mm hr^{-1} would be quite intense! The infiltration rate decreases in time as infiltrating water moistens the surface layers and reduces the gradient in capillary-pressure head. Ultimately, the near-surface portion of the soil approaches saturation, the capillary-pressure head gradient approaches zero and Equation 8.4 indicates that the infiltration rate approaches the saturated conductivity (Figure 8.9). The infiltration-capacity curve for any given soil, of course, would depend on the soil properties because the capillary-pressure

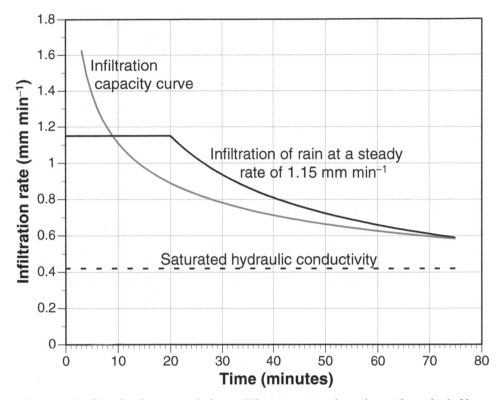

Figure 8.9 Infiltration into a sandy loam. When water ponds at the surface, the infiltration rate (*blue curve*) starts out at a very high value because of the large downward gradients in capillary-pressure head and decreases steadily toward the limiting value of the saturated hydraulic conductivity of the soil, a rate that would be achieved when the downward gradient in hydraulic head was due only to the elevation head. When water is sprinkled on a soil surface at a relatively slow rate, the infiltration rate will equal the sprinkling rate initially. If the rainfall rate exceeds the hydraulic conductivity of the soil and if the rainfall persists for long enough, the ability of the soil to infiltrate water will drop below the rainfall rate after hydraulic gradients near the surface become reduced by the wetting process. The time at which the infiltration rate starts to drop below the rainfall rate is known as the time to ponding. This occurs to the right of the infiltration capacity curve because the initial rates are much lower for the sprinkling case than for the ponding case; approximately the same total volume of water must infiltrate in the sprinkling case as in the ponding case before the infiltration rate starts to drop for the sprinkling case.

head gradient and unsaturated values of hydraulic conductivity govern the time progress of infiltration.

Now consider the case of a steady rainfall beginning at time $t = 0$. In this case, the rate of delivery of water to the surface is usually insufficient to cause initial ponding of water, that is, the infiltration capacity exceeds the rate of supply. Initially, all of the water reaching the surface infiltrates. Provided the rainfall rate is larger than the saturated hydraulic conductivity and that rainfall continues for a long enough time, the infiltration capacity will decrease continually as the soil becomes wet. The infiltration capacity ultimately will be reduced below the rainfall rate. Subsequently, the rate of infiltration will be controlled by the rate at which the soil can transmit water and "excess water" will accumulate at the surface (Figure 8.9). This excess water will flow rapidly over the surface to streams and can be an important process in generating stormflow in streams and rivers, as is discussed in Chapter 10.

8.8.1 The Green-Ampt equation

Observation of infiltration into dry soils, especially sandy soils, indicates that the water tends to progress downward as a "slug." In other words, a sharp wetting front separates the unsaturated soil below from the saturated soil above and it is this front that progresses downward as infiltration proceeds. Green and Ampt (1911) presented an analysis of the infiltration problem assuming that this wetting front was infinitely sharp, that is, horizontal (Figure 8.10). In this case, the flux everywhere in the saturated upper portion must equal the infiltration rate and so the hydraulic gradient is uniform. Darcy's law then gives:

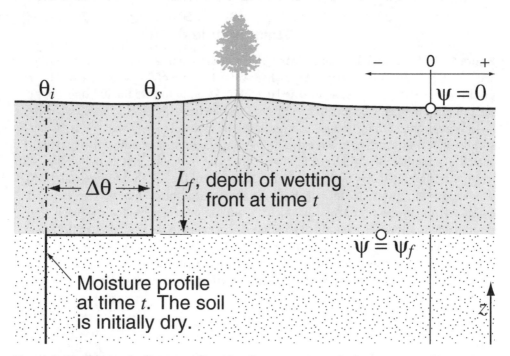

Figure 8.10 Schematic diagram of wetting front movement in the unsaturated zone, as conceptualized in the Green-Ampt model. The basic concept is that the soil surface is held at saturation and that water moves downward through the soil as a sharp "wetting front." Darcy's law can be applied under these idealized conditions to give an equation for calculating the rate of infiltration with time.

$$i = -K_S \left(\frac{-\psi_f + 0}{L_f} + 1 \right),$$ (8.10)

where i = infiltration rate (equal to the specific discharge), ψ_f = the capillary-pressure head at the wetting front, L_f = the depth to the wetting front, and K_s = saturated conductivity (Figure 8.10). Equation 8.10 can be written:

$$i = -K_S \left(\frac{-\psi_f + L_f}{L_f} \right).$$ (8.11)

Furthermore, for the type of flow envisioned by Green and Ampt (often referred to as *plug* or *piston flow*), the total cumulative amount of water infiltrated, I, is equal to the product of L_f and $\Delta\theta$, the difference between the saturated moisture content, θ_s, and the initial moisture content, θ_i. Then, because the infiltration rate, i, must equal the negative of the time derivative of I (negative flux means downward motion),

$$i = -\frac{dI}{dt} = -\Delta\theta \frac{dL_f}{dt}$$ (8.12)

and Equation 8.11 becomes:

$$\Delta\theta \frac{dL_f}{dt} = K_S \left(\frac{-\psi_f + L_f}{L_f} \right).$$ (8.13)

This equation can be integrated from $t = 0$ to an arbitrary time, t, to give:

$$\frac{K_S t}{\Delta\theta} = L_f + \psi_f \ln \left[1 + \frac{L_f}{(-\psi_f)} \right],$$ (8.14)

or equivalently

$$t = \frac{I}{K_S} + \psi_f \frac{\Delta\theta}{K_S} \ln \left[1 + \frac{I}{(-\psi_f \Delta\theta)} \right].$$ (8.15)

The above equation is an implicit relationship between the cumulative infiltration, I, and time, t. Thus, from known soil parameters (K_s, $\Delta\theta$ and ψ_f) an infiltration capacity curve similar to that shown in Figure 8.9 can be constructed. For example, for $K_s = 7.0 \times 10^{-6}$ m s^{-1}, $\Delta\theta = 0.04$ and $\psi_f = -0.106$ m, the Green-Ampt calculation produces the infiltration curve shown in Figure 8.9 (Clapp, 1977). These parameters are appropriate for a sandy loam soil (for which the assumption of a piston-like wetting front should be reasonable). Comparison of the Green-Ampt results with results from a more complete theory indicates that the calculated values of infiltration are sound from a theoretical standpoint.

Even for non-sandy soils, the form of the Green-Ampt infiltration curve can reproduce observations fairly well. Consider data from measurements using an **infiltrometer** made on a loamy soil in central Virginia. A ring infiltrometer is a simple device to measure infiltration rates. A metal ring is inserted into the ground and water is added to maintain a level pool of water in the ring. The amount of water required to maintain the level is the rate at which water enters the soil surface, that is, the infiltration rate. Data recorded are typically time (t) and cumulative depth of water infiltrated (I). Equation 8.13 can be used to calculate an $I - t$ curve for given values of K_s, ψ_f, and $\Delta\theta$. The computation is accomplished by selecting a suite of values of I for which calculations are to be done and then using 8.13 directly to calculate a corresponding value for t. For the infiltrometer data, the Green-Ampt equation fits the measured data reasonably well with $K_s = 0.8$ mm min^{-1}, $\psi_f = -250$ mm, and $\Delta\theta = 0.1$ (Figure 8.11).

The Green-Ampt method assumes that the soil column is homogeneous and the water table is deep in a way that the wetting front does not reach the water table during a rainfall event. In the presence of a shallow water table or of a bedrock at a depth L_s, infiltration water may saturate the soil column. Rainfall onto saturated soil consequently

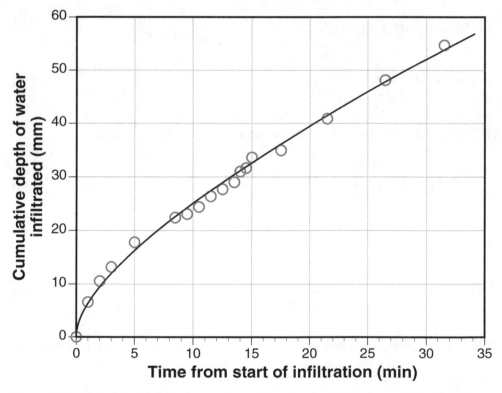

Figure 8.11 Cumulative infiltration measured for a loamy soil in central Virginia (*blue circles*) and calculated using the Green-Ampt equation (*black line*).

generates runoff, which is not due to limited soil infiltration capacity but to limited soil storage capacity, as discussed in Chapter 10. The soil column becomes completely saturated when the cumulative infiltration I is equal to the initial soil storage capacity $\Delta\theta\, L_s$. Thus, the time (since the beginning of the rainstorm or snowmelt event) when the soil becomes saturated—and runoff is produced regardless of the soil infiltration capacity—can be calculated using Equation 8.15 with $I = \Delta\theta\, L_s$.

8.9 Field Measurements in a Soil

Except in the most intense rainstorms, calculations of infiltration based on an assumption of saturation at the soil surface (e.g., the Green-Ampt approach) are not considered sound. Numerical solutions to the Richards' equation (Equation 8.6) can be used to gain insight into how moisture will infiltrate a soil under various rainfall rates. The physical principles developed so far in this chapter serve as a base for understanding processes that occur in the field, but added complications do arise.

Let's look at some data that were collected in the soil of a forested catchment near Orono, Maine, USA. A sprinkler was set up at the soil surface to simulate rainfall. Capillary-pressure heads were measured using tensiometers (see Section 8.7) and moisture contents were measured using the time domain reflectometry (TDR) technique (Figure 8.12).

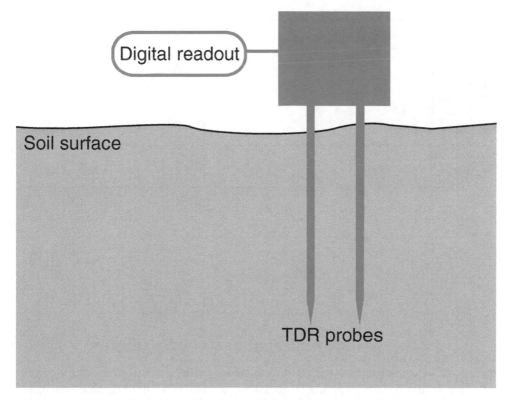

Figure 8.12 Schematic diagram of a time domain reflectometry (TDR) device for measuring soil moisture.

Measurements were taken in an organic **soil horizon** near the soil surface (approximately 50 mm below the surface) for a period of 9 hours. The sprinkler delivered water to the soil surface at a rate of 11 mm hr^{-1} for the first 4.5 hours and was subsequently shut off.

The soil started at a capillary-pressure head of about −0.40 m (Figure 8.13). Between 15 minutes and 45 minutes after the initiation of sprinkling, capillary-pressure heads rose rapidly to approximately −0.15 m, indicating passage of a sharp front. The capillary-pressure head continued to increase slowly for the rest of the duration of sprinkling. This makes sense given the principles that we have developed in this chapter. As water infiltrates the soil and moves downward, it fills pores that had been occupied by air and capillary pressures increase.

Measurements of soil moisture confirm this general picture (Figure 8.13). The moisture content is initially about 26% and rose steadily from $t = 0.5$ hr to $t = 4.5$ hr. The moisture content did not rise as rapidly as did the capillary-pressure head, however, but after sprinkling ceased, the moisture content and the capillary-pressure head appeared to change together in a more direct fashion. This is not what we would expect from the previously postulated straightforward relationship between moisture content and capillary pressure, the moisture characteristic. What does the moisture characteristic look like for the soil in the sprinkling experiment? We can answer this question directly by

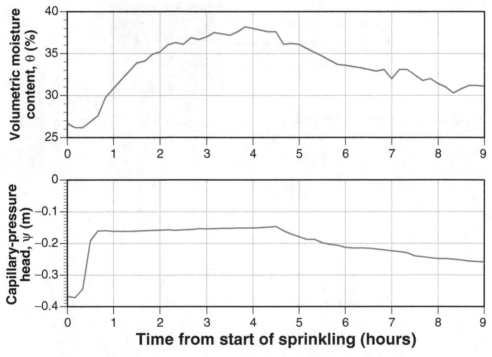

Figure 8.13 Capillary-pressure head (*bottom*) and volumetric moisture content (*top*) in a soil during a sprinkling experiment. Sprinkling at a rate of 11 mm hr^{-1} was started at time zero and continued for 4.5 hours. Subsequent to 4.5 hours, there was essentially no flux at the soil surface and water drained vertically downward.

plotting the measured values of moisture content against the measured values of capillary-pressure head.

The moisture characteristic for the Maine soil shows a **hysteresis** loop (Figure 8.14). Initially, the soil is relatively dry (ψ is approximately -0.40 m, θ is approximately 26%). As the soil wets up from the sprinkling, ψ rapidly increases to about -0.16 m. As this change takes place, θ increases relatively little, from 26% to 27%. The next phase of change is a relatively slower increase in ψ to about -0.15 m, but a correspondingly large increase in θ to about 38%. After sprinkling ceases, the moisture content and capillary-pressure head covary along a different characteristic curve, a branch above the curve defined by the wetting of the soil. Hysteresis refers to a phenomenon whereby different paths are followed on the $\psi - \theta$ plot depending on whether the soil is wetting or drying.

A conceptual appreciation of how the hysteresis phenomenon arises can be gained by extending our capillary tube analogy. Consider a capillary tube with a bulb-shaped expansion in it (Figure 8.15). (This capillary-tube analogy to explain hysteresis is often called the "ink-bottle effect" because the shape of the capillary tube is reminiscent of the tube that was used to transfer ink from a bottle to a pen in times past.) Suppose that the small-diameter portion of the tube can hold water against an applied capillary-pressure

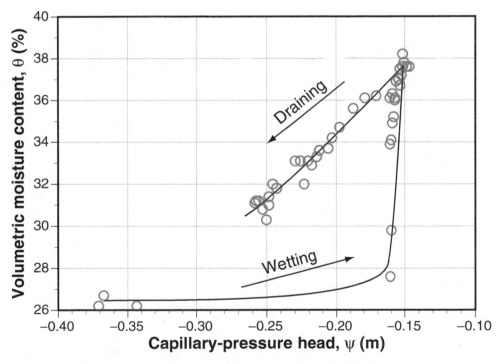

Figure 8.14 The moisture characteristic of the organic horizon of the Maine soil determined from a sprinkler experiment. The hysteresis loop in the curve is evident. As the soil imbibes water during the sprinkling, ψ and θ covary along the bottom part of the loop. After sprinkling ceases and the soil is draining, ψ and θ covary along the top part of the loop.

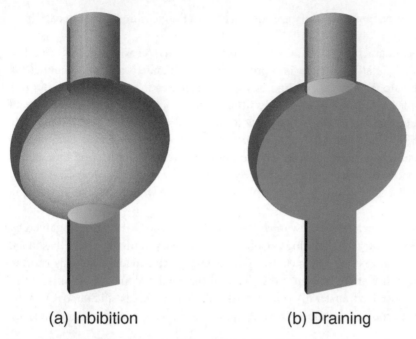

(a) Inbibition   (b) Draining

Figure 8.15 Hysteresis in a capillary tube with a bulb-shaped expansion. As shown in *a*, the "moisture content" of the tube remains low during "imbibition" of water, because water cannot enter the large-diameter portion of the tube until the capillary-pressure head increases to the value associated with the large diameter. On the other hand, as shown in *b*, during "draining" the "moisture content" remains high until the capillary-pressure head drops to the critical value for the small-diameter tube.

head of −0.40 m and that the large-diameter portion of the tube can retain water against a pressure head of −0.15 m. Consider the "wetting" of the tube from below depicted in Figure 8.15a. The water cannot cross the large-diameter portion of the tube, which holds much of the water in the tube, until the capillary-pressure head at the base of the bulb-shaped expansion reaches −0.15 m. Note that the "moisture content" in the tube will remain low until the bulb is filled. So with this analogy, as water enters the tube from below, the pressure head increases from −0.40 m to −0.16 m with relatively little change in "moisture content." The "moisture content" increases suddenly as capillary-pressure head reaches −0.15 m at the bottom of the bulb, however, because water can then enter the bulb-shaped portion of the tube. Conversely, if the tube is full (Figure 8.15b; high "moisture content") and capillary pressure at the base of the tube is reduced to drain the tube, the bulb-shaped portion of the tube will not drain until the pressure head decreases to −0.40 m at the top of the bulb. This is because water is held in the small-diameter part of the tube above the bulb and water can be retained in this upper part of the tube. This analogy indicates that for this particular capillary tube, the "moisture content" will be higher for a given applied capillary-pressure head when the tube is *draining* than it will be for the same capillary-pressure head when the tube is *imbibing* water. This is exactly the form of the hysteresis loop observed for the Maine soil.

Hysteresis is not the only complication that arises in considering flows in the field. Real soils are heterogeneous. As one example, we can consider vertical changes in soil properties. Different soil horizons (loosely speaking, "layers") typically have different hydraulic properties. For the sprinkling experiment in Maine, capillary-pressure heads were measured at four different depths in the soil, in four different horizons. Considering the observations, we see that the smooth progress of a wetting front downward through the soil as envisioned in the Green-Ampt model is interrupted (Figure 8.16). In Figures 8.13 and 8.14 above, we looked at the Oa horizon near the soil surface. The Bhs horizon is the next deepest. The capillary-pressure head builds up in the Bhs horizon from about −0.45 m to slightly *positive* values. Positive pressure heads indicate that a saturated layer had to build in this horizon before water "pushed through" into the next lower horizon, Bs2. Clearly there was an impediment to flow into the Bs2 horizon that caused water to "back up" until positive pressures increased the hydraulic gradient (and hence the flow) into the layer. In the C horizon, the positive pressure heads indicate that a water

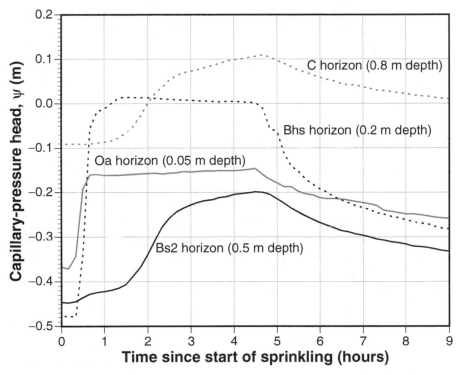

Figure 8.16 Measured pressure heads in four horizons of a forest soil in Maine for a sprinkling experiment. Sprinkling began at time zero at a constant rate of 11 mm hr^{-1} and stopped after 4.5 hours. The lag in the pressure response at the various depths is due to the time of travel of moisture vertically downward in the soil. The pressure builds to slightly positive values in the Bhs horizon, indicating that moisture flow into the underlying Bs2 horizon is impeded. A water table builds up in the C horizon because the base material is a glacial till with very low permeability.

table developed with a saturated depth of 100 mm above the tensiometer. The changes in hydraulic properties with depth influence the dynamics of vertical soil-water movement in layered soils. Interestingly, the "backing up" of the water above the Bs2 horizon is probably because the Bs2 horizon is *coarser* than the Bhs horizon. The resistance to flow in unsaturated materials provided by coarse layers of sediments (described further in the next section) is an important but somewhat non-intuitive phenomenon.

Another example of soil heterogeneity is the nearly universal presence of **macropores**. Beven and Germann (1982) considered macropores to be those openings in a soil that are large in relation to those in the surrounding soil, such that the movement of water, *once initiated*, may be much faster than within the surrounding soil matrix. The statement above regarding the initiation of flow in macropores deserves further explanation. These large pores may be thought of as large capillary tubes, and therefore will tend to drain at (negative) pressures close to zero. If we imagine rainfall encountering a dry soil, water will tend to be drawn by capillary forces into the smallest pores first. The flow of water in macropores begins only when saturation occurs locally at portions of the soil surface. Furthermore, water may flow quickly through a macropore, and encounter drier soil at some depth below the general wetting front. At that point, capillary forces may once again cause water from the macropores to be drawn into the surrounding small pores.

Soil macropores may be produced by both biotic (animal burrows, earthworm channels, plant roots) and non-biotic (soil pipes, shrinkage cracks) processes. Their importance in the movement of water within hillslopes continues to be an important area of study in catchment hydrology.

8.10 Evapotranspiration from the Unsaturated Zone

As indicated above, moisture content in the unsaturated zone can be quite variable in time. Infiltration of water during precipitation or snowmelt events increases the soil moisture. Afterward, drainage of water downward toward the water table and movement of water upward out of the soil due to evaporation and transpiration decrease soil moisture. The amount of water stored in the unsaturated zone therefore changes with time. A detailed water balance for the land surface requires an accounting of inflows, outflows, and rates of change of storage in the unsaturated zone.

Measurements that are used to keep track of moisture dynamics in surface soils include several that we have discussed. Capillary-pressure head can be measured with tensiometers (Section 8.7) and moisture content can be measured using time domain reflectometry. It is quite difficult to separate drainage rates from evapotranspiration rates using only measurements made in the soil. One experimental technique for measuring evapotranspiration involves the use of weighing lysimeters. A weighing lysimeter is essentially a large pot, filled with soil, and mounted flush with the soil surface. A device to record the total weight of the lysimeter is placed beneath the lysimeter. Plants are allowed to grow in the lysimeter. The amount of precipitation that infiltrates through the surface is recorded carefully as is the amount of water that drains from the base of the lysimeter. Changes in the water content within the lysimeter are determined by changes in the total weight. All of these measurements on a lysimeter can be done with high accuracy.

Evapotranspiration then can be computed directly from Equation 1.7 because all terms except *et* are known.

Rates of removal of water from the unsaturated zone by evapotranspiration are controlled by a number of factors. One of these controlling factors is the wetness of the soil itself (Figure 2.11). Typically, if a vegetated surface is supplied with plenty of water (e.g., a well-watered lawn), evapotranspiration will be controlled by atmospheric conditions. That is, evapotranspiration will proceed at the potential rate, which is a function of solar radiation, wind speed, humidity, and so forth. If the lawn is not watered and if there is a prolonged period without rain, soil moisture will decrease (outflows exceed inflows). Evapotranspiration will proceed at the potential rate for some time, but ultimately the rate will drop. Why does the rate drop when the atmosphere is capable of taking up water at the potential rate? The explanation stems from the soil physical principles that have been covered in this chapter. As water is pulled from the soil near a plant root, the moisture content in the soil surrounding the root decreases. Decreases in θ lead to large decreases in K. Therefore, by virtue of Darcy's law, to maintain a steady flow of water to the plant root, the plant must exert ever greater suctions (ever more negative capillary-pressure heads) so that increases in the hydraulic gradient counterbalance the decreases in K. At some point, the plant cannot sustain this battle with a drying soil and the transpiration rate falls below the potential rate, as discussed in Chapter 9. The plants regulate the transpiration rate by adjusting the opening of the stomata on the leaves (see Section 2.4).

8.11 Capillary Barriers

Flow of fluids through unsaturated media often results in counterintuitive behavior. One very important example is that flow through coarse material is impeded relative to flow through fine material under unsaturated conditions. That is, under unsaturated conditions, water flows more readily through clayey soil than through gravel! The key to understanding how gravel layers can be **capillary barriers** to flow of water in the unsaturated zone is appreciating that gravel has almost all "large" pores. Under negative capillary-pressure heads, these large pores fill with air and essentially stop the transport of water. (Water actually does move by vapor diffusion, but this is an exceedingly slow process.) Clays, on the other hand, have almost all "small" pores. Thus, under anything but extreme low (negative) capillary-pressure heads, many of the pores will be filled with water and will conduct water, albeit at slow rates (relative to saturated gravel, but at very fast rates relative to vapor diffusion). The situation can be appreciated by looking at the hydraulic conductivity curves for gravel versus a clayey material (Figure 8.17). At saturation, $K_{\text{gravel}} \gg K_{\text{clay}}$, but at moisture contents not too far below saturation, $K_{\text{clay}} \gg K_{\text{gravel}}$. That is, flows are impeded in the gravel relative to the clay at intermediate to low values of the moisture content.

The recognition that coarse materials serve as barriers to flow in unsaturated regimes has several practical applications. In desert regions, cobbles can be used to mulch agricultural fields. During intense thunderstorms (the usual method of delivery of water to the surface in semiarid regions), the gravel layer is wet and readily allows water to infiltrate. Once the surface gravel dries out in the following fair weather, however, the evaporative flux of water from the surface is essentially prevented because the hydraulic conductivity

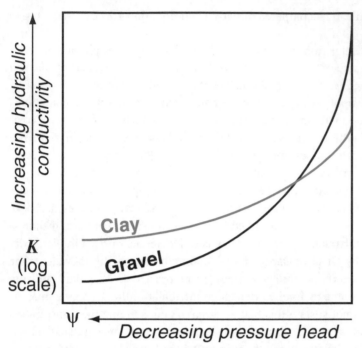

Figure 8.17 The hydraulic conductivity of clay is orders of magnitude less than that of gravel at saturation but is orders of magnitude larger at low values of the moisture content.

of dry gravel is very low. The infiltrated water is available for use by plants for transpiration. Another application is the construction of barriers to flow into waste disposal trenches. If wastes are placed in the vadose zone, a gravel cap is placed over the trench, and fine material is filled in over the gravel, the gravel will act as a capillary barrier and deflect the flow of infiltrating water through the fine material and away from the waste trench (under unsaturated conditions).

8.12 Concluding Remarks

Understanding the mechanics of flow of soil moisture is prerequisite for dealing with many problems in physical hydrology. The process of redistribution of moisture following an infiltration event, percolation of water to a groundwater table, movement of soil moisture in response to an evaporation demand, etc., are all important aspects of the field water cycle. Furthermore, solutions to "applied" problems, such as determination of optimal irrigation rates or siting of septic fields, also depend on a knowledge of flow in the unsaturated zone. As one example, let's reconsider the proposed LLRW disposal site in Ward Valley, California. (Note that the Ward Valley site no longer is under consideration as the California legislature cancelled the project in 2002.)

Ward Valley is in the arid Mojave Desert. Annual rainfall averages less than 150 mm. The site is on a broad alluvial surface some 17 m above the calculated level of the 100-year flood in nearby Homer Wash. The alluvial and basin-fill deposits at the site are about 600 m thick with the water table about 225 m below ground surface.

In 1993, after contractors for the state of California had spent more than two years collecting data to characterize the proposed disposal site and were ready to apply for licensing, several geologists questioned the safety of the site for disposal of radioactive wastes. One of the concerns was the potential for infiltrating waters to dissolve the wastes and recharge the groundwater beneath the site at rates that would be too fast to allow enough time for radioactive decay to render the wastes harmless. How can the claims of (fairly rapid) water movement in the unsaturated zone be evaluated? One set of measurements that would be useful is the vertical distribution of capillary-pressure heads. Recall our discussion of the equilibrium moisture profile above a water table. We concluded that, *if capillary-pressure head declined with height above datum to exactly balance the increase in head due to gravity*, there would be no movement of water. That is, if the vertical gradient in hydraulic head is zero, Darcy's law indicates that flow is zero. Conversely, if the gradient in hydraulic head is in the downward direction, recharge is indicated and if the gradient is upward, discharge (due to evapotranspiration) is indicated. A committee of the National Research Council concluded that the data for Ward Valley were insufficient to estimate the gradient, but that data from similar arid sites show upward gradients (NRC, 1995c). The committee concluded that the dry state of the materials at the Ward Valley site did indicate that vertical flow rates would be very slow. Detailed assessments of potential waste-disposal sites in the vadose zone require careful measurements and interpretation in light of the theory outlined in this chapter.

8.13 Key Points

- Pore spaces in the unsaturated zone may be partly filled with water and partly filled with air. The volumetric moisture content, θ, is a measure of the amount of water held in a soil or rock. {Section 8.1}

- Moisture content in the unsaturated (or vadose) zone is quite variable near the surface where interactions with the atmosphere are strong. In particular, plants take up water from this zone of soil water. Where unsaturated zones are deep, below about 1 to 3 meters, the temporal variability in moisture content caused by atmospheric changes are dampened. Water flows in this intermediate zone are approximately steady. Near the water table, a saturated zone develops due to capillary rise of water in the pores of the soil or rock. This saturated zone above the water table is known as the capillary fringe. {Section 8.1}

- In unsaturated soils, negative pressure heads develop because of capillary forces acting on curved air-water interfaces. {Section 8.2}

- One of the important relationships needed to understand moisture dynamics in unsaturated materials is that between capillary-pressure head and moisture content. The moisture characteristic defines this relationship. {Section 8.3}

- The gradient in hydraulic head drives flow in the unsaturated zone, just as it does in the saturated zone. In the unsaturated zone the hydraulic head is defined by $h = \psi + z$, where ψ is the (negative) capillary-pressure head. {Section 8.4}

- Darcy's law for flow in the unsaturated zone has the same form as it does for flow in the saturated zone. For unsaturated flow, hydraulic conductivity is a function of moisture content, so $q = -K(\theta)(dh/dl)$. {Section 8.4}

- Darcy's law can be used to infer the direction of water flow and the magnitude of the specific discharge. If the gradient in hydraulic head is positive, flow is downward. If the gradient in hydraulic head is negative, flow is upward. The specific discharge is the product of K and the gradient in hydraulic head. {Section 8.5}

- For the case of zero specific discharge in the vertical in the unsaturated zone above a static water table, the gradient in elevation head must be exactly balanced by the gradient in capillary-pressure head. Thus, $\psi = -z$ under these conditions. This indicates that the equilibrium profile of pressure head above a water table is a straight line upward to the left at 45°. The equilibrium profile of moisture content has the same shape as the moisture characteristic. {Section 8.6}

- Infiltration into a dry soil is very rapid initially because hydraulic gradients can be very large. The rate of infiltration decreases with time if the surface conditions remain constant, and over time approaches a constant value equal to the saturated hydraulic conductivity. {Section 8.8}

- The Green-Ampt equation can be used to calculate the infiltration-time curve if several soil parameters can be estimated. These parameters are the saturated hydraulic conductivity, the wetting-front pressure head, and the change in moisture content between saturation and the underlying soil. {Section 8.8.1}

- Field conditions rarely conform to the ideal. Hysteresis in the hydraulic relationships for a soil occur, meaning that the relationship between ψ and θ, for example, is different when the soil is imbibing water than when it is draining. Heterogeneities such as macropores are common in soils, which may be responsible for rapid movement of water in the unsaturated zone under certain conditions. {Section 8.9}

- Coarse material can act as a *barrier* to flow of water in unsaturated soils because of the relationships among ψ, θ, and K. {Section 8.10}

8.14 Example Problems

Problem 1. Tensiometers are installed at 0.4 m and 0.5 m above the water table in a uniform sandy soil with the moisture characteristic and hydraulic conductivity curves given in Figures 8.4 and 8.5, respectively. One set of tensiometer readings indicates that the capillary-pressure head at the first of these tensiometers is −0.45 m and at the second is −0.6 m.

A. What is the direction of water movement between the two tensiometers?

B. Estimate the magnitude of the specific discharge between the two tensiometers.

Problem 2. Heidmann et al. (1990) report that the moisture characteristic for a sandy loam soil in Arizona can be represented accurately by the equation $\psi = -963.7\theta^{-4.659}$, where ψ is in meters and θ is expressed as a percentage (e.g, 26% rather than 0.26). The saturation value of moisture content for this soil is estimated to be 27%.

A. Plot the moisture characteristic for this soil.

B. What is the equilibrium profile of (a) capillary-pressure head and (b) moisture content above a static water table in this sand?

C. Suppose that moisture content measurements from 10 m below the soil surface in this sandy loam to a depth of 50 m show a constant moisture content of 3%. The water table is at a depth of 80 m. Over the interval of the moisture measurements, how do capillary pressure head and hydraulic head vary?

Problem 3. Assume that a wetting front moves into the sandy loam soil described in Problem 2. The moisture content at the surface of the soil is held constant at 27%. The underlying moisture content is constant at 6%. If the saturated hydraulic conductivity for this soil is 3×10^{-6} m s^{-1}, estimate how long it will take the wetting front to move 1 m into the soil.

8.15 Suggested Reading

Fetter, C.W. 2000. *Applied hydrogeology*, 4th ed. Upper Saddle River, NJ: Prentice Hall. Chapter 6.

9 Ecohydrology: Interactions between Hydrological Processes and the Biota

9.1 Introduction

Hydrological processes play an important role in the dynamics of terrestrial ecosystems because they determine the availability of freshwater resources that sustain life on Earth. At the same time, the biota affect fundamental hydrological processes, such as evapotranspiration, precipitation, runoff and infiltration, which have a major impact on the water cycle at global, regional, and local scales. In recent years hydrologists have paid increasing attention to the coupling between ecological and hydrological processes and have become aware that a better understanding of the dynamics underlying changes in water fluxes and stocks in continental land masses can be achieved by investigating interactions between water and the biota. The study of these interactions has led to the emergence of ecohydrology, a subdiscipline at the confluence between ecology and hydrology. Research in this field is investigating both the controls exerted by hydrologic processes on the dynamics of terrestrial ecosystems (e.g., carbon sequestration, biogeochemical cycles, plant and microbial stress), and the effect of the biota on water flows and stocks. Understanding these hydrologic-biotic interactions is crucial to the prediction of the effects of changes in water availability and climate on ecosystems, as well as the evaluation of land use change impacts on the water cycle.

9.2 Hydrologic Controls on Plants and Soil Microorganisms

Most plants take up water through their roots. Water then moves through the **xylem**, capillary tubes within the plant, to the leaves and is lost by the plant as water vapor fluxes (i.e., transpiration) through the **stomata**, small cavities typically located beneath the leaves (Figure 9.1). Plants also assimilate CO_2 from the atmosphere through the stomata during photosynthesis. Plants can regulate the size of opening of the stomata to control the rates of transpiration and carbon assimilation. Thus, the water and carbon cycles are strongly coupled through stomatal physiology.

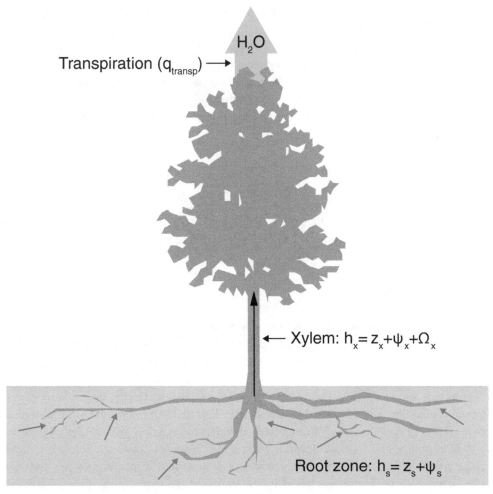

Figure 9.1 Schematic representation of water flow through the soil-plant-atmosphere system. Plants take up water from the soil through their roots. This water is transported through the xylem and released as water vapor fluxes (transpiration) to the atmosphere through the stomata.

9.2.1 Plant-water relations

Classic theories of water transport in soil, roots, and xylem assume that the flow is laminar. Soil pores and plant vessels are so tiny and the flow velocities are so small that the Reynolds numbers are definitely within the laminar range (see Chapter 3). Therefore, plant physiologists investigate water flow in plants using Darcy's law, which assumes the flow to be laminar (see Chapter 6). According to Darcy's law, the flow rate is proportional to gradients of hydraulic head or—in the jargon of plant physiologists—gradients of water potential, h [L]. The hydraulic head of soil water (or **soil water potential**, h_s) is the sum of elevation head (z_s) and soil capillary-pressure head (ψ_s), which in turn depends on the soil water content through the moisture characteristic curves (Figure 8.4). Water flow within plants is partly driven by concentration gradients across semipermeable membranes; known as **osmosis**, this process entails the transport of water molecules through semipermeable membranes from lower to higher solute concentration areas (see Box 9.1). To account for this phenomenon, the **xylem water potential** (h_x) is expressed as the sum of elevation head (z_x), xylem capillary-pressure head (ψ_x), and the osmotic potential (Ω), which is zero in pure water and negative in the presence of solutes (Ω decreases with increasing solute concentrations). In a steady flow the rate of root uptake is equal to the transpiration rate, q_{transp} [L T^{-1}] (see Figure 9.1), and can be expressed as

$$q_{transp} = -c(h_x - h_s), \tag{9.1}$$

where $h_x = z_x + \psi_x + \Omega$; $h_s = z_s + \psi_s$; and c [T^{-1}] is a constant representing the conductivity of the soil-root-xylem system. The capillary-pressure head in the xylem can be thus expressed as

$$\psi_x = \psi_s - \Delta z - \Omega - q_{transp}/c, \tag{9.2}$$

where $\Delta z = z_x - z_s$ is the height above the rooting zone (Figure 9.1).

Based on Equation 9.2, the xylem capillary-pressure head, ψ_x, decreases with increasing plant height, transpiration rate, and osmotic potential. Low values of ψ_x cause conditions of plant water stress in two ways. First, to maintain turgidity, the "stiffness" of cells that allows the non-woody parts of plants to stand vertically, the pressure of the water inside plant cells needs to be positive. Because the water potential of intracellular water (h_c) is in equilibrium with that of xylem water ($h_c \approx h_x$), low values of ψ_x may cause loss of turgidity and wilting. Second, when water pressure decreases below a critical value air bubbles form in the xylem. Known as **cavitation**, this phenomenon is undesirable because air bubbles cause the hydraulic failure of the xylem by clogging and damaging the conduits (embolisms). As a result of cavitation, fewer vessels remain available for xylem flow, thereby decreasing the conductivity, c, of the soil-root-xylem system and further reducing ψ_x (Equation 9.2). To maintain turgidity and avoid cavitation, plants can either modify (decrease) the osmotic potential by changing solute concentration in xylem water—a process known as **osmotic compensation**—or reduce the transpiration rates (see Equation 9.2). We recall from Chapter 8 that the relationship between ψ_s and the water content is expressed by the moisture characteristic curve (Figure 8.4): the capillary-pressure head ψ_s of soil water decreases as the soil becomes drier. During a dry period, plants might use osmotic compensation to prevent turgidity loss and hydraulic failure, but below a critical soil moisture value, θ^*, they eventually have to decrease transpiration by closing their stomata. As a result of stomata closure, photosynthetic uptake is also reduced because CO_2 fluxes occur through the same stomata. Under these conditions, if the rate of carbon assimilation is insufficient to meet the metabolic demand, plants need to use their carbohydrate reserves. Prolonged droughts may lead to the depletion of these reserves, thereby, causing plant mortality by carbon starvation (McDowell et al., 2008). Overall, as soil moisture decreases below the critical value, θ^*, plant transpiration is reduced, eventually becoming zero at a soil moisture level, θ_w, termed **permanent wilting point** (Figure 9.2).

The discussion above suggests that plants exposed to droughts need to walk a fine line of using stomatal closure to prevent desiccation and hydraulic failure while maintaining adequate rates of photosynthetic uptake to avoid carbon starvation. Two different stomatal regulation strategies may be used by plants (Figure 9.2): **isohydric plants** tend to close the stomata and reduce photosynthesis as the soil becomes drier, while **anisohydric plants** prefer maintaining higher photosynthetic uptakes even when soil moisture is low. The isohydric strategy avoids hydraulic failure at the cost of exposing the plant to carbon starvation. Conversely, anisohydric plants are less prone to carbon starvation but more susceptible to cavitation. An interesting example of the different drought sensitivity of isohydric and anisohydric plants has been observed in piñon-juniper woodlands across the southwestern United States. A severe drought in 2000–2002 caused about 95% mortality in piñon trees and 25% mortality in junipers. Scientists

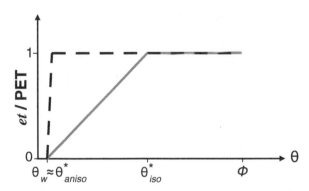

Figure 9.2 Relationship between the ratio of evapotranspiration (*et*) to potential evapotranspiration (PET) and soil moisture (θ) for isohydric (blue, solid) and anisohydric (black, dashed) plants. The value of θ is maximum (and equal to the porosity, ϕ) at saturation. For soil moisture values greater than θ^*, the stomata are wide open and transpiration occurs at the potential rate (PET). As soil moisture decreases below θ^* the plant starts closing the stomata thereby reducing *et*. At the permanent wilting point (θ_w), *et* is zero. Isohydric plants have a value of θ^* greater than anisohydric plants.

BOX 9.1 Osmotic Pressure

When ions (charged particles) in aqueous solution are not free to diffuse through-out a medium in which concentration gradients are maintained, an **osmotic pressure** develops. For example, in a device in which a semi-permeable membrane (a membrane with pores that allow water molecules to fit through, but which filter ions so they cannot pass) maintains a concentration gradient, osmotic pressure causes water to flow from the pure water into the solution until the hydrostatic pressure in the pure water phase balances the osmotic pressure (see figure below). The device sketched in the figure, a simple **osmometer**, can be used to measure osmotic pressure of the solution. Notice how this device resembles a plant that maintains solute concentrations in the xylem higher than in soil water (osmotic compensation). The existence of such a concentration gradient across semipermeable membranes allows plants to enhance root up-take, while reducing the risk of hydraulic failure due to low xylem pressure values.

Interestingly, even though no semi-permeable membranes exist in soils, in some clays there is a mechanism for segregating a relatively ion-rich phase from a relatively ion-deficient phase. The surface of clay particles tends to have a negative charge due to the structure of the mineral. Because of the attraction of particles having opposite electrical charges, an increased concentration of cations (positively charged ions) is found in the space between clay particles.

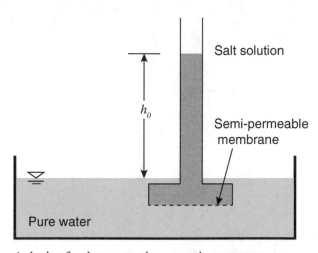

A device for demonstrating osmotic pressure.

BOX 9.1   *(continued)*

If there is very dilute water (very low cation concentration) in the pore spaces exterior to the clay plates, the increased ion concentration between any two clay plates sustains an osmotic-pressure head gradient that tends to force water in the direction of *greater* ion concentration; that is, into the space between clay plates. In a fashion similar to the osmometer (see figure), water flow will continue until the osmotic pressure is balanced by capillary pressure. This phenomenon explains the expansion (shrinking) of some clays upon imbibition (drying).

have explained these different mortality rates as an effect of the drought response strategies of these two species. Junipers exhibit anisohydric behavior (i.e., more susceptible to hydraulic failure), while piñons are isohydric plants (more prone to carbon starvation). There is some evidence that during the 2000–2002 drought, piñon trees had a negative carbon budget and were likely affected by high mortality rates due to carbon starvation (McDowell et al., 2008).

9.2.2 Waterlogging

Most plants need oxygen at their roots for metabolic processes. Because the oxygen supply to the root zone occurs by diffusion through air-filled soil pores, oxygen availability becomes restricted as soils become saturated. Lack of oxygen makes saturated soils unsuitable for the growth of many terrestrial plant species. Thus water stress in vegetation can be caused by both a water deficit and by an excess of water (**waterlogging**). Plants are sometime classified on the basis of their tolerance of waterlogging conditions. While most plants take up water from the unsaturated zone, **phreatophytes** place part of their root system below the water table and take up water from the groundwater. These plants rely on water uptake from the saturated zone but may still have only a limited tolerance to waterlogging. **Hydrophytes** withstand prolonged periods in which the ground surface is flooded.

The different sensitivity of plant species to flooding typically explains their distribution along topographic gradients in flood prone areas. Changes in hydrologic conditions may alter these distributions. For instance, in the Florida Everglades, shifts in plant community composition have been observed in response to altered hydrologic conditions. The Everglades are an extensive wetland with a seasonal 60 km wide sheet-flow delivering freshwater from Lake Okeechobee to Florida Bay over a distance of about 150 km. This slow shallow flow takes place on a limestone substrate partly vegetated by herbaceous plants. The landscape exhibits a number of islands that emerge from the water for most of the year. Because of their higher elevation the islands are more infrequently flooded than the surrounding marshes and are, therefore, densely covered by trees and other woody plants with limited flood tolerance. In 1948 the U.S. Congress approved a project that led to the construction of canals and levies that partitioned the Everglades into basins used for agriculture, water storage (Water Conservation Areas, WCAs), and biological conservation (the Everglades National Park). The WCAs were established to retain some of this freshwater and make it available for drinking and agricultural uses during the dry season. The consequent increase in flooding frequency in portions of the water conservation areas caused a die-off of trees in the lower elevation parts of the islands and a complete die-off in low islands.

9.2.3 Hydraulic redistribution

Most plants keep their stomata closed at night and no transpiration occurs. When vertical hydraulic head gradients exist along the soil profile, water is redistributed through the root system from shallow to deep soil or vice versa, depending on the direction of decreasing hydraulic head (see Section 8.5). Because of their higher hydraulic conductivity with respect to unsaturated soils, plant roots provide a preferential pathway for the nocturnal redistribution of water (Caldwell and Richards, 1989). If the soil is much drier at the surface than at depth (e.g., during a dry spell), then the capillary-pressure head at the surface is much lower than in the deep soil; this gradient in pressure head can overcome the opposite elevation head gradient and bring deeper water up through the roots to the surface (Figure 9.3). Known as **hydraulic lift**, this process may facilitate the establishment and survival of shallow-rooted plants and enhance the uptake of nutrients stored in the shallow soil. The opposite process can also occur: right after a rainfall

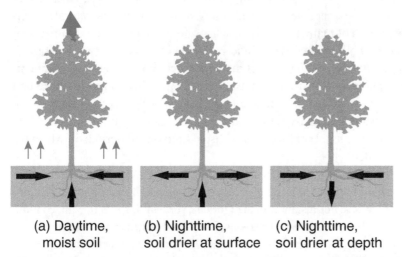

(a) Daytime,　　　(b) Nighttime,　　　(c) Nighttime,
  moist soil　　soil drier at surface　soil drier at depth

Figure 9.3 The role of roots in hydraulic redistribution. During the day water is taken up from the soil by the roots and transported to the leaves where it is transpired (*a*). At night, if the soil is much drier at the surface than at depth, water is taken up by deeper roots and released into the ground by shallow roots (*b*). Conversely, if the deeper soil is much drier than the shallow soil, at night water is drawn from the shallow soil and released into the ground by the deeper roots (*c*).

event, if the deeper soil is sufficiently dry, the hydraulic gradient from the shallow to the deep soil could draw water downward through the root system (Figure 9.3). This phenomenon of **hydraulic descent** can be used by deep-rooted plants to bring soil moisture out of the reach of shallow rooted plants (Burgess et al., 1998). The occurrence of hydraulic lift has been well documented in several forest ecosystems worldwide, including the Amazon. Model simulations indicate that hydraulic lift may enhance root uptake from deeper soil layers, and induce a substantial increase in dry season transpiration and photosynthesis with important impacts on the regional and global climate (Lee et al., 2005).

9.2.4 Soil moisture control on microbial activity

Soil moisture affects microbial processes in multiple ways. Low soil moisture inhibits microbial activity because the substrate (carbon and nutrients) used by microbes is transported in the water phase and its mobility is strongly reduced in dry soils. Moreover, in hyperarid (extremely dry) conditions microbes are affected by dehydration. Overall the water stress tolerance of soil micro-organisms is much higher than for plants; that is, the wilting point is much lower for microbes than for most types of vegetation. As water content increases above the wilting point, microbial activity increases because availability of substrate increases. The activity increases until effects of waterlogging become important after which increases in moisture lead to decreases in aerobic microbial activity (Figure 9.4).

The sensitivity of microbes to water logging depends on the microbial metabolism. While anaerobic microorganisms thrive in waterlogged soils and anoxic conditions, the activity of aerobic communities relies on the supply of oxygen and requires soil aeration.

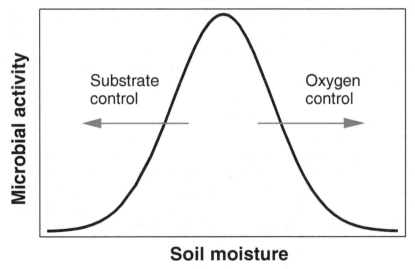

Figure 9.4 A conceptual representation of soil moisture control on aerobic microbial activity.

Modified after Skopp et al. (1990).

For example, the decomposition of soil organic matter is performed by aerobic microorganisms that convert litter and other organic residuals into inorganic compounds while releasing CO_2 into the atmosphere. Water logging conditions hinder decomposition, and provide an ideal environment for the accumulation of organic matter. For instance, peat deposits are commonly found in wetlands where the rate of organic matter production exceeds the decomposition rate. It has been estimated that a large fraction of global carbon stocks is in peat deposits. In boreal regions, defined by very cold winters and relatively short and cool summers, peatlands are prone to drainage as an effect of climate warming. This phenomenon is expected to increase soil aeration and soil organic matter decomposition, thereby causing peat loss, decreases in ground surface elevation (subsidence), and increases in atmospheric CO_2 emissions.

At the beginning of this section we indicated that the carbon and water cycles are strongly coupled. At this point it is clear that the coupling acts both through the process of atmospheric carbon assimilation by plants (hydrologic controls on stomatal regulation of photosynthetic uptake) and through soil moisture controls on organic matter decomposition.

9.3 Biotic Controls on Hydrological Processes

Hydrological processes are affected by the biota in various ways. At the plot scale, plant roots and soil organisms enhance infiltration by providing preferential pathways for water flow through the soil column, while plant uptake strongly modifies soil moisture dynamics. At the watershed scale, vegetation affects the water balance by enhancing evapotranspiration and reducing runoff. At the global scale, the terrestrial biosphere sustains a faster water cycle, that is, in the presence of vegetation, water cycles faster between the Earth's surface and the atmosphere than it would in the absence of vegetation. Using model simulations, Fraedrich et al. (1999) compared the water cycle in a desert planet (no vegetation) and in a green planet (with all the non-glaciated land covered by forest vegetation) and found that in the vegetated scenario evapotranspiration from continental land masses increased by 243% and precipitation by 93%, while runoff decreased by 24 % with respect to the unvegetated case. These two (unrealistic) extreme cases give us an idea of the important role played by vegetation in the global water cycle.

In Chapter 1 we explained how the water received at the Earth's surface as precipitation leaves continental landmasses either as evapotranspiration or runoff. Because most of the evapotranspiration from terrestrial ecosystems is contributed by plants, this flux is sometimes termed "green water flow," while surface and groundwater runoff are referred to as "blue water flows" (Figure 9.5) (Falkenmark et al., 2006). To investigate the impact of the biosphere on the water balance we can consider how vegetation affects the partitioning of precipitation over land into evapotranspiration and runoff (i.e., into green and blue water flows).

9.3.1 Impact of vegetation on evapotranspiration

Evapotranspiration from a vegetated landscape occurs at a faster rate than evaporation from bare soil. Thus, removal of vegetation reduces green water flows and increases

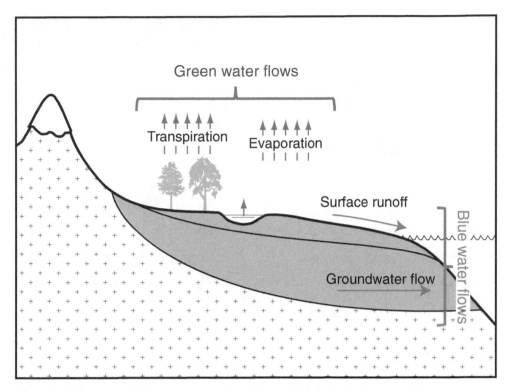

Figure 9.5 Green and blue water out of continental landmasses.

runoff. Deforestation is one of the major contributors to land cover change: it has been estimated that forests currently cover about 31% of the global land surface and that they are currently lost at the rate of 13 million ha yr^{-1} (FAO, 2010). Central and South America, Africa, and South-East Asia are particularly affected by deforestation. At the same time other regions of the world (e.g., the eastern U.S.) are undergoing reforestation as a result of abandonment of agricultural land, changes in climate, and land management. In most cases deforestation does not entail the replacement of a vegetated landscape with bare soil but with pasture or cropland. This change in land cover is still typically associated with a decrease in evapotranspiration (Figure 9.6) because, when compared to crops or pastures, forests have (a) a greater leaf area index (LAI, total foliage area per unit ground area), (b) deeper root systems that give them access to deeper soil moisture, and (c) greater land surface roughness that enhances evapotranspiration (see Chapter 2). Moreover, forests typically have a lower albedo (i.e., they are darker), which allows them to receive a higher net solar radiation than croplands and pastures. In other words, more energy is available to sustain evapotranspiration (see Equation 2.19) in a forest than in a pasture.

While deforestation is expected to reduce evapotranspiration and increase runoff, other trends in land use change may have the opposite effect. For instance, irrigation takes water from streams and aquifers and makes it available for crop evapotranspiration. As a result of irrigation, blue water flows are strongly reduced while green water flows

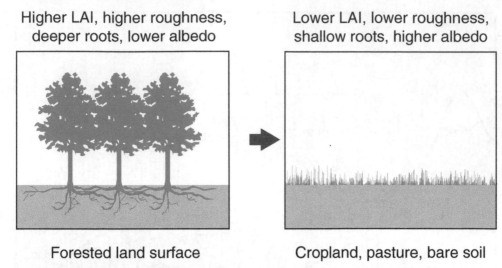

Higher LAI, higher roughness,
deeper roots, lower albedo

Lower LAI, lower roughness,
shallow roots, higher albedo

Forested land surface   Cropland, pasture, bare soil

Figure 9.6 Major land surface changes associated with deforestation, with forested conditions (*left*) contrasted with cropland (*right*).

increase. Many rivers on earth (e.g., the Rio Grande or the Colorado River) do not reach the ocean anymore as a result of water withdrawal for agriculture and other uses. Even though only 20% of the global agricultural land is currently irrigated, irrigation is expected to increase over the next few decades (to meet the escalating food demand) thereby further reducing river flows and increasing green water flows (Falkenmark et al., 2006).

9.3.2 Impact of vegetation on infiltration and runoff

Research in forest hydrology has clarified the effect of vegetation on the hydrologic response of catchments to precipitation. Forest soils are very permeable and allow high infiltration rates because roots provide preferential pathways for water flow through the soil matrix, particularly through the formation of macropores (i.e., relatively large cavities; see Chapter 10) resulting from the decay of dead roots. Moreover, plant canopies and litter shelter the soil surface against rainsplash compaction (the compaction induced by raindrop impacts onto the soil surface), thereby further enhancing the soil infiltration capacity. As a result, flow of water across the land surface in forested watersheds tends to be restricted to areas where soils are more easily saturated, typically in sites with convergent topography located close to the channel network (see Chapter 10). The removal of forest vegetation disturbs the soil surface and exposes it to compaction by rainsplash; it also reduces soil bioturbation and macropore formation by plant roots and other soil organisms. As a result, deforestation causes a decrease in soil infiltration capacity and an increase in surface runoff (Figure 9.7).

Forested watersheds exhibit higher evapotranspiration rates than their deforested counterparts: the replacement of a forest with pasture or bare soil reduces both interception and transpiration. The amount of water that flows through the watershed as surface and groundwater runoff is consequently greater in clearcut than in forested landscapes.

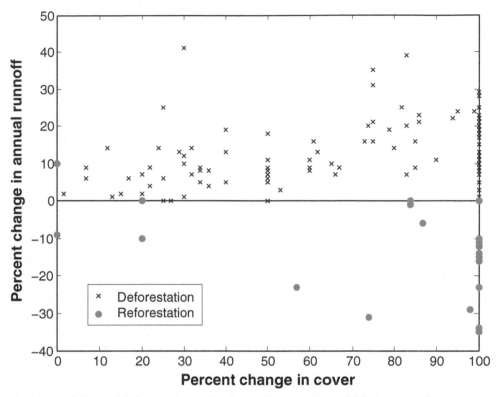

Figure 9.7 Effect of deforestation and reforestation on water yields (expressed as a percentage of mean annual precipitation). A decrease in forest cover is generally associated with an increase in runoff, while most of the reforestation experiments exhibit a decrease in runoff.
Data from Andreassian (2004).

These effects are in general more noticeable during the growing season, when vegetation plays a more active role in the water balance.

The impact of forest vegetation on runoff is typically investigated in **paired watershed** experiments, which compare the hydrologic response of two small watersheds located in the same region with similar topographic and hydrogeological features but different land cover conditions. These studies have shown that the increase in runoff induced by forest cutting affects streamflow in both fair weather periods and during storms. Overall a denuded landscape has a more limited capacity to retain and store water and tends to respond more quickly to rainfall events, in some cases with abrupt flash-floods. Snowmelt floods are also faster and occur earlier in the year. While in some regions the increase in runoff resulting from deforestation has the undesirable effect of increasing the potential for flooding and soil erosion, in others the decrease in runoff resulting from afforestation by bush encroachment or tree plantations is recognized as a serious threat to water yields and water availability, especially in semiarid catchments.

9.3.3 Effect of vegetation on precipitation

The impact of terrestrial vegetation on precipitation is more difficult to assess than its more direct effect on evapotranspiration and runoff. Because the effect on precipitation is expected to be detectable at scales that are much larger than those of the paired watershed studies discussed in the previous section (<10–100 km^2), direct experimental evidence is not available. Therefore scientists rely on process-based models that account for the coupling between hydrological and atmospheric processes. These model-based studies are in overall agreement: regional-scale deforestation is expected to cause a decrease both in evapotranspiration and in regional precipitation (e.g., Bonan, 2008). The decline in precipitation can be explained as a direct effect of the decrease in evapotranspiration (Section 9.3.1), and as a result of the altered surface energy balance resulting from changes in land cover conditions. A decrease in evapotranspiration entails a reduction in the terrestrial supply of atmospheric water vapor that contributes to precipitation recycling (Section 1.3). Vegetation removal is typically associated with an increase in albedo (Figure 9.6), hence a decrease in the net solar radiation that is received at the land surface. As a result, surface heating is reduced and convection-induced uplift—one of the mechanisms causing precipitation (Chapter 1)—is weakened. It has also been reported that vegetation can affect precipitation by altering the terrestrial supply of aerosols suitable as cloud condensation nuclei (e.g., aerosols resulting from biogenic gas emissions, smoke from forest fires, or mineral dust emitted by deforested landscapes). A quantitative assessment of these effects is a subject of ongoing research.

An interesting mechanism contributing to the supply of water to some terrestrial ecosystems is associated with the interaction of vegetated canopies with moist air. In Chapter 1 we stressed how condensation requires a surface on which water molecules can condense. Close to the ground, this surface may be provided by plant canopies, thereby leading to canopy condensation and dew formation. Alternatively, small droplets may already be present in fog or low clouds; as this damp air moves through the canopy, some of droplets are deposited onto leaf surfaces. Water eventually drips from wet leaves to the ground surface, a phenomenon known as **occult precipitation** (Chapter 2). In some coastal foggy regions occult precipitation may contribute to a substantial fraction of the total water input and these inputs may play an important ecological role, particularly if they occur during the dry season (e.g., the west coast of the U.S.) or in deserts (e.g., the Namibian or the Atacama Desert). It is worth noticing that occult precipitation would not occur (or would be severely reduced) in the absence of vegetation canopies (Runyan et al., 2012). To enhance their water supply, some communities living in "fog deserts" have built systems of nets and gutters that trap atmospheric moisture imitating the way plant canopies function in these environments.

9.4 Concluding Remarks

In the previous sections we have seen both how plants depend on water (Section 9.2) and how the rate of major hydrological processes can be modified by vegetation (Section

9.3). We can now consider the case of ecosystems in which, by altering hydrological processes, plants are able to establish conditions that reduce their exposure to water stress caused either by water scarcity or by waterlogging (Runyan et al., 2012).

Fog deposition can be crucial to the survival of plant communities in arid and semiarid ecosystems. This is an example of systems that rely on mutual interactions between vegetation and hydrological processes. In fact, the decrease in occult precipitation that would occur in response to the removal of plant canopies could impede vegetation re-establishment. Similarly, regional climate models have shown that vegetation removal from Mongolia, the Sahel region of Africa, or other desert margins would result in a reduction of precipitation, thereby preventing vegetation re-establishment (Bonan, 2008). Similar impacts of vegetation on precipitation are often invoked to explain major climate and vegetation changes in the history of the Earth. For example, fossil records indicate that up to the mid-Holocene (i.e., about 5000 years ago) large parts of the Sahara used to be vegetated. It has been argued that the decline of the "Green Sahara" was sustained by a decrease in precipitation induced by the loss of vegetation cover. Although changes in orbital conditions appear to have played a major role in the shift to the desert state, paleoclimate models indicate that the decrease in precipitation was likely enhanced by vegetation losses (Brovkin et al., 1998).

Analogous conditions may result from plant-water table interactions. The typical effect of vegetation on the underlying shallow unconfined aquifer is to lower the water table by taking up water from the overlying unsaturated zone or through direct depletion by phreatophytic vegetation. In landscapes with shallow water tables vegetation may play a crucial role in draining the shallow soil, thereby preventing the emergence of water logging conditions that would be detrimental to the survival of the existing plant species. Some ecosystems may strongly depend on these positive interactions between vegetation and water table dynamics. In some cases, the removal of native vegetation may result in a dramatic increase in water table elevation. For example, the removal of eucalyptus trees from the Murray-Darling basin (Australia) has allowed the water table to rise so close to the ground surface that direct evaporation of shallow groundwater is causing salt accumulation at the soil surface, thereby making soils unsuitable for most plants (Runyan et al., 2012).

These are just some examples of ecosystems that strongly depend on mutual interactions between hydrological processes and the biota. Understanding these interactions is crucial the study of ecosystem sensitivity to changes in climate and hydrologic conditions.

9.5 Key Points

- Water transport within plant vessels occurs as a laminar flow. To take up water from the ground, plants need to exert a suction that is strong enough to establish a hydraulic gradient between root zone and leaves. There are, however, some hydraulic constraints to the maximum suction tolerated by plants: strong suctions (i.e., low negative pressures) may induce loss of turgidity and cavitation, thereby causing the failure of the hydraulic function of the xylem. {Section 9.2.1}

- The water characteristic curves discussed in Chapter 8 (Figure 8.4) indicate that drier soils exhibit a lower (negative) capillary-pressure head. Thus in dry periods plants need to maintain a stronger suction to sustain uptake. To avoid damage caused by cavitation, plants tend to reduce transpiration by closing the leaf stomata, small cavities through which water vapor (transpiration) and CO_2 (photosynthesis) are exchanged with the atmosphere. {Section 9.2.1}

- As a result of stomata regulation, photosynthetic carbon uptake by plants is also reduced. Thus, prolonged droughts may induce plant mortality by carbon starvation. {Section 9.2.1}

- Water stress conditions can emerge not only when the ground is dry but also in saturated soils (waterlogging). Oxygen is needed in the root zone to sustain plant metabolic processes. Because oxygen is supplied by diffusion through air-filled soil pores, its availability is restricted in saturated soils. Thus, most plant species do not tolerate waterlogging. Similarly microbial reactions that require oxygen (e.g., decomposition) are restricted in saturated soils. {Sections 9.2.2 and 9.2.4}

- At night plant roots may redistribute soil water along decreasing gradients of soil moisture. If the soil is much drier at the surface then at depth, hydraulic redistribution entails an upward water flow through the root system ("hydraulic lift"). Conversely, if soil moisture is much greater in the shallow soil then at depth, downward redistribution through the roots may occur. {Section 9.2.3}

- Vegetation enhances evapotranspiration and precipitation in continental landmasses, thereby intensifying the terrestrial portion of the water cycle. {Section 9.3}

- Forest removal (deforestation) reduces evapotranspiration (or "green water flows") and increases runoff (or "blue water flows"). Conversely, irrigation reduces blue water flows and increases green water flows. {Section 9.3}

- The ability of forests to enhance evapotranspiration is due to the higher leaf area index (hence greater transpiration and interception), deeper roots, and greater roughness of forest vegetation with respect to croplands or pastures. {Section 9.3.1}

- Vegetation has the effect of increasing soil infiltration capacity because plant canopies shelter the soil surface against rainsplash compaction, while soil bioturbation by plant roots creates macropores, thereby increasing the soil permeability. Overall, the impact of vegetation on the hydrologic regime is to increase infiltration and decrease runoff. {Section 9.3.2}

- At the regional scale, forest vegetation increases precipitation. In fact, vegetation enhances evapotranspiration, thereby intensifying precipitation recycling. Moreover, because of their lower albedo, plants modify the surface energy balance thereby enhancing surface heating and convection. {Section 9.3.3}

- Occult precipitation (i.e., fog deposition and canopy condensation) can be an important contributor to the water balance in regions affected by fog and low clouds. {Section 9.3.3}

- The ability of vegetation to modify the local or regional water balance may be crucial to the survival of some plant communities in water scarce environments. In these ecosystems vegetation removal can dramatically alter the water budget to the point of preventing the re-establishment of the same plant species. {Section 9.4}

9.6 Example Problems

Problem 1. Irrigation is used in areas in which agricultural productivity can be enhanced by maintaining the soil water content well above the permanent wilting point (e.g., $\theta \geq 2\theta_w$) throughout the growing season. When irrigation water is applied, root zone soil moisture should not exceed field capacity θ_{fc}, because, if $\theta > \theta_{fc}$, water and fertilizer are lost from the root zone by drainage. With this management approach soil moisture remains below field capacity and the only losses of soil water from the root zone are by evapotranspiration. Assuming that no rainfall occurs, the temporal dynamics of depth-average soil moisture θ in the root zone can be expressed by the soil water balance equation

$$Z\frac{d\theta}{dt} = -et,$$

where Z is the depth of the root zone. In other words, changes in the water stored in the control volume (i.e., the root zone) are due to the difference between water inputs and outputs (see Chapter 1). In this specific case there are no rainfall-induced inputs, while the only outflow is due to et. The irrigation period (i.e., the time between two consecutive applications of irrigation water) is calculated by integrating the soil water balance equation between time $t=0$ when $\theta = \theta_{fc}$ (i.e., right after irrigation) and time t_i when $\theta = \theta_0$ (and it is time to irrigate again), assuming that the evapotranspiration rate linearly decreases from the potential rate PET at $\theta = \theta_{fc}$ to zero at $\theta = \theta_w$ (i.e., $et = \text{PET}(\theta - \theta_w)/(\theta_{fc} - \theta_w)$)

$$t_i = \int_{\theta_{fc}}^{\theta_0} -\frac{Z(\theta_{fc} - \theta_w)}{PET} \frac{d\theta}{\theta - \theta_w} = \frac{Z(\theta_{fc} - \theta_w)}{PET} \ln\frac{(\theta_{fc} - \theta_w)}{(\theta_0 - \theta_w)}.$$

Consider the case of a crop grown on a sandy soil with $\theta_{fc} = 0.19$ and a crop-specific wilting point, $\theta_w = 0.05$. The root zone is 40 cm deep. After each application of irrigation water, soil moisture is equal to θ_{fc}. Assuming that PET $= 6$ mm d^{-1}, calculate the irrigation period, t_i, i.e., how long it would take for soil moisture to decrease from field capacity to the value $\theta_0 = 2\theta_w$ in the absence of any rainfall input.

Problem 2. With the irrigation scheme described in the previous problem, during each application of irrigation water soil moisture is increased from $\theta_0 = 2\theta_w$ to θ_{fc}. The volume of water (per unit area) required to make this change is $Z(\theta_{fc} - \theta_0)$. Calculate the amount of water per unit area that should be applied every time the area is irrigated.

9.7 Suggested Readings

Bonan, G. 2008. *Ecological climatology*. Cambridge, UK: Cambridge University Press.

D'Odorico, P., and A. Porporato. 2006. *Dryland ecohydrology*. Dordrecht, the Netherlands: Springer.

Rodriguez-Iturbe, I., and A. Porporato. 2005. *Ecohydrology of water-controlled ecosystems: Soil moisture and plant dynamics*. New York: Cambridge University Press.

10 Catchment Hydrology: The Hillslope-Stream Continuum

10.1 Introduction

One of the problems that has occupied hydrologists for many decades is the identification and quantification of the pathways by which rainfall and snowmelt move through catchments and ultimately produce runoff. In Chapters 5, 7, and 8 we described the individual flow paths over and beneath the Earth's surface that precipitation might take through a catchment and provided the physical basis for understanding each. To describe the catchment hydrological cycle, we have to integrate all the individual processes of surface and subsurface flow of water. This is a formidable problem to solve in a quantitative fashion. In this chapter we introduce some of the ideas and tools used in catchment hydrology.

There are many reasons why hydrologists are interested in understanding flow paths within catchments. One reason is that understanding the routing of rainfall and snowmelt is required for understanding how water interacts chemically with rocks, sediments, and biota within a catchment. For example, hydrologists are interested in understanding the effects of acid rain on the chemical composition of streams around the world. The pH of rainfall and snowfall in many parts of North America and Europe in the latter part of the twentieth century was typically below 4.5, which is well below the level at which water would be considered to be acidic (<5.65). "Acid rain" is known to result from the chemical interaction of water in the atmosphere with gaseous emissions from power plants, factories, and automobiles. (Although the trend in the acidity of rain has been positive over the past decade or so and the extent of areas with rainfall pH below 4.5 has decreased considerably, concerns remain about the impact of deposition of acidic compounds on catchments (Burns et al., 2011).) In some catchments, the stream pH shows an episodic depression (known as acidification) that at first glance might appear to be caused by rain falling directly on the stream surface. However, for many streams this explanation cannot be correct because the pH of the stream remains low even after the rainfall has ceased (for example, Bear Brook in Maine, Figure 10.1). How, then, is the composition of the rain altered when it comes into contact with soils and rocks? Answers to such questions require knowledge of the flow paths through and residence times within various portions of the catchment, and are relevant to a variety of environmental issues. For example, depressions of pH below about 5 may cause damage to certain types of fish.

Another reason for desiring an understanding of runoff generation in catchments is to be able to predict the effects that humans will have on the hydrological cycle through their land-based activities. It is now known from experimental studies at research sites throughout the world (for example, the Hubbard Brook Experimental Forest in New Hampshire) that various silvicultural practices (such as selective cutting, clear-cutting, and others) can have dramatic effects on stream-water discharge and quality for many years after the particular management activity is performed. Hydrological changes that have been demonstrated experimentally include increases in annual water yields, decreases in evapotranspiration, increases in nitrate concentrations, and increases in stream temperature and sediment loads.

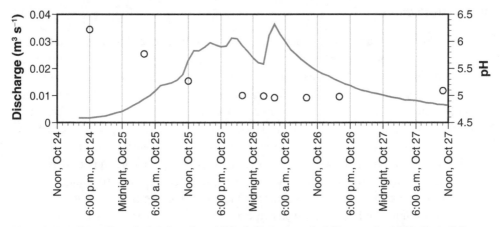

Figure 10.1 Bear Brook, Maine, is acidified during a rainfall event in 1992. Rainfall produces a rising hydrograph (*blue line*) and a decrease in the stream pH (*circles*).

Hydrologists often conceptualize a catchment as functioning like a group of "reservoirs" that store and release water in much the same way as a simple flood-control reservoir (see Chapter 5). Precipitation and snowmelt are the "inputs" to the catchment reservoir and evapotranspiration and runoff are the "outputs." This conceptual model of a catchment can be useful in trying to integrate the various processes that affect runoff. Much of our treatment of catchment hydrology in this chapter will make use of the idea of identifying and quantifying flow in various "reservoirs."

10.2 Streamflow Hydrographs

Consider the hydrograph for the stream draining Watershed 34 (W34) at the Coweeta Hydrologic Laboratory in North Carolina (Figure 10.2). Watershed 34 (area = 0.33 km^2) is a mid-elevation catchment with relatively deep soils (for the mountains of western North Carolina). Consideration of the hydrograph for W34 leads to speculation that some of the flow during precipitation events occurs immediately after the onset of precipitation, while other portions of the flow occur for some time after the end of precipitation. These two components of a streamflow hydrograph are known as **quickflow** and **baseflow.**

The separation of a hydrograph into two components suggests that water is being routed through two different storage "reservoirs" (Figure 10.3). That is, considering the input to a reservoir to be precipitation and the outflow from the reservoir to be either the quickflow *or* the baseflow, we observe the typical effects of flow routing through a

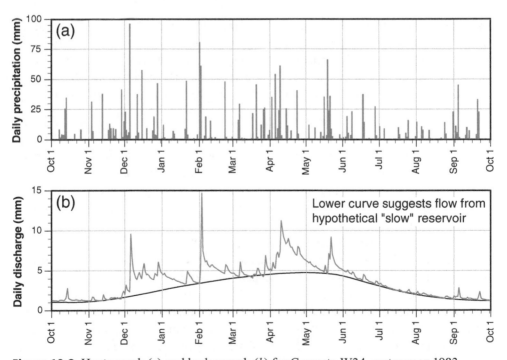

Figure 10.2 Hyetograph (*a*) and hydrograph (*b*) for Coweeta W34, water year 1983.

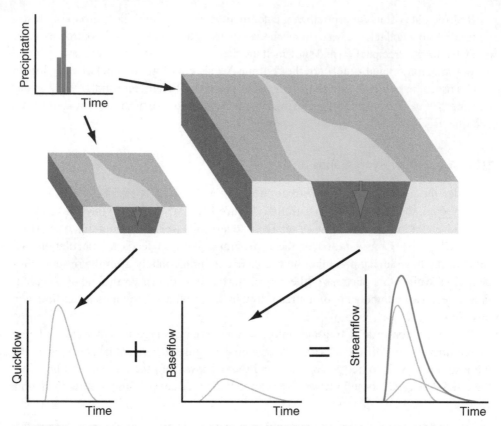

Figure 10.3 Rainfall routed through two conceptual "reservoirs." Although depicted here as surface-water reservoirs, in this chapter the term reservoir refers to portions of the catchment hydrological system with differing dynamic behavior. For example, the baseflow reservoir may be groundwater; because groundwater recharge and discharge are relatively slow processes, this reservoir behaves like the one at the right above.

reservoir: peaks are delayed and attenuated. From our discussion of flow routing through surface impoundments (Chapter 5.4), we can infer by analogy that the baseflow reservoir has characteristics of storage and release that are quite different from those of the quick-flow reservoir. Remember that these catchment "reservoirs" are conceptual, meant to represent portions of the hydrological system that behave *similarly* to surface-water reservoirs, by storing and releasing water.

The reservoirs shown in Figure 10.3 operate simultaneously. This makes the job of separating the hydrograph into contributions from the two (or more) reservoirs a difficult one. The observation that precipitation appears to be routed through "quickflow" and "baseflow" reservoirs in many catchments has led to a great deal of research on flow paths through catchments. What makes up the quickflow and baseflow reservoirs? What are the characteristics of catchments that determine the relative amounts of quickflow and base-flow? Does the water that flows through the different reservoirs acquire different chemical signatures from their contact with different geological materials?

10.3 Hydrograph Separation

Questions such as those posed in the previous section have not been answered completely despite many decades of research. One of the difficulties lies in the determination of the components of the streamflow hydrograph. After all, we don't measure two hydrographs (quickflow and baseflow); we measure only the total flow. The question then arises: how can we "separate" the measured hydrograph? The smooth line in Figure 10.2 showing the baseflow was drawn "by eye." The separation seems reasonable, but has no scientific basis. Hydrograph separation techniques have been used by hydrologists for many years, but all of the methods are empirical. These empirical methods are used widely to enable the calculation of quickflow from catchments. Once the hydrograph has been separated into quickflow and baseflow, a very simple computation, known as the unit hydrograph method, can be used to route the portion of the precipitation that goes through the quickflow "reservoir" to the stream, a procedure that has application in engineering hydrology where the size of pipes to carry stormwater is decided on the basis of such routing calculations.

For many years, a standard practice was to *assume* that the components derived from empirical hydrograph separation methods were physically "real"—that the quickflow arose from the flow of water overland where precipitation rates exceeded infiltration rates and that the baseflow originated as water that had infiltrated the soil surface, percolated to the groundwater zone, and discharged into a stream channel over prolonged periods of time. The "reservoirs" of Figure 10.3 under this view would be the hillslope surfaces (the quickflow reservoir) and the unconfined aquifer (the baseflow reservoir). Unfortunately, this simple picture is not valid in many instances. The surface-water-groundwater-reservoir interpretation of the empirical results turns out to be counter to observations made by field hydrologists in forested catchments, for example.

An alternate method for separating hydrographs is based on differences in the chemical composition of waters in the "reservoirs" thought to store and release water (e.g., see Sklash and Farvolden, 1979). The assumption behind this alternate method for separating hydrographs is that streamflow during a storm event is a mixture of water that has been precipitated into the catchment during the particular storm (i.e., event rainfall or "new" water) and water that was stored in the catchment prior to the onset of the storm (i.e., pre-event or "old" water). Provided that the concentration of some chemical component (i.e., tracer) of these two waters naturally differs (and does not vary during the storm or vary spatially across the catchment), measurements of the concentration of the tracer in the two components and in the mixture through time can be used to back-calculate the contributions from the "old" and "new" reservoirs. For example, suppose we make measurements of the chloride concentration in the rain falling on the catchment, in the groundwater within the catchment, and in streamflow (the mixture) during the course of a storm. Let Q_t be the total measured streamflow (measured at a standard stream gage). Let Q_n and Q_o be the flow contributions to the hydrograph associated with the "new" water (rain) and the "old" water, respectively. Because we assume that there are no other flow components, at any time t we know that

$$Q_t = Q_o + Q_n. \tag{10.1}$$

Note that the streamflow (Q_t) and the flow components (Q_n and Q_o) vary with time.

If the concentration of chloride in precipitation (rain or snowmelt) is relatively constant over the storm and if the concentration of chloride in "old" water in the catchment is relatively constant in both time and space, we can represent these concentrations as C_o and C_n (note that we have measurements of these concentrations). Because the stream water is a mixture of the old and the new waters, the concentration of chloride in the stream will change over the course of the storm. We represent this temporally changing concentration as C_t. Because we assume that the mass of chloride observed in the stream is derived from only these two reservoirs, we have:

$$Q_t C_t = Q_o C_o + Q_n C_n. \tag{10.2}$$

Again note that we assume that we have measurements of C_t (i.e., stream chloride concentrations) at several times over the course of the hydrograph. Given measurements of: (1) Q_t, (2) C_t (at several times), (3) C_n (assumed constant), and (4) C_o (assumed constant), Equations 10.1 and 10.2 can be solved for Q_n and Q_o at every time for which measurements of C_t are available:

$$Q_n = \left[\frac{(C_t - C_o)}{(C_n - C_o)} \right] Q_t. \tag{10.3}$$

This calculation yields a hydrograph separation in terms of "old" and "new" water.

Consider results for a storm that occurred in June 1992, in the Shaver Hollow catchment, Shenandoah National Park, Virginia. Bazemore (1993) made hydrological measurements and chemical observations of a 0.082-km^2 sub-catchment at Shaver Hollow. The pre-event groundwater had a chloride concentration of about 27 mol L^{-1} and the event water that fell through the forest canopy had a chloride concentration of about 4 mol L^{-1}. Chloride concentrations in the stream, measured over the storm hydrograph, were used to separate the hydrograph into "old" and "new" water (Figure 10.4). Note that approximately 90% of the water under the storm hydrograph was determined to be "old" (pre-event) water and at the time of peak discharge more than 80% of the total discharge was determined to be old water (Bazemore, 1993). The use of chemical composition of throughfall, soil water, and ground water to separate hydrographs into components requires a number of assumptions about the temporal and spatial variability in the various "components." The method can be criticized because the assumptions may be poor for natural catchments (e.g., see Harris et al., 1995). Despite controversy about the details of chemical hydrograph separations, the frequent observation from such work that a large fraction of stormflow in forested, temperate-zone catchments is "old water" is generally agreed to be valid. This approach of using chemical signatures to identify contributions

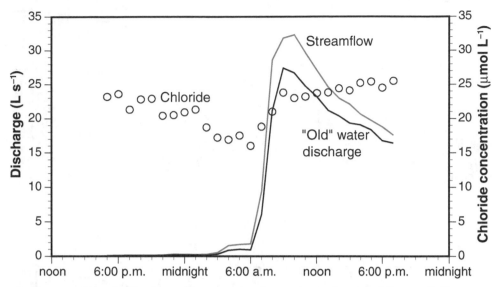

Figure 10.4 Chemical hydrograph separation for a sub-catchment of Shaver Hollow, Shenandoah National Park, Virginia. The data cover June 4 and 5, 1992.
Data courtesy of David Bazemore.

to streamflow from several sources can be useful for a variety of problems and is generally termed end-member mixing analysis.

The main conclusion from the application of the chemical-hydrograph-separation method to this and other forested, upland catchments located in humid, temperate regions is that observed streamflow is mostly old water. Quickflow is not synonymous with "new" water and baseflow is not synonymous with "old" water. An empirical separation of the hydrograph into quickflow and baseflow is insufficient to describe actual flow paths within a catchment, which are necessary, for example, to account for chemical reactions in a catchment. Describing the transport of solutes within a catchment requires a detailed understanding of the physical processes that govern the flow of water in catchments.

10.4 Runoff Processes

There are four paths by which water precipitated onto a catchment ultimately can be discharged into a stream channel. The four flow paths (Figure 10.5) are: direct precipitation (or throughfall) onto an active stream channel, overland flow, shallow subsurface stormflow, and groundwater flow. Snowfall can be considered "delayed rainfall" once snowmelt takes place, and can be assumed to follow the same flow paths to the channel.

10.4.1 Direct precipitation onto stream channels

The contribution of direct precipitation (or throughfall) to an active stream channel is usually quite small because the surface area of the perennial channel system in most

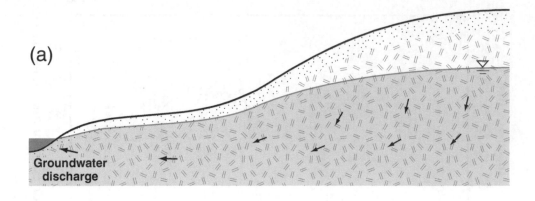

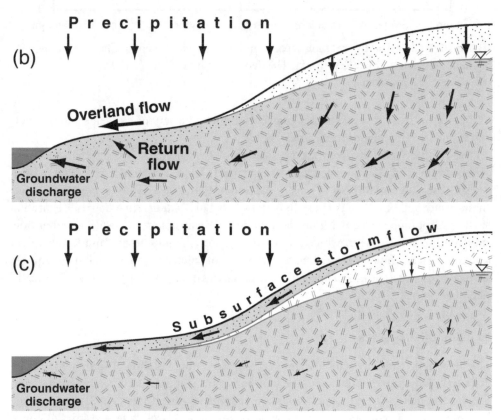

Figure 10.5 Runoff processes. Between precipitation or snowmelt events (*a*), streamflow is maintained by groundwater discharge (baseflow). Runoff is enhanced during and after precipitation events. Direct precipitation onto stream channels or overland flow (*b*) contributes directly to enhanced surface runoff. Infiltrating rainwater or snowmelt that reaches the groundwater system can enhance baseflow or provide return flow in saturated areas (*b*). Saturation of permeable soil horizons also can produce shallow lateral flow called subsurface stormflow (*c*).

catchments is a small percentage of the catchment area. However, in many catchments dramatic expansion of the channel system can occur during wet seasons and during periods of precipitation. In upland catchments, a significant fraction of storm runoff can result from precipitation falling on this expanded channel network (Dunne and Black, 1970). Overland flow caused by precipitation on saturated areas near the stream is part of the variable source mechanism for runoff production described below.

10.4.2 Overland flow

Overland flow is water that flows across the ground surface and discharges into a stream channel. If overland flow is to occur, water must accumulate at the surface rather than infiltrate into the soil. Overland flow can occur on a catchment for a number of reasons: (a) the catchment surface may be nearly impermeable due to the presence of exposed bedrock or paving over the surface; (b) the instantaneous rate of infiltration through the pervious surface may be exceeded by the instantaneous rate of rainfall (or snowmelt) onto the catchment surface, causing ponding of water at the surface; and (c) the catchment soil upon which the rainfall is precipitated may be saturated to the soil surface, causing ponding because the precipitated water cannot infiltrate into an already saturated soil. Overland flow in catchments is one of the most rapid paths that rainfall can follow to the stream channel; therefore, overland flow is expected to contribute to quickflow.

Overland flow dominates the response of many urbanized catchments with large expanses of directly connected, impervious surfaces. Because urbanization increases the fraction of precipitated water that follows the rapid overland flow path through a catchment, increased frequency of floods is one result of development. Leopold (1994) reports that after suburban development of a small catchment in Maryland, flood frequencies on Watts Branch, the stream draining the catchment, increased dramatically. Overbank flows in Watts Branch occurred about 1.4 times per year before urbanization and about 6.5 times per year afterwards.

Overland flow of water from a catchment that occurs when the rainfall or snowmelt rate exceeds the ability of the soil to allow water to infiltrate is referred to as **infiltration-excess overland flow**, or **Hortonian overland flow**, thereby crediting the American hydrologist Robert Horton, who reported occurrences of this runoff mechanism. Horton (1933) argued that, over the time course of a storm, the rainfall rate eventually would exceed the rate at which water could infiltrate into the soils of a catchment (as in Figure 8.9), resulting in the accumulation of excess water at the ground surface and overland flow in sheets. As discussed in Chapter 8, infiltration into soils depends on hydraulic conductivity of the soil profile. The hydraulic conductivity of the surface soil is conditioned strongly by vegetation as well as by the texture of the mineral soil. For example, hydrologists have observed that forest soils typically have a high infiltration capacity due to the presence of above-ground vegetation (and decomposing vegetation on the surface), which protects the surface from compaction caused by raindrop impact, and of below-ground vegetation, which provides roots that keep the soil porous and highly permeable (Chapter 9.3.2). Infiltration-excess overland flow is not a significant runoff-generating process in such catchments. For example, up to 150 mm of water per hour was sprinkled for several hours on an isolated block of soil in a forested catchment in Maine with essentially no overland flow (Hornberger et al., 1990). Conversely, when vegetation is absent from the surface of a catchment, the surface soil tends to develop a "crust" with relatively low hydraulic conductivity. Infiltration-excess overland flow can be the dominant runoff-generating mechanism in catchments where the land surface has been strongly disturbed (e.g., a plowed agricultural field or construction site where vegetation was removed and the soil surface left exposed) and in arid and semiarid regions with sparse vegetative cover. Infiltration-excess overland flow generally does not occur at a constant rate over entire catchments due to spatial differences in soil properties. Some areas of the catchment may produce overland flow during virtually all storms, while other areas may rarely, if ever, do so.

In many forested catchments in the temperate zone, the soil is capable of allowing essentially all of the incident precipitation to infiltrate. In such catchments overland flow develops not because precipitation intensity exceeds the infiltration rate, but because it falls on temporarily or permanently saturated areas (wetlands) with no capacity for water to infiltrate. Flow developing under these conditions is known as **saturation-excess overland flow** (Figure 10.5b). The areas of a catchment that are prone to saturation tend to be near the stream channels, or where groundwater discharge areas occur. These areas grow in size during a storm and shrink during extended dry periods (Dunne et al., 1975; Beven, 1978). The areas on which saturation-excess overland flow develops, therefore, change with time over a storm. Because the expansion of saturated areas occurs in a similar fashion as the expansion of the channel network, this mechanism of runoff generation is often referred to as the **variable contributing area concept**.

10.4.3 Shallow subsurface stormflow

Water that has infiltrated the soil surface continues to be influenced by gravity, causing water to percolate downward through the soil profile. In general, the ability of the soils and rocks of a catchment to conduct water (hydraulic conductivity) decreases with depth; water percolating downward has thus been observed to "pile up," causing localized areas of saturation in the soil. In these instances, water may move laterally toward a stream by a process known as **shallow subsurface stormflow** (Figure 10.5c). Some of the water in subsurface stormflow moves at a relatively slow pace through the soil and contributes to the baseflow of streams, particularly during wetter winter and spring periods. Subsurface stormflow also may occur along preferred flow pathways (e.g., soil cracks, old animal burrows, decayed root channels, etc.) called macropores (Chapter 8) and the flow to the stream channel along these pathways may be quite rapid. Thus, subsurface stormflow can contribute significantly to quickflow. A portion of the infiltrated water does not become subsurface stormflow; it percolates downward until it reaches the water table and becomes groundwater.

10.4.4 Groundwater flow

In Chapter 7 we presented several examples of groundwater movement and discharge to topographic lows. In general, groundwater flow is the slowest of all flow paths through a catchment. Baseflow in low-flow periods is composed almost entirely of groundwater discharge. As such, baseflow tends to vary quite slowly over long time periods in response to changing inputs of water through net recharge. In catchments, the groundwater discharge to a stream channel may lag the occurrence of precipitation by days, weeks, or even years. When subsurface water flowing downslope to the stream enters the saturated area near the stream, some of the water is forced to reemerge onto the ground surface because the capacity of the soils and rocks to transmit all of the flowing water downslope is insufficient. This reemerging subsurface water is known as **return flow** (Figure 10.5b). Like saturation-excess overland flow, return flow can be quite rapid in some catchments with shallow water tables where groundwater mounds may form which dramatically increase local hydraulic gradients toward the stream (Freeze, 1972a, 1972b; Winter, 1983). As was the case with subsurface stormflow, groundwater flow can contribute to either the quickflow or baseflow components of a stream hydrograph, although in most instances the contribution to baseflow is dominant.

10.4.5 "Old" and "new" water

As we have noted (Section 10.3), for streams in forested upland catchments in temperate climates, the bulk of quickflow (even at the peak of a hydrograph) has the chemical composition of water that was resident in the subsurface prior to the storm event. That is, the bulk of quickflow is "old" water, indicating that water contained in soils and rocks in a catchment must be able to reach the stream quickly under storm conditions. This observation may seem at first glance to be at odds with the physical processes described above in which "old" groundwater is viewed as contributing primarily to the (slow) baseflow portion of the hydrograph. There are a number of physical processes that can be called upon to explain the observation that a large fraction of quickflow is "old" water. Consideration of the physics of soil water suggests that processes that cause "new" water delivered to the surface of a catchment to displace "old" water and force it into the stream are common. For example, even if only a small amount of infiltrated water reaches the capillary fringe quickly (e.g., through flow in macropores, especially in the riparian, or near-stream, area), it can change the negative capillary-pressure head in the capillary fringe to a positive pressure head. Such rapid increases in positive pressure head can force "old" groundwater into the stream rapidly. Another mechanism that contributes to the delivery of "old" water to a stream under storm conditions is the entrainment of "old" soil water by overland flow. Overland flow in riparian areas occurs in patches, with surface water flowing into the soil, mixing with shallow soil water, and then reemerging on the surface. The resulting overland flow contains a portion of the "old" soil water with which the "new" water has mixed. The conceptual explanations offered above appear to be consistent with observations, but by themselves do not provide quantitative estimates for the various flows. Quantification requires consideration of inflow, outflow, and storage of water within a catchment.

10.5 Contributing Area and Topographic Controls on Saturation

As we saw in Chapter 7, the topography of the landscape exerts an enormous influence on the movement of water in the subsurface. Topography likewise should control the development of areas of surface saturation and runoff. If we could break a catchment up into blocks ("reservoirs"), we might be able to use the conservation of mass equation to determine the degree of saturation and potential for runoff generation for each one. Each block would differ in its position along the hillslope and in the slope of the land surface (and probably the water table) through the block. Consideration of inflows, outflows, and runoff potential for all of the blocks in a catchment could provide the starting point for routing water through the catchment.

10.5.1 Contributing area

The degree of saturation of each catchment "block" depends on its water balance. If the inflow to the upslope face of the block from higher portions of the catchment is greater than the outflow from the downslope face of the block, the water table within the block will increase. The inflow rate for a catchment block depends on the **contributing area**, the area of catchment upslope from a given block that contributes inflow to that block. Contributing area, A, depends on the distance to the divide above the block as well as whether it is convergent, divergent, or planar (Figure 10.6).

To define A, elevation contours are drawn at a specified contour interval (for example, 10 m) for the catchment. Beginning at the base of the catchment, lines are drawn perpendicular to each contour they cross, forming a network of curves similar to the flow nets described in Chapter 6. The lines perpendicular to the contour lines represent flow lines. In areas of the catchment with a uniform slope (planar sections), the flow lines will have a constant spacing (Figure 10.7). In areas where the hillslope is concave, the flow lines tend to converge as one follows them downslope. The surface soil layer between two flow lines on a concave, or convergent, slope is something like a converging channel. As the upstream subsurface flow gets funneled into a smaller area, its depth increases and, with sufficient supply, can saturate the soil. The opposite happens in a divergent section. The increasing distance between flow lines allows the subsurface flow to spread out and thin.

While inflow to a catchment block is proportional to contributing area, local slope, $\tan \beta$, controls outflow from the block. For example, if the topographic and water table slopes of a block are relatively flat, the hydraulic gradient is small and Darcy's law indicates that water movement will be relatively slow in the absence of changes in hydraulic conductivity. Therefore, we might expect a small outflow and an increase in water storage through time within that block depending on the volume of inflow. The increase in storage is even greater if the block is at the base of a convergent hillslope, such that a great deal of upslope flow into the block occurs. If the water table reaches the surface, the block is completely saturated and any additional water supplied to the block will run off as saturation excess overland flow.

10.5.2 Topographic index

As described above, the important characteristics of a hillslope that influence the likelihood of areas of saturation and runoff developing are the upslope contributing area per unit contour length, $a = A/c$, and the local slope of the block, $\tan \beta$. These can be related to each other as discussed in Section 10.5.3, or they can be combined into a single variable that quantitatively captures the effect of topography, such as the **topographic index**:

$$TI = \ln(a/\tan \beta). \tag{10.5}$$

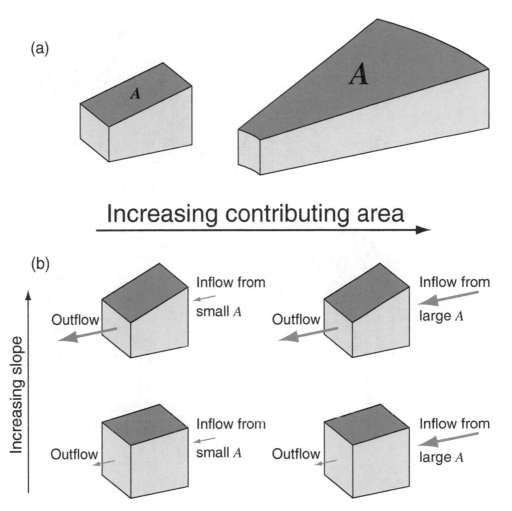

Figure 10.6 Local slope and contributing area control the water balance for a catchment "block." The inflow rate is proportional to the contributing area A, which depends on how long the hillslope is as well as whether it is convergent, divergent, or planar (*a*). The local slope controls the outflow from the blocks (*b*). If inflow is smaller than outflow (*upper left in b*), the water table declines. Conversely, if inflow is greater than outflow (*lower right in b*), the water table will rise and surface saturation may occur.

Specific contributing area, $a = A/c$, can be defined for each point within a catchment if the topography of a catchment is known. Although it would be possible to estimate a from a high-resolution topographic map, most studies of this sort use digital elevation data that can be used in conjunction with geographic information system (GIS) software to determine the specific contributing area for each point in the catchment. It is important to note that the results of an analysis such as this are highly dependent on the quality

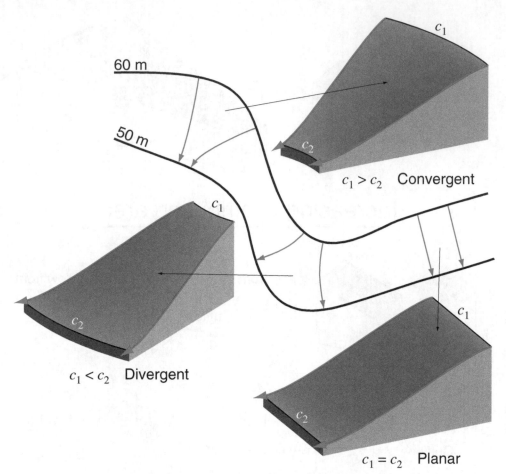

Figure 10.7 Influence of topography on contributing area.

and resolution of the digital elevation data. Accurate identification of the channel network, in particular, depends on using high-resolution elevation data. The current state of the art technique for generating high-resolution topographic data is light detection and ranging (lidar). In fact, the critical need for good topographic data to define channels and channel networks has led to calls for lidar maps to be produced to allow accurate mapping of flood plains (NRC, 2007).

A map of topographic indices for a catchment reveals areas where runoff processes such as saturation-excess overland flow are likely to occur (Figure 10.8). High values of the topographic index indicate areas with large contributing areas and relatively flat slopes, typically at the base of hillslopes and near the stream. These areas also correspond with expected groundwater discharge areas (Chapter 7). Low TI values are found at the tops of hills, where there is relatively little upslope contributing area and slopes are steep. These areas correspond generally with groundwater recharge areas.

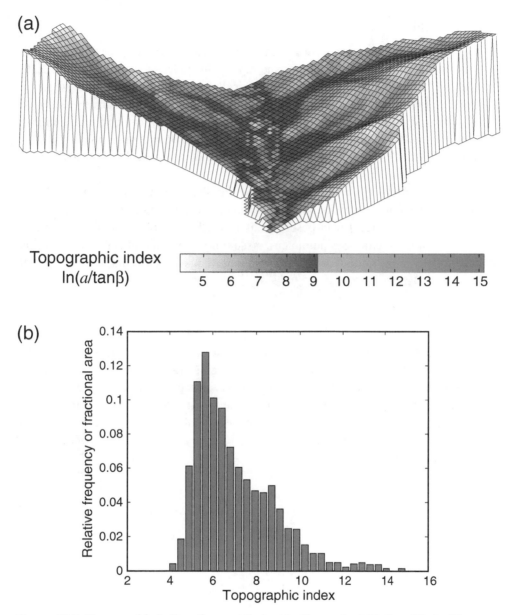

Figure 10.8 Topographic indices for a catchment in Shenandoah National Park. The spatial pattern (*a*) indicates a likelihood of saturation in the central valley of the catchment. The distribution of values (*b*) is used in TOPMODEL calculations as described in Section 10.6.

10.5.3 Hillslope stability

The hillslopes of catchments are not static features of the landscape. Over time, hillslopes change slowly due to erosion by runoff and rapidly due to landslides (also known as debris flows). Landsliding on steep hillslopes typically occurs in localized areas where the soils are saturated and the water pressure in the pore spaces is high. Surface runoff most often occurs when soils become saturated so that precipitation can no longer infiltrate (saturation-excess overland flow). If the surface runoff is deep enough and/or the slope steep enough, the flow can dislodge and carry soil particles from the hillslope to the channel, resulting in erosion of the hillslope.

Dietrich et al. (1992) combined simple expressions describing the thresholds of saturation-excess overland flow, landsliding, and hillslope erosion with detailed digital elevation data and careful field observations to predict locations within a catchment where each of these processes dominates. Assuming a constant transmissivity T of the surface soil layer, the saturated subsurface soil discharge across a contour line of length c is given by Darcy's law as

$$Q_{subsurface} = Tc\tan\beta. \tag{10.6}$$

The water-table slope is assumed to be equal to the surface slope. The total amount of water reaching the length of contour (c) over a specified period of time is Aq_{total}, where $q_{total} = R$ (the recharge rate, [L T^{-1}]) and A is the upslope contributing area. In other words, q_{total} is the volume of water per unit surface area (or depth) that is moving through the hillslope per unit time. The difference between total runoff past a contour interval (q_{total}) and saturated subsurface discharge ($q_{subsurface}$) is saturation-excess overland flow. Thus, overland flow occurs when:

$$Aq_{total} > Tc\tan\beta \tag{10.7}$$

or

$$\frac{A}{c} > \frac{T}{q_{total}}\tan\beta \tag{10.8}$$

Erosion by overland flow will only occur in the parts of the catchment where the overland flow is deep enough (large specific contributing area A/c) or the slope is steep enough (large $\tan\beta$) for the flow to dislodge the soil grains. Dietrich et al. (1992) proposed the following expression for the erosion threshold:

$$\frac{A}{c} > \frac{T}{q_{total}}\tan\beta + \frac{\alpha}{q_{total}(\tan\beta)^2}, \tag{10.9}$$

where α [L^2 T^{-1}] characterizes the resistance of the soil to erosion.

Cohesionless material on a sloping surface becomes unstable, leading to shallow landsliding, when the slope of the surface exceeds a critical value dependent on the soil and water properties and the degree of saturation described by:

$$\tan\beta > \left(\frac{\rho_s - \sigma\rho}{\rho_s}\right)\tan\phi_f, \tag{10.10}$$

where ρ is water density (i.e., 1000 kg m^{-3}), ρ_s is soil density at saturation (of the order of about 2000 kg m^{-3}), σ is the degree of saturation of the soil mantle, and ϕ_f is the internal angle of friction (a parameter expressing the shear strength due to friction among soil grains). When the soil is saturated ($\sigma = 1$), this reduces to $\tan\beta > 0.5\tan\phi_f$ for typical values of soil and water density. When the soil is unsaturated ($Q_{total} < Q_{subsurface}$), $\sigma = Q_{total}/Q_{subsurface} = Aq_{total}/(Tc\,\tan\beta)$.

These expressions for thresholds of saturation overland flow (Equation 10.8), erosion (Equation 10.9), and landsliding (Equation 10.10) all depend on the specific contributing area A/c and the slope $\tan\beta$. A plot of the curves defining each threshold in terms of these parameters shows their relationship to each other and the topographic parameters. In Figure 10.9, these threshold curves are plotted for a total runoff $q_{total} = 50$ mm day^{-1}, assuming $T = 10^{-4}$ m^2 s^{-1}, $\phi_f = 35°$, and $\alpha = 8 \times 10^{-6}$ m^2 s^{-1}, which produces good agreement between predictions and observations of hillslope hydrologic characteristics in a small (1.2 km^2) northern California catchment studied by Dietrich et al. (1992).

A diagram such as Figure 10.9 can be used to determine areas of a catchment that are susceptible to erosion and landsliding, which may serve a variety of purposes including guiding land-use decisions. The threshold most susceptible to land-use practices is the erosion threshold. The value of the parameter α characterizing the resistance of the soil to erosion decreases rapidly with removal of vegetative soil covers and soil disturbance. As the value decreases, the erosion threshold shifts to the left in Figure 10.9, resulting in a larger portion of the catchment that is prone to erosion.

10.6 Routing Water through a Catchment Using Catchment Models

The question that we address in this section is how precipitation can be routed through a catchment to calculate streamflow. The most straightforward way to use the theory developed in previous chapters to solve the catchment routing problem is to link together equations for overland flow (e.g., Manning's equation), for flow in the unsaturated zone (e.g., Richards' equation), and for flow in the water-table aquifer using an equation for groundwater flow. The earliest application of such a model was presented by Freeze (1971, 1972a, 1972b), who examined the runoff responses of a hypothetical hillslope to precipitation inputs. More recently with advances in computational power, this approach has been used successfully to simulate flows through catchments in great detail. For example, the Penn State Integrated Hydrologic Model (PIHM, Qu and Duffy, 2007; www.pihm.psu .edu) has been applied at several catchments to study coupled hydrological-biogeochemical processes.

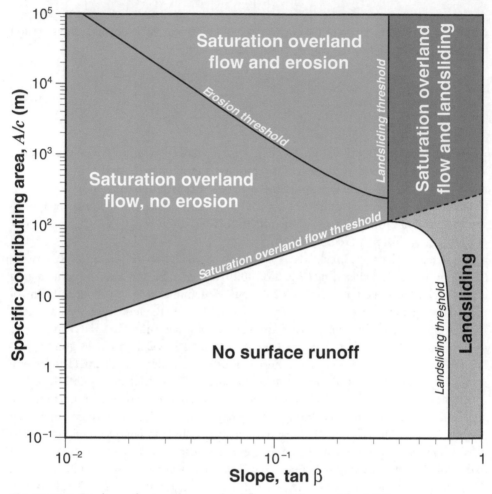

Figure 10.9 Regions of saturation overland flow, erosion, and landsliding.

Here we will consider a slightly different and simpler approach using the concepts of catchment "blocks" and the topographic index described in Section 10.5. One framework that uses this approach is known as **TOPMODEL**, a catchment model that is based on the idea that topography exerts a dominant control on flow routing through upland catchments (Beven and Kirkby, 1979). TOPMODEL uses the equation for conservation of mass ("inflow rate minus outflow rate equals rate of change of storage") for several "reservoirs" in a catchment–for example, an "interception reservoir" and a "soil reservoir" (Figure 10.10). Rainfall provides the input to the interception reservoir, which is taken to have a capacity of a few mm of water depending on the vegetation type (see discussion of interception in Chapter 2). The outputs from the interception reservoir are evaporation, calculated using an evaporation formula (see Chapter 2), and throughfall, which then forms the input to the soil reservoir. The conservation of mass equation again provides a method for calculating the water balance for the soil reservoir. By linking together the water

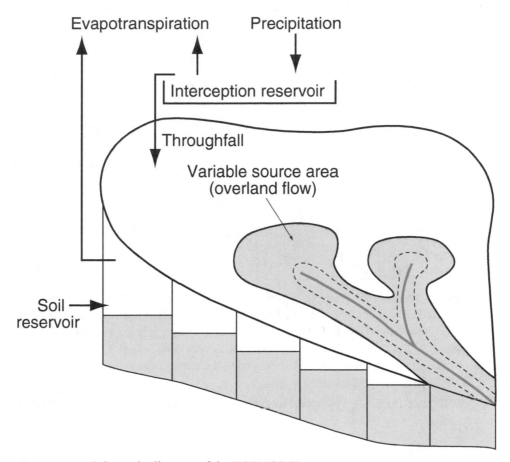

Figure 10.10 Schematic diagram of the TOPMODEL concept.

balance equations for all of the hypothetical reservoirs in the catchment, a routing computation can be completed.

TOPMODEL performs the bookkeeping for the water balance computations in the framework of topographically defined elements and uses Darcy's law to calculate flow rates through the soil. Consider a segment of a catchment defined by a cut along an elevation contour line at the bottom, and "sides" running perpendicular to contours up to the catchment divide (Figure 10.11). Recall our assumption that flow is driven by topography; hence, this segment is just a portion of a flow net for the catchment. The flow of subsurface water is conditioned strongly by the local topography. The degree of convergence of "flow lines" (lines perpendicular to the contours) determines how much upslope area drains to a unit length of contour at any given point. The local slope, the thickness of the soil, and the hydraulic conductivity of the soil determine the "ability" of the soil to move water farther down the slope once it has arrived at the given point. Source areas for surface runoff occur where subsurface water accumulates—points to which large upslope areas drain (such as convergent hillslopes or "hollows") and where the capacity to drain the water downslope is limited (where slopes flatten at the base of hollows).

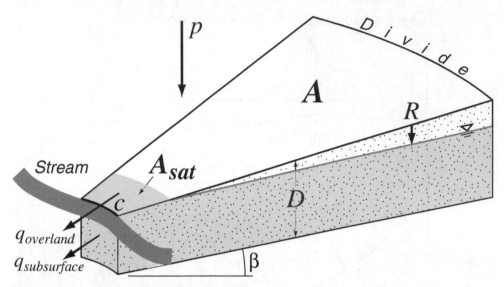

Figure 10.11 The water balance for a catchment hillslope segment. Throughfall at rate p falls on the segment of area A and thickness D. A portion, R, of this recharges the subsurface. Subsurface flow from the segment occurs at rate $q_{subsurface}$. Surface flow, $q_{overland}$, occurs from saturated areas (saturation-excess overland flow). The local slope at the outflow point, β, is considered to be equal to the slope of the water table.

Conservation of mass can be applied to the segment depicted in Figure 10.11 to determine the fluxes.

10.6.1 TOPMODEL calculations

Streamflow is the sum of subsurface flow and of overland flow from saturated contributing areas:

$$q_{total} = q_{subsurface} + q_{overland}, \tag{10.11}$$

where q_{total} is total streamflow. It has dimensions of $[L\ T^{-1}]$ (discharge $[L^3\ T^{-1}]$ divided by area $[L^2]$); all of the flow quantities in TOPMODEL have these units. The surface flow contribution is $q_{overland}$ and $q_{subsurface}$ is the subsurface flow contribution (Figure 10.11).

Surface flow is generated when precipitation falls on a saturated area and from return flow, so:

$$q_{overland} = \frac{A_{sat}}{A} p + q_{return}, \tag{10.12}$$

where A_{sat}/A is the fraction of the hillslope area that is saturated (Figure 10.11), $p\ [L\ T^{-1}]$ is the throughfall or snowmelt rate, and $q_{return}\ [L\ T^{-1}]$ is the return flow.

We calculate $q_{subsurface}\ [L\ T^{-1}]$ from total subsurface discharge $Q_{subsurface} = Tc\ \tan\beta$ $[L^3\ T^{-1}]$ (Equation 10.6), where T is the transmissivity of the soil $[L^2\ T^{-1}]$, $c\ [L]$ is the contour width (length perpendicular to the flow direction), and $\tan\beta$ is the slope. The transmissivity is equal to the soil depth multiplied by the soil hydraulic conductivity. Note that the slope of the water table is assumed to be the same as that of the land surface.

We assume that the saturated hydraulic conductivity of the soil decreases with soil depth exponentially, a situation often observed:

$$K(z) = K_0 e^{-fz},\qquad(10.13)$$

where $K(z)$ [L T^{-1}] is the hydraulic conductivity at depth z (measured positively in the downward direction), K_0 is the hydraulic conductivity at the surface, and f [L^{-1}] is a parameter that governs the rate of decrease of K with depth. To determine the transmissivity of a saturated zone of a given thickness (from a depth of water table z to a depth to bedrock D) Equation 10.13 is integrated to obtain:

$$T = \frac{K_0}{f}(e^{-fz} - e^{-fD})\qquad(10.14)$$

The term e^{-fD} is generally much smaller than the term e^{-fz}, so Equation 10.14 can be simplified:

$$T = \frac{K_0}{f}e^{-fz}.\qquad(10.15)$$

Combining Equations 10.6 and 10.15 gives the following equation for subsurface flow:

$$Q_{subsurface} = \frac{K_0}{f}e^{-fz}c\tan\beta.\qquad(10.16)$$

TOPMODEL does water-balance accounting by keeping track of the "saturation deficit," the amount of water that one would have to add to the soil at a given point to bring the water table to the surface. Because one has to track saturated areas if saturation-excess overland flow is to be computed, this makes sense. To implement computations in terms of s, the saturation deficit, z is replaced by s/ϕ where s [L] is the saturation deficit and ϕ is the porosity of the soil. Substituting for z in Equation 10.16 gives:

$$Q_{subsurface} = \frac{K_0}{f}e^{-f\frac{s}{\phi}}c\tan\beta.\qquad(10.17)$$

To make things "neater," we introduce some simplifying notation. We can replace K_0/f with T_{max}, because this term is the transmissivity when the soil is completely saturated ($s=0$). We can also replace f/ϕ with $1/m$, a soil parameter inasmuch as f defines the decrease of K with depth and ϕ is porosity. Equation 10.17 can then be written:

$$Q_{subsurface} = T_{max}e^{-\frac{s}{m}}c\tan\beta.\qquad(10.18)$$

We may now proceed with a calculation of the water balance for a hillslope slice (Figure 10.11). This will lead, in conjunction with Equation 10.18, to expressions for A_{sat}/A and $q_{subsurface}$. Equation 10.18 gives the subsurface flow being transmitted downslope at any point. The flow coming into the slice at any time is:

$$Q_R = RA, \tag{10.19}$$

where R [L T^{-1}] is the recharge rate and A is the area of the hillslope slice—the section of hillslope that drains past the section of contour (c) in question. The "great leap" of TOPMODEL is to assume steady-state conditions. Then, $Q_{subsurface} = Q_R$, or

$$RA = T_{max} e^{-\frac{s}{m}} c \tan \beta. \tag{10.20}$$

This equation can be solved for s:

$$s = -m \ln\left(\frac{R}{T_{max}}\right) - m \ln\left(\frac{a}{\tan \beta}\right), \tag{10.21}$$

where $a = A/c$, the specific contributing area. The second term on the right of Equation 10.21 describes the way in which topography controls the propensity for every point in the catchment to reach saturation (i.e., the propensity of each point to generate saturation-excess overland flow during storms). If s is less than or equal to zero, the soil is saturated. From Equation 10.21 we see that this occurs most easily for points within the catchment where the topographic index (TI = $\ln(a/\tan \beta)$) is large.

Until this point in the discussion, we have been referring to an individual catchment hillslope, or hillslope segment, defined by a pair of streamlines and extending from the stream to the catchment divide (Figure 10.11). However, we could also consider any point in the catchment and calculate the upslope contributing area and the local slope. In this way, we can compute the distribution of topographic indices for the entire catchment. In practice, the computations are often done for "blocks" delineated on the basis of DEM (Digital Elevation Model) or surveying data. The topographic index, and therefore the contributions of surface and subsurface flow to streamflow, can be calculated for each block. The saturation deficit can also be calculated for each block, using Equation 10.21. Furthermore, these quantities will be identical for two blocks with the same topographic index, as long as R and T_{max} are spatially constant.

To solve for the catchment-average saturation deficit ($\bar{s}$), we can integrate Equation 10.21 over the catchment and divide by the area. Here we assume that R and T_{max} are constant over the catchment:

$$\bar{s} = -m \ln\left(\frac{R}{T_{max}}\right) - m\lambda, \tag{10.22}$$

where λ is the mean $\ln(a/\tan \beta)$ for the catchment. Combining Equations 10.21 and 10.22 gives:

$$s = \overline{s} + m\left[\lambda - \ln\left(\frac{a}{\tan\beta}\right)\right].$$ (10.23)

This equation states that the saturation deficit at any point in a catchment is equal to the average saturation deficit for the catchment plus a soil parameter, m, times the difference between the average topographic index and the local topographic index.

Now we have a way to calculate A_{sat}/A—compute s at any point and check to see if it is less than or equal to zero. We can estimate m from soil characteristics, calculate λ and $\ln(a/\tan\beta)$ from a topographic map, and can keep track of $\overline{s}$ by water-balance accounting (p, interception, et, subsurface flow, and overland flow). If $s < 0$, the soil is completely saturated and any rain on the surface will become overland flow. The rate of flow produced by this mechanism is determined using the throughfall intensity and fractional catchment area that is saturated (e.g., Equation 10.12). Return flow occurs where $s < 0$, and the rate of return flow is equal to $|s|A_{sat}/A$.

Next we develop an expression for the mean subsurface discharge, $\overline{q}_{subsurface}$. Integrating Equation 10.18 over the catchment area and dividing the result by the catchment area yields:

$$\overline{q}_{subsurface} = T_{max}e^{-\lambda}e^{-\overline{s}/m}.$$ (10.24)

Thus, in TOPMODEL, subsurface flow is controlled by soil characteristics (T_{max} and m), topography (λ), and the average saturation deficit of the catchment.

This is not all of TOPMODEL, but it is a summary of the main conceptual points. Added to the formulations above are the other standard components of the water budget—evapotranspiration, snowmelt, channel routing (e.g., a simple reservoir routing method). All of the water-balance accounting parts of the model are simple applications of the conservation of mass. A fuller description of TOPMODEL is available in Wolock (1993).

10.6.2 Hydrograph simulation using TOPMODEL

TOPMODEL simulates the runoff response of a catchment to precipitation events. By tracking the change in storage within each "block" defined by a value of TI, the model not only routes water through the catchment, but enables predictions of areas that will become saturated during a storm. The soils at the base of hillslopes (blocks with high TIs) tend to become saturated as a storm progresses and saturation-excess overland flow is produced. Streamflow is taken as the sum of subsurface flow (net outflow from the soil reservoir) and the overland flow. Evapotranspiration removes water from the soil store according to a rate calculated with an evapotranspiration formula.

To illustrate the use of TOPMODEL to simulate streamflow from a catchment, we examine the Snake River (Hornberger et al., 1994). The catchment of the Snake River near Montezuma, Colorado, is 12 km² in area and is mountainous, ranging in elevation from about 3350 m to 4120 m. The Continental Divide bounds the catchment on the south and east. Approximately half of the catchment is above the tree line. The data necessary to implement the model were taken from nearby stations. A snowmelt model that takes

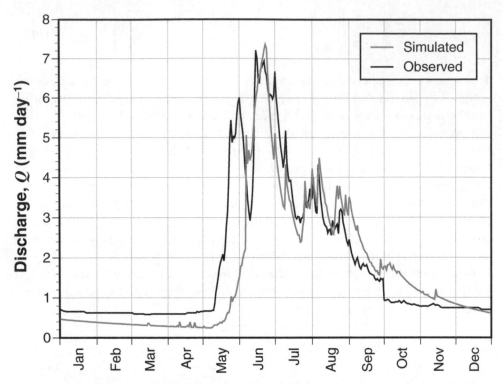

Figure 10.12 Observed discharge (1984) in the Snake River, Colorado, compared with TOPMODEL calculations.

Data courtesy of Ken Bencala and Diane McKnight.

melt to be proportional to temperature was used in the water-balance calculations within TOPMODEL.

The general shape of the snowmelt hydrograph for 1984 simulated using TOPMODEL is in accord with the measured hydrograph (Figure 10.12). The initial timing of the hydrograph rise is late in the simulated hydrograph; presumably, the temperature index model calculates the initiation of significant snowmelt to be later than when the melt is actually initiated. This likely is due to errors in the simple model that was used for snowmelt. The remainder of the simulation follows the observed hydrograph quite well although the late part of the recession is overpredicted. Overall this simulation is quite reasonable given the limitations in the data (e.g., the need to extrapolate the discharge, precipitation, and temperature data from downstream stations).

10.7 Concluding Remarks

At the beginning of the chapter we suggested that solving the catchment routing problem was important for a number of reasons. The material presented above introduced some of the ideas used in routing flows through a catchment and discussed some of the complexities in the processes that give rise to observed streamflow dynamics. The advances that have been made in catchment modeling have been substantial, but the state of the science is still inadequate to allow acceptable routing for some of the most serious problems (Hornberger and Boyer, 1995). One example is predicting the flow paths of natural solutes and contaminants in catchments, an example of which is the acidification of streams during storms observed in the northeastern United States (Figure 10.1) and elsewhere in the world.

Models such as TOPMODEL are necessary for calculation of acidification and of the movement of other chemicals through a catchment. A mass-balance model is required to do an accounting of chemicals, in addition to an accounting of water. Robson et al. (1992) combined TOPMODEL with a simple chemical mixing model to describe the episodic acidification observed in a stream in Wales. This is an example of the approaches that are being used to study a variety of pollution problems in catchments. The transport of chemicals through a catchment is controlled to a great extent by interactions between the water and the soil and rocks in the catchment. Determination of the flows through the soil and groundwater reservoirs in the catchment is critical for realistic predictions of the fate of pollutants.

TOPMODEL calculates not only the stream hydrograph but information that is useful for linking hydrological calculations to hydrochemical models (Cosby et al., 1987). For example, TOPMODEL calculates subsurface flow, overland flow, and saturation deficit (depth to water table), quantities that can be used to determine how waters from different parts of a catchment mix to produce the chemical composition observed in streams. As a final illustration, we look once again at the Snake River in Colorado.

In many lake and stream systems, the concentration and composition of dissolved organic material (DOM) is a critical water quality characteristic. One example of a process controlled by DOM is the formation of trihalomethanes, compounds that are known carcinogens, in a drinking water supply as a result of interactions between chlorine (used to treat drinking water in public supplies) and components of the DOM during water treatment. The DOM also can have indirect effects on water quality by influencing internal processes of aquatic ecosystems, such as photosynthesis and heterotrophic activity. Therefore, one question that hydrologists and ecologists are interested in answering is how hydrological processes transport dissolved organic carbon (DOC) to streams. An example is the Snake River catchment, where DOC builds up in near-surface soils beneath the snowpack due to microbiological activity (Boyer et al., 1997) and is released during snowmelt in the spring. Quantifying the "flushing" of DOC from near-surface soils requires knowledge of the extent of saturation of the soils and the rates of water movement through them. TOPMODEL can provide a computation of these hydrological items. For example, based on the TOPMODEL simulation shown in Figure 10.12, the concentrations of DOC in the stream can be calculated (Figure 10.13). The calculated values early in the melt period do not rise to the measured levels because the snowmelt was not calculated in TOPMODEL correctly at early times (Figure 10.12), but the general timing and magnitude of calculated DOC levels are consistent with measurements. These results suggest that such calculations could be useful to water managers who might want

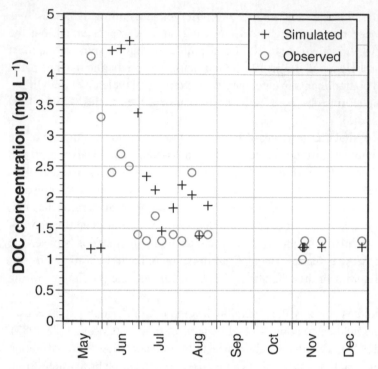

Figure 10.13 Measured and simulated dissolved organic carbon (DOC) concentrations (1984) in the Snake River, Colorado.

Data courtesy of Ken Bencala and Diane McKnight.

to avoid taking water into the water-supply system during times when DOC concentrations in the stream are high. Forecasts of flushing of DOC would enable managers to plan ahead to store treated water so they could avoid undue production of trihalomethanes in the drinking water supply.

10.8 Key Points

- Streamflow hydrographs appear to have two temporal components (quickflow and baseflow). Separation of hydrographs into these components is done by "eye" or by purely empirical methods. {Section 10.2}

- Streamflow hydrographs can also be separated into two components that attempt to describe the source of stream water (e.g., old and new water). This "chemical" separation is based on the mixing of chemically different waters. By solving mass balance equations for water and a conservative solute, estimates of Q_n and Q_o can be computed for any time for which a measurement of C_t is available by: $Q_n/Q_t = [(C_t - C_o)/C_n - C_o)]$ and $Q_o/Q_t = 1 - (Q_n/Q_t)$. {Section 10.3}

- The application of the chemical-hydrograph-separation method to many forested, upland catchments located in humid, temperate regions shows that observed streamflow is mostly old water. "Quickflow" is not synonymous with "new" water and baseflow is not synonymous with "old" water. {Section 10.3}

- Runoff from a catchment can follow four different flow paths: direct precipitation onto stream channels, overland flow, shallow subsurface stormflow, and groundwater flow. {Section 10.4}

- There are two different mechanisms by which overland flow is produced in catchments: infiltration-excess overland flow and saturation-excess overland flow. Infiltration-excess overland flow is considered dominant in systems where the soil profile or soil surface has been radically disturbed (e.g., agricultural catchments), in arid and semiarid regions where vegetation density is low, and in urban areas where the surface is made essentially impermeable by paving or other construction. Saturation-excess overland flow is most significant in humid regions with dense vegetation and topographic conditions (concave slopes with flat valley bottoms) that cause the water table to be located relatively close to the surface. {Section 10.4.2}

- Shallow subsurface stormflow may occur when permeable surficial soils become saturated. Water may then flow to the stream through these soils. {Section 10.4.3}

- Discharge of groundwater to streams is responsible for baseflow to streams during fair weather periods. Also, groundwater can discharge to saturated surface areas (return flow) or to a stream or river during precipitation events, and thereby contribute to increased streamflow during storm events. {Section 10.4.4}

- Inflow of water to a segment of a hillslope is proportional to upslope contributing area, which depends on the distance to the divide and the shape of the hillslope (e.g., planar, convergent or divergent.) {Section 10.5.1}

- The ratio of contributing area per unit contour length to the local slope of the topography, termed the topographic index, provides a useful measure of the likelihood of saturation in a section of a hillslope. {Section 10.5.2}

- Relationships between the ratio of contributing area per unit contour length and local slope can be used to identify regions of a hillslope prone to surface runoff, erosion, and landsliding. {Section 10.5.3}

- Routing of water through a catchment can be done using computer simulation models. One such model, TOPMODEL, is based on use of a steady-state flow net to compute subsurface flow. The model also calculates overland flow from saturated areas by linking topography with water-balance computations. {Section 10.6}

10.9 Example Problems

Problem 1. A hydrologist studying runoff generation in a catchment measures the following chloride concentrations during the peak of a rainstorm event: $C_n = 4.5$ µmol L^{-1}, $C_o = 40.5$ µmol L^{-1}, and $C_t = 36.0$ µmol L^{-1}. What fractions of total streamflow are contributed by new water and old water?

Problem 2. Data on hydraulic conductivity versus depth in a soil on a forested slope are presented by Harr (1977). Examine the data to determine whether the assumption that K decreases exponentially with depth is reasonable, and, if it is, estimate K_0 and f.

Depth in soil (m)	Hydraulic conductivity, K (m s^{-1})
0.10	9.8×10^{-4}
0.30	1.1×10^{-3}
0.70	4.5×10^{-4}
1.10	4.9×10^{-4}
1.30	4.4×10^{-5}
1.50	6.1×10^{-5}

Problem 3. Two "streamtube" segments of a catchment (i.e., portions of a flow net based on topography) have properties shown below:

	Segment 1	Segment 2
Upslope area, A (m²)	500	500
Length of contour at base of segment, c (m)	3.5	25
Slope at base of segment, $\tan \beta$	0.02	0.08

Calculate the topographic index for each segment and indicate which segment is more likely to produce saturation-excess overland flow.

Problem 4. The soil of a given catchment has a porosity (ϕ) of 0.4 and hydraulic conductivity as described in Problem 2. The catchment has an average value of the topographic index of 3.5. Under conditions where the average saturation deficit of the catchment is 100 mm, calculate the subsurface flow to the stream, $q_{subsurface}$. If the throughfall rate at this time is 4×10^{-3} mm s^{-1} (14 mm hr^{-1}), then what fraction of the catchment would have to be saturated if saturation-excess overland flow were to be equal to one-half of the subsurface flow?

10.10 Suggested Readings

Dunne, T., and L.B. Leopold. 1978. *Water in environmental planning.* San Francisco: W. H. Freeman, Chapter 9, pp. 255–278.

Wolock, D.M. 1993. *Simulating the variable-source-area concept of streamflow generation with the Watershed Model TOPMODEL.* U.S. Geological Survey Water-Resources Investigations Report 93-4124.

11 Water, Climate, Energy, and Food

11.1 The Role of Hydrology in a Changing Planet

As noted in Chapter 1, the study of hydrology began long ago. As human populations have grown and redistributed, the pressing questions for which an understanding of hydrological processes is valuable have shifted and expanded. Of course the core question remains how to provide the quantity and quality of water needed to support human populations and the environment on which we depend. But with a growing population and changing climate, new questions have emerged that require hydrological research and applications to move in new directions. In fact, although the hydrological sciences are critical to addressing new and existing water-related issues, the complex interactions and feedbacks among hydrology, geomorphology, biogeochemistry, ecology, and climate mean that understanding and solving problems related to water is an increasingly interdisciplinary enterprise.

A recent study by the National Academy of Sciences (NRC, 2012a) examined the challenges and opportunities in the hydrological sciences in the early 21st century. The report notes:

> Fundamental new drivers of hydrologic sciences in the 21st century rest on the realization that: (a) humans are a dominant influence on water sustainability both at the global and local scales, (b) the world is becoming exceedingly "flat" not only with respect to rapid dissemination of scientific knowledge, but also with respect to learning from distant environments currently undergoing rapid change (e.g., deforestation, drought, agricultural expansion, etc.) and predicting future water scenarios in other parts of the world, and (c) the natural world is a highly non-linear system of interacting parts at multiple scales prone to abrupt changes, tipping points, and surprises much more often than previously thought possible.

Among the key challenges identified in the report are impacts of humans and climate on the water cycle, interactions between hydrological processes and terrestrial ecosystems (see Chapter 9), and the ways in which hydrological sciences can inform solutions to the problems of ensuring the quantity and quality of water needed for drinking, agriculture and energy (NRC, 2012a). Here we consider how hydrological research and understanding can contribute to the first and last of these.

11.2 Water, Energy, and Food: Complex Interdependencies

Several years ago, Richard Smalley of Rice University posed the question to a variety of audiences, "What will be the top ten problems facing humanity over the next fifty years?" The answers consistently showed *energy*, *water*, and *food* as top items (http://cnst.rice .edu). Without a doubt, these resources are of critical importance and will only become more so in the future as global population increases toward 9 billion and as the effects of anthropogenic climate change become manifest. As we seek to manage these resources to meet demands, we are confronted by the realization that the development and use of each has impacts on and is impacted by the development and use of the others (Perrone and Hornberger, 2013).

11.2.1 The water-energy nexus

As one example, consider the generation of electricity. A large fraction of the generation is performed in thermoelectric plants where water is converted to steam (using heat from burning coal or natural gas or from nuclear reactions), the steam is run through turbines to generate electricity, and then the steam is condensed by cooling. The cooling process requires water, most often fresh water, and lots of it. According to the U.S. Geological Survey (Barber, 2009) water withdrawals in the United States for thermoelectric power generation in 2005 represented 49% of total water withdrawals and exceeded the sum of withdrawals for irrigation (31%), public water supplies (11%), and industry (4%). Much of the water withdrawn for cooling is returned to streams, albeit at a higher temperature, but some of the water is consumed by evaporation. In 2008, U.S. power plants withdrew between about 225 and 650 million cubic meters and consumed from 11 to 23 million cubic meters of water (Averyt et al., 2011).

Thermoelectric plants are not the only form of power generation contributing to the withdrawal and consumption of large volumes of water. For instance, solar power generation requires substantial amounts of water to clean the solar panels from the dust deposited on their surfaces. Needless to say, solar panels are often deployed in predominantly clear sky regions with prevailing cloudless conditions and consequently arid climate. Thus, the supply of the water required to clean solar panels and sustain solar energy production can be problematic. Although hydropower often is considered to be fairly non-consumptive in terms of water, an analysis of large hydroelectric dams worldwide indicates that the blue water consumption from them amounts to about 10% of the blue water consumed by crops (Mekonnen and Hoekstra, 2012). Also, in some instances the release of water from dams to run the turbines can preclude its diversion into irrigation canals causing a conflict between water for crops and water for energy. Consider the case of Sri Lanka, where about 40% of the electricity is generated through hydropower. In times of plenty, there may be enough water to supply all uses, but under drought, the competition for water between agriculture and electricity generation can create problems, and decisions have to be made about how to apportion the releases. The issues involved in making decisions are complex and there is no "right answer" (e.g., see Molle et al., 2008).

Conversely, water distribution, treatment, and end-use require quite a significant amount of energy. A 2005 study for California indicated that "water-related energy use consumes 19 percent of the state's electricity, 30 percent of its natural gas, and 88 billion gallons [330 billion liters] of diesel fuel every year—and this demand is growing" (Klein, 2005). This is a tremendous amount of energy and one conclusion of the study was that the state could most readily meet its short-term (5 years) energy conservation goals by conserving water.

The water-energy linkage is further complicated when we consider the third of Smalley's top problems. The link between water and food is obvious. In rain-fed agriculture, crops are part of the green water portion of the water cycle whereas irrigation converts substantial portions of blue water into green water (see Chapter 9). There also are links of food to energy, perhaps most notably with respect to the increasing demand for biofuels which leads to displacement of land and water resources away from growing food. Globally withdrawal of fresh water from streams and aquifers is dominated by irrigated agriculture, accounting for some 70% of the total (Rosegrant et al., 2009). Considering population growth, the escalating demand for biofuels, and shifts toward diets with more meat protein, which requires more water than a diet dominated by grains (see Section 11.2), the demand for water for irrigation is likely to grow globally.

11.2.2 The water-food nexus

To explain the relationship between water and food security, we need to stress three important points and stay away from some common misconceptions. First, about 86% of the human appropriation of water resources, when both blue and green water (i.e., both irrigated and rain-fed agriculture) are accounted for, is used by agriculture in the production of food, fibers, and biofuels (Figure 1.2). Most of the per capita water consumption (or water footprint) is contributed by food production (600–1800 m^3 per person each year) while a much smaller amount of water is used for drinking ($\approx$1 m^3/person/y) and household needs (25–360 m^3/person/y; e.g., Falkenmark and Rockström, 2004). Thus, the global water crisis is about a hungry, rather than a thirsty, humanity. Second, even though most of agriculture (about 80% of the cultivated land) is rain fed, the debate about water crisis often tends to misleadingly concentrate on drinking and irrigation water (i.e., blue water) without regard for our greater reliance on green water. Third, when we look at the use of blue water resources, we need to recognize the difference between water withdrawal and consumption: while withdrawal refers to the extraction from surface water bodies and aquifers, consumption refers to losses by evaporation, transpiration, or incorporation in crops, organisms, and products. As noticed in the previous section, much of the water withdrawn for hydropower production is not consumed but returned to water bodies with a lower gravitational potential energy. Conversely, part of the water withdrawn by cooling plants for thermo-power generation is evaporated (i.e., consumed), while the rest is returned to water bodies at higher temperatures. Similarly, only part of blue water withdrawals for irrigation is consumed, while the rest is eventually returned to water bodies. Because transpiration sustains the productivity of natural and agro ecosystems, it is often referred to as a productive consumptive use of water, while evaporation is defined as an unproductive form of water consumption. Thus, water consumption in irrigated crops is partly contributed by uptake and transpiration by plants (productive consumption) and partly lost in evaporation (unproductive consumption); the relative importance of these productive and unproductive forms of consumption determines the efficiency of an irrigation technique.

The estimation of the blue water footprint of crops or other commodities is based on consumption and not withdrawal rates (Table 11.1). The calculation of the overall water footprint of these commodities should account for both blue and green water consumption. Such a calculation requires an analysis of all the water consumed in the course of the production cycle of that good (a life cycle analysis). The results of this analysis are available for all the major agricultural and industrial commodities. Some examples are summarized in Table 11.1.

Animal products have in general a much bigger water footprint than vegetables. The production of the same amount of food calories requires on average 8 times more water in the case of meat (4 m^3/1000 kcal) than for plant products (0.5 m^3/1000 kcal; Falkenmark and Rockström, 2004). Thus, the water footprint of food varies across societies (600–1800 m^3/person/y), depending on the prevalent diet, particularly on the reliance on meat. Falkenmark and Rockström (2004) estimated that the water footprint of a balanced diet (20% of kcal intake from meat) is about 1300 m^3/person/y. Typical patterns of economic development exhibit an increase in meat consumption as nations become wealthier. This shift in diet and the escalating demographic growth are both responsible for the ongoing increase in human demand for water resources. Thus, mankind's appropriation of water resources poses some important ethical questions on how far humanity should go in subtracting water (and land) from natural ecosystems and depleting groundwater stores (i.e., aquifers), and how water should be shared among different societies.

11.2.3 Globalization of water

While climate controls the supply of freshwater resources (e.g., see Figure 1.1), population size and the type of diet determine the demand (Figure 11.1). In regions where the demand exceeds the supply, societies might either have to sustain smaller consumption rates or import food products from other, more water-rich regions (Figure 11.1). Exporting and importing food can be thought of as an indirect trade in water; this water is referred to as virtual water (Allan, 1998).

Virtual water trade allows nations to virtually share and exchange water resources. Often known as the globalization of water (Hoekstra and Chapagain, 2008) this phenomenon can be visualized by mapping the global network of virtual water trade (Figure 11.2). In 2010 the total volume of virtual water traded in food commodities (2.8×10^{12} m^3 y^{-1})

Table 11.1. The water footprint (in m³ kg⁻¹) of some major food commodities

Product		Australia	Brazil	China	USA	Global Average
		\multicolumn{4}{c}{Producing Country}		
Maize	*Green*	0.75	1.62	0.79	0.52	0.95
	Blue	0.68	0.00	0.07	0.06	0.08
	Total	**1.43**	**1.62**	**0.86**	**0.59**	**1.03**
Rice, paddy	*Green*	0.25	1.94	0.55	0.42	1.15
	Blue	1.15	0.36	0.25	0.85	0.34
	Total	**1.40**	**2.30**	**0.79**	**1.27**	**1.49**
Wheat	*Green*	2.00	1.99	0.82	1.87	1.28
	Blue	0.02	0.00	0.47	0.09	0.34
	Total	**2.01**	**1.99**	**1.29**	**1.96**	**1.62**
Soy beans	*Green*	1.89	2.18	2.55	1.56	2.04
	Blue	0.00	0.00	0.25	0.09	0.07
	Total	**1.89**	**2.18**	**2.80**	**1.65**	**2.11**
Tomatoes	*Green*	0.04	0.07	0.18	0.03	0.11
	Blue	0.07	0.02	0.00	0.07	0.06
	Total	**0.11**	**0.09**	**0.19**	**0.11**	**0.17**
Swine meat	*Green*	3.86	4.44	3.69	2.99	3.58
	Blue	0.88	0.54	0.29	0.46	0.33
	Total	**4.74**	**4.97**	**3.97**	**3.46**	**3.91**
Bovine meat	*Green*	10.30	13.65	9.08	9.18	10.23
	Blue	0.43	0.12	0.35	0.37	0.39
	Total	**10.73**	**13.78**	**9.43**	**9.55**	**10.62**

Source: Based on Mekonnen and Hoekstra (2010).

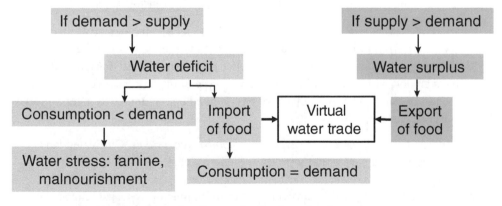

Figure 11.1 Demand and supply of water for food production.

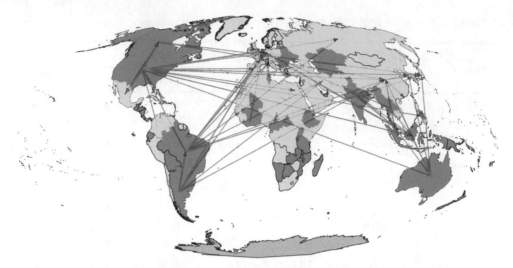

Figure 11.2 The global network of virtual water trade for the year 2010, based on data from FAOSTAT. Only the top links, which transfer 50% of the total virtual water flux, are shown. Net exporters are shown in dark blue; net importers in light blue.
Courtesy of Joel A. Carr.

accounted for about 24% of the average global freshwater resources used for food production (1.18×10^{13} m^3 y^{-1}; Carr et al., 2012). Virtual water trade is crucial to the food security of many countries. Although the trade in virtual water can be viewed as benevolent in that it can reduce malnourishment in countries where local food supplies become limited by water availability, there is some concern that in the long run, the ability of countries to cope with drought might be compromised by an over-reliance on virtual water (D'Odorico et al., 2010).

There is no obvious solution to the likely water shortages facing populations living in regions of chronic water scarcity or temporary drought. Overall, there are only a few options to meet the water needs of a constantly growing global population while reducing malnourishment and famine (e.g., Falkenmark and Rockström, 2004):

- *Increase the use of blue water for irrigation.* Crop yields can be greatly enhanced by switching from rain fed to irrigated agriculture. However, most of the suitable freshwater resources are already overexploited and overcommitted. Many water courses are left with only minimal flows, and some of them do not even make it to the ocean anymore. Therefore there are only limited opportunities to increase water withdrawals for irrigation.

- *Expand agricultural land at the expense of natural ecosystems.* Cultivated land area is not likely to substantially increase (Fedoroff et al., 2010). The clearing of most tropical forests results in agricultural soils that are unsuitable to sustain good long-term crop yields. Considering the environmental costs associated with CO_2 emissions and biodiversity loss, the benefits of new conversions to agriculture are not sufficient to justify further expansion of cultivated lands (Foley et al., 2011).

• *"More crop per drop."* Food production can be enhanced by using genetically modified crops with higher water use efficiency and "climate smart" agriculture.

• *Reduce non-productive water consumption.* The use of more efficient irrigation techniques (e.g., drip irrigation) and of new soil and water management methods should allow for a decrease in evaporation while sustaining productive water consumption (transpiration).

These approaches are not mutually exclusive. The combined adoption of a variety of methods aimed at increasing agricultural yields through an integrated use of new crop and irrigation technology while enhancing food awareness through educational programs appears to be the most promising strategy to meet the increasing demand for water and food (e.g., Tilman et al., 2011).

11.3 Impacts of Changing Climate on the Water Cycle

It is clear from Section 11.2 and elsewhere in this text that humans are directly affecting the water cycle in dramatic ways through agriculture, energy production, groundwater extraction, deforestation (and reforestation), and urbanization, to mention just a few. Humans are also affecting the water cycle more indirectly by increasing atmospheric levels of carbon dioxide (CO_2), which alters air temperature and transpiration. It has been suggested that these alterations may cause an acceleration of the water cycle (see Box 11.1) and global redistribution of water, leading to more extreme weather events that may increase the number of floods in some regions while making other regions are more drought-prone. Because the water cycle is tightly linked to biogeochemical cycles (e.g., carbon and nitrogen) and vegetation distributions, changes in the water cycle due to climate change are likely to have far-reaching effects on ecosystems and landscapes.

11.3.1 Effects of increasing temperature on precipitation and runoff

As atmospheric CO_2 levels rise, so will air temperature. Climate models help to quantitatively project future global surface warming in response to specific scenarios of future emissions of CO_2 and other greenhouse gases. Low to high emission scenarios lead to predicted increases in global mean surface air temperature of 1.8°C to 3.4°C by 2100 (Meehl et al., 2007). These increases will not be uniformly distributed across the globe, with some areas, such as portions of the southwest U.S., experiencing a large increase in the number of days with a heat index >100°F (37.8°C) by the end of the century (NRC, 2012b).

Changes in air temperature will affect precipitation, evaporation and transpiration. Higher temperatures will increase evaporation rates of surface water (from oceans,

BOX 11.1   Intensification of the Global Water Cycle

As noted in Chapter 1, water is conserved in the global hydrological cycle. Therefore, globally and over long time scales, precipitation (p) equals evapotranspiration (et). Using equation (2.8) we can then express precipitation as

$$p = et = K' e_{sat} (1 - e / e_{sat}),$$

where K' is a factor expressing the (global) conductance of the Earth's surface (see Equation 2.9), e_{sat} is the saturation vapor pressure, and e is the actual water vapor pressure of the air, which, as noted in Chapter 2, is proportional to the water vapor density and is a good indicator of atmospheric humidity. An increase in global temperatures is expected to increase both the atmospheric humidity (or e)—because of the more intense evapotranspiration—and the saturated vapor density (see Figure 2.1), which measures the amount of moisture the air can hold in unsaturated conditions. Therefore, it has been argued that the ratio e/e_{sat}, also known as **relative humidity**, will not substantially change as an effect of climate warming. Thus, only the term e_{sat} on the right-hand side of the above equation is expected to change with T. The dependency of e_{sat} on temperature is shown in Figure 2.1: because e_{sat} is an increasing function of T, climate warming is expected to increase global precipitation, an effect known as the **intensification of the global water cycle**. The dependency between e_{sat} and T suggests that global precipitation should increase by 6.8% per each degree Celsius of global warming. Climate change models and observations, however, indicate that the increase in precipitation associated with 1°C of warming is more in the 2–3% range, presumably owing to other factors affecting the global conductance parameter K' (e.g., Katul et al., 2012). This analysis is based on global mean values of temperature and precipitation. Climate change research has shown that changes in both temperature and precipitation are not expected to be uniform around the globe (see Figure 11.3).

lakes, or soil). In addition, warmer air can hold more water (see Figure 2.1). An increase in evaporation will decrease soil moisture, which could reduce transpiration, but changes in air temperature and vapor pressure will also affect transpiration rates. Because some of these changes can be offsetting, sorting out how precipitation and evapotranspiration respond to increases in air temperature is challenging. Hydrologists and other environmental scientists are combining theory, models and measurements (in the lab and field) to make the best forecasts of future change in precipitation, evapotranspiration and runoff given our current understanding (see Box 11.1).

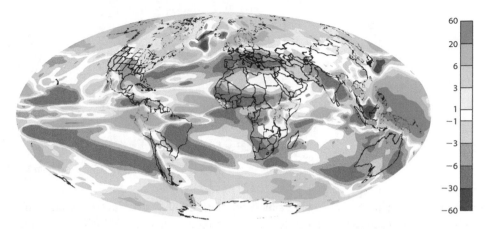

Figure 11.3 Annual mean change in precipitation (in inches of water per year) for 2081–2100 relative to 1950–2000 calculated based on the Geophysical Fluid Dynamics Laboratory (GFDL) CM2.1 model in which carbon dioxide was increased from 370 to 717 ppm. Positive values correspond to wetter conditions.
Redrawn from an image developed by NOAA GFDL (www.gfdl.noaa.gov).

Climate models have been used to predict the change in amounts and global distribution of precipitation as the air warms. Figure 11.3 shows the distribution of predicted changes in annual mean precipitation in 2081–2100 relative to 1950–2000 calculated based on the Geophysical Fluid Dynamics Laboratory (GFDL) CM2.1 model in which carbon dioxide was increased from 370 to 717 ppm. Whereas high latitudes are likely to experience more precipitation (as rain or snow), low latitudes, except in the immediate vicinity of the equator, tend to become drier.

Not surprisingly, these patterns are generally mirrored in predicted changes in runoff in the 21st century (Figure 11.4; Milly et al., 2005). Increases in evaporation caused by increasing temperatures coupled with decreases in precipitation lead to decreases in runoff in many parts of the world. In other parts, increases in precipitation are leading to increases in runoff. For example, reconstruction of monthly discharge from the world's largest rivers indicates that global continental runoff increased during the last century (Labat et al., 2004), most likely due to increases in precipitation in the watersheds feeding these rivers. Regardless of whether precipitation is increasing or decreasing, observations suggest an increase in the intensity of the largest rainfall events.

11.3.2 Effects of increased temperature and CO_2 on evapotranspiration

Understanding the ways in which climate change will affect evapotranspiration is more difficult than for precipitation and runoff because of feedbacks between soil moisture and evapotranspiration: evapotranspiration dries soils but drier soils slow evapotranspiration

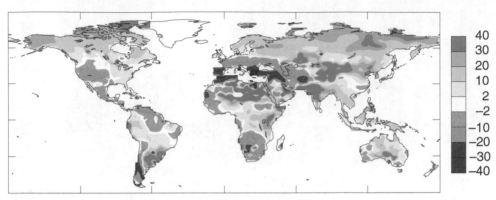

Figure 11.4 Annual mean change in runoff for 2041–2060 as a percentage relative to 1900–1970 calculated as the ensemble mean of 35 model runs using 12 different models. Positive values correspond to higher runoff.

Adapted by permission from Macmillan Publishers Ltd: Nature, from Milly et al. (2005).

(see Figure 2.11). In addition, increases in atmospheric CO_2 have been found to decrease stomatal conductance (Leakey et al., 2009), which could decrease evapotranspiration rates and thereby increase soil moisture and runoff (Gedney et al., 2006), although the magnitude of this effect is uncertain. The challenge of predicting how evapotranspiration will change as climate changes is compounded by the lack of relatively long-term, globally distributed measurements such as those available for precipitation and runoff. Even the records that do exist are complicated by changes in land use during the last century, such as deforestation, that could overwhelm climate signals at many sites (e.g., see Figure 9.7).

11.4 Challenges for the Future

As the Danish physicist Niels Bohr once said, "Prediction is very difficult, especially about the future." Nevertheless, it is clear that we face serious issues with regard to what are called social-ecological systems in the coming decades, and that many of these issues will be framed by the topics above—ensuring water and food security in the face of a changing climate. Whether predictions (or forecasts or scenarios) about the future are borne out will depend on actions taken by individuals and collectives of individuals, including nation states. We believe that students of environmental science should have a broad understanding of the fundamentals of physical hydrology so they can help shape the future in ways that are informed by science. The NRC (2102a) report noted at the beginning of this chapter concludes that: "Compelling challenges and opportunities lie ahead in understanding, quantifying, and predicting water cycle dynamics, the interaction of water and life, and how to build a path to the sustained provision of clean water for people and ecosystems." The hydrological underpinnings for meeting the challenges and taking advantage of the opportunities form the substance of this book.

11.5 Suggested Readings

Falkenmark, M., and J. Rockström. 2004. *Balancing water for humans and nature.* London: Earthscan.

National Research Council (NRC). 2012a. *Challenges and opportunities in the hydrologic sciences.* Washington, DC: The National Academies Press.

National Research Council (NRC). 2012b. *Climate change: Evidence, impacts, and choices.* Washington, DC: The National Academies Press.

APPENDIX 1

Units, Dimensions, and Conversions

Hydrological Quantities

We can separate the quantities that we encounter in hydrology into two classes. *Basic measurements* constitute the first class. For example, we might measure the height of water in a manometer, piezometer, or water well, the velocity in a stream, the mass of water in a precipitation collector, or the temperature of water in a lake. The second class includes the discharge in a stream (calculated from the mean velocity and stream cross-sectional area), or the total volume of precipitation that falls on a catchment during a storm (calculated from the catchment area and the mean precipitation depth). These are *derived quantities*, because they are not measured directly but are calculated from measured variables using an equation representing a relationship between variables. Regardless of their type, hydrological quantities have an associated unit, such as centimeters per second, and dimension, such as length per time (although some quantities may be unitless and dimensionless). In recording basic measurements or manipulating these values to derive other quantities, we need to be concerned with several things, such as *precision* and appropriate units and dimensions, which are the subjects of this appendix.

Units and Dimensions

Consider a simple example. A field investigation in a small catchment records some basic information, including the depth of water in a single precipitation collector placed within the catchment. We would like to know, as *precisely* and *accurately* as we can with the available data, the total amount (volume) of water delivered to the catchment during a single event:

Volume = Depth × Area.

The measure of precipitation has an associated *dimension*, in this case length or [L], and *unit*, for example, centimeters or inches. We use as fundamental dimensions length [L], mass [M], time [T], and temperature [Θ]. Many quantities have a dimension that is some combination of these fundamental dimensions (Table A1.1). For example, the volume of precipitation falling on a catchment has dimensions of length cubed [L^3]. To continue with our calculation, we can first check to see if the relationship between depth and volume is *dimensionally homogeneous*, that is, whether the dimensions on both sides of the above equation are the same:

$$[L^3] = [L] \times [L^2] = [L^3].$$

As we would expect, this equation is dimensionally homogeneous. Dimensional homogeneity does not guarantee that the equation will give an accurate or correct result; for example, what if the length of the stream were used erroneously in place of the precipitation depth in the equation? However, an equation that completely and accurately describes a physical relation must be dimensionally homogeneous. It is always useful to check that any equation you use is dimensionally homogeneous.

There are a variety of units that correspond to each dimensional quantity. Length might be given in meters, centimeters, inches, or even gallons per square foot. Therefore,

Table A1.1 Base and derived units relevant to hydrology in SI measurement

Quantity	Dimension	Unit	SI Symbol	Formula
Base units				
Length	$[L]$	meter	m	
Mass	$[M]$	kilogram	kg	
Temperature	$[\Theta]$	kelvin	K	
Time	$[T]$	second	s	
Derived units				
Area	$[L^2]$	square meter		m^2
Volume	$[L^3]$	cubic meter		m^3
Velocity	$[L\,T^{-1}]$	meter per second		$m\ s^{-1}$
Acceleration	$[L\,T^{-2}]$	meter per second squared		$m\ s^{-2}$
Density	$[M\,L^{-3}]$	kilogram per cubic meter		$kg\ m^{-3}$
Force	$[M\,L\,T^{-2}]$	newton	N	$kg\ m\ s^{-2}$
Pressure	$[M\,L^{-1}\,T^{-2}]$	pascal	Pa	$N\ m^{-2}$
Stress	$[M\,L^{-1}\,T^{-2}]$	pascal	Pa	$N\ m^{-2}$
Energy	$[M\,L^2\,T^{-2}]$	joule	J	$N{\cdot}m$
Quantity of heat	$[M\,L^2\,T^{-2}]$	joule	J	$N{\cdot}m$
Work	$[M\,L^2\,T^{-2}]$	joule	J	$N{\cdot}m$
Power	$[M\,L^2\,T^{-3}]$	watt	W	$J\ s^{-1}$
Viscosity, dynamic	$[M\,L^{-1}\,T^{-1}]$	pascal-second		$Pa{\cdot}s$
Viscosity, kinematic	$[L^2\,T^{-1}]$	square meter per second		$m^2\ s^{-1}$
Specific heat	$[L^2\,\Theta^{-1}T^{-2}]$	joule per kilogram-kelvin		$J\ kg^{-1}\ K^{-1}$

Table A1.2 Prefixes used in the SI system

Prefix	SI symbol	Multiplication factor
tera	T	$1\,000\,000\,000\,000 = 10^{12}$
giga	G	$1\,000\,000\,000 = 10^{9}$
mega	M	$1\,000\,000 = 10^{6}$
kilo	k	$1\,000 = 10^{3}$
hecto[a]	h	$100 = 10^{2}$
deka[a]	da	$10 = 10^{1}$
deci[a]	d	$0.1 = 10^{-1}$
centi[a]	c	$0.01 = 10^{-2}$
milli	m	$0.001 = 10^{-3}$
micro	μ	$0.000\,001 = 10^{-6}$
nano	n	$0.000\,000\,001 = 10^{-9}$
pico	p	$0.000\,000\,000\,001 = 10^{-12}$
femto	f	$0.000\,000\,000\,000\,001 = 10^{-15}$
atto	a	$0.000\,000\,000\,000\,000\,001 = 10^{-18}$

[a]Avoid use of this prefix where possible.

care must be taken to make sure any calculation is based on a *unitarily homogeneous* form of an equation. For the example of precipitation over a catchment,

Volume (m^3) = Depth (m) × Area (m^2).

Needless to say, a wrong answer will result if the units on one side of an equation are different from those on the other side.

The most common system of units employed today is the SI (*System International d'Unites*; Table A1.1). Other widely used systems of units include the *English system* (foot, pound) and the *cgs system* (centimeters, grams, seconds). The SI system includes a sequence of standard prefixes to indicate magnitude (Table A1.2). For example, a kilogram is equal to 1000 grams, and a nanogram is equal to 10^{-9} grams.

Significant Figures and Precision

To complete the simple example with which we began, we will use measured quantities to derive the volume of precipitation (in m^3) given a precipitation depth of 13 mm (0.013 m) and a catchment area of 2.065×10^5 m^2. The volume of water received by the catchment is $0.013 \text{ m} \times 2.065 \times 10^5 \text{ m}^2 = 2684.5 \text{ m}^3$. This answer is correct mathematically, but it is expressed *with greater relative precision* than is justified by the measured values. There are too many significant figures in the answer. This requires some explanation. First,

Table A1.3 Examples of quantities and their significant figures

Quantity	Significant figures	Scientific notation
650,000	2[a]	6.5×10^5
30	1[a]	3×10^1
30.	2	3.0×10^1
30.0	3	3.00×10^1
30.01	4	3.001×10^1
.01	1	1×10^{-2}
0.01	1	1×10^{-2}
0.010	2	1.0×10^{-2}
0.00500	3	5.00×10^{-3}
1,000.0010	8	1.0000010×10^3

[a]Trailing zeroes without a decimal point are ambiguous.

the measured quantities have a certain *absolute precision*. For the measured precipitation depth, the absolute precision is 1 mm. It is likely that the rain gage could only measure precipitation depth *to the nearest mm*. The absolute precision for the area (206,500 m^2) is 100 m^2. This is apparent because we wrote the value initially using scientific notation. For a number written as 206,500, it is not clear whether the absolute precision is 100 m, 10 m, or 1 m. The *relative precision* is best thought of in terms of *significant figures*. In a given quantity, a significant figure is any given digit, except for zeros to the left of the first nonzero digit, which serves only to fix the position of the decimal point. Some examples are given in Table A1.3.

There are a number of rules for dealing with derived quantities to ensure that answers are expressed with the correct relative precision or number of significant figures. When multiplying or dividing numbers, the answer should be expressed using the same number of significant figures as *the least relatively precise number involved in the calculation*, that is, the one with the fewest significant figures. In our example, the precipitation depth (13 mm) has two significant figures and the catchment area (2.065×10^5 m^2) has four. Therefore, our answer should have only two significant figures, 2.7×10^3 m^3.

When adding or subtracting numbers, the rule is that the answer should be expressed using the same number of significant figures as *the number with the fewest decimal places*. Finally, numbers should not be rounded to the appropriate precision in a calculation with several steps *until the very end*.

Unit Conversions

Measurements that describe hydrological quantities may be expressed in a variety of different units. As a result, one often has to convert a quantity from one unit to another. For example, suppose that a value of hydraulic conductivity (see Chapter 6) found in a research report is given as 6.2 feet per day. We would like to express this value in the SI units of meters per second. One approach is to multiply the value by ratios of equivalent units. The ratios are formed such that the old units cancel, leaving the new units. The procedure is illustrated below:

$$6.2 \frac{\text{ft}}{\text{day}} \times \frac{1 \text{ m}}{3.281 \text{ ft}} \times \frac{1 \text{ day}}{86,400 \text{ s}} = 2.2 \times 10^{-5} \frac{\text{m}}{\text{s}}.$$

Note that the answer is expressed in scientific notation, using the same number of significant figures (2) as the least relatively precise number, which in this case is the hydraulic conductivity value. It should also be obvious that the multiplication has resulted in the original units (ft, day) being "canceled."

Tables A1.4 through A1.8 provide equivalent values for many quantities used frequently in hydrology. To use the tables, look down the left column to find the unit you are interested in using to express a quantity. In the example above, we wanted to convert feet to meters. Looking down the first column of Table A1.4 you will find a row beginning with "meter." Scanning across this row, you will find that 1 meter is equivalent to 3.281 feet, which is the value used to construct the ratio in the example above.

Table A1.4 Equivalent units for length

Unit	Equivalent[a]					
	millimeter	inch	foot	*meter*	kilometer	mile
millimeter	1	0.03937	0.003281	0.001000	1.0×10^{-6}	0.6214×10^{-6}
inch	25.4	1	0.0833	0.02540	25.4×10^{-6}	15.78×10^{-6}
foot	304.8	12	1	0.3048	304.6×10^{-6}	189.4×10^{-6}
meter	1000	39.37	3.281	1	0.001	621.4×10^{-6}
kilometer	1,000,000	39,370	3281	1000	1	0.6214
mile	1,609,000	63,360	5280	1609	1.609	1

[a] In Tables A1.4 through A1.8, values are shown to four significant figures, and the SI expression, in base or derived units, is in italics.

Table A1.5 Equivalent units for area

Unit	Equivalent						
	inch²	foot²	*meter²*	acre	hectare	kilometer²	mile²
inch²	1	0.006944	645.2×10^{-6}	15.94×10^{-8}	64.52×10^{-9}	645.2×10^{-12}	249.1×10^{-12}
foot²	144	1	929.0×10^{-4}	22.96×10^{-6}	9.290×10^{-9}	92.90×10^{-9}	35.87×10^{-9}
meter²	1550	10.76	1	247.1×10^{-6}	10^{-4}	10^{-6}	386.1×10^{-9}
acre	6.273×10^{6}	4.356×10^{4}	4047	1	0.4047	0.004047	0.001563
hectare	1.550×10^{7}	1.076×10^{5}	10^{4}	2.471	1	0.01	0.003861
kilometer²	1.550×10^{9}	1.076×10^{7}	10^{6}	247.1	100	1	0.3861
mile²	4.014×10^{9}	2.788×10^{7}	2.590×10^{6}	640	259	2.590	1

Table A1.6 Equivalent units for volume

Unit	Equivalent						
	$inch^3$	liter	U.S. gallon	$foot^3$	$yard^3$	$meter^3$	acre-foot
$inch^3$	1	0.01639	0.004329	578.7×10^{-6}	21.43×10^{-6}	16.39×10^{-6}	13.29×10^{-9}
liter	61.02	1	0.2642	0.03531	0.001308	10^{-3}	810.6×10^{-9}
U.S. gallon	231.0	3.785	1	0.1337	0.004951	0.003785	3.068×10^{-6}
$foot^3$	1728	28.32	7.481	1	0.03704	0.02832	22.96×10^{-6}
$yard^3$	46660	764.6	202.0	27	1	0.7646	619.8×10^{-6}
$meter^3$	61020	1000	264.2	35.31	1.308	1	810.6×10^{-6}
acre-foot	75.27×10^6	1.233×10^6	3.259×10^5	4.356×10^5	1613	1233	1

Table A1.7 Equivalent units for discharge

Unit	Equivalent				
	gallon $minute^{-1}$	liter $second^{-1}$	acre-foot day^{-1}	$foot^3$ $second^{-1}$	$meter^3$ $second^{-1}$
gallon $minute^{-1}$	1	0.06309	0.004419	0.002228	63.09×10^{-6}
liter $second^{-1}$	15.85	1	0.07005	0.03531	10^{-3}
acre-foot day^{-1}	226.3	14.28	1	0.5042	0.01428
$foot^3$ $second^{-1}$	448.8	28.32	1.983	1	0.02832
$meter^3$ $second^{-1}$	15,850	1000	70.04	35.31	1

Table A1.8 Equivalent values for velocity

Unit	Equivalent				
	foot day^{-1}	km hr^{-1}	foot s^{-1}	mile hr^{-1}	meter s^{-1}
foot day^{-1}	1	12.70×10^{-6}	11.57×10^{-6}	7.891×10^{-6}	3.528×10^{-6}
km hr^{-1}	78,740	1	0.9113	0.6214	0.2778
foot s^{-1}	86,400	1.097	1	0.6818	0.3048
mile hr^{-1}	126,700	1.609	1.467	1	0.4470
meter s^{-1}	283,500	3.600	3.281	2.237	1

APPENDIX 2

Properties of Water

General Properties

The Earth's hydrosphere contains almost 1.4 billion km^3 of water. In nearly all of its physical properties, water is either unique or at the extreme end of the range of a property. Water remains a liquid within the temperature range most suited to life processes.

In basic structure, the water molecule, composed of one atom of oxygen and two of hydrogen, has a small dipole moment. Water will dissolve almost anything to some degree, although the extent may be very small for some substances. Once dissolved, substances tend to remain dissolved because of one of water's exceptional properties. The *dielectric constant* for water is greater than that for any other substance. The dielectric constant (multiplied by the square of the distance separating ions) indicates the magnitude of the force of attraction between positive and negative ions. Ionic substances in solution are therefore tightly held by the surrounding water dipoles. As a result, water found in nature contains some amount of dissolved material.

Water also has the greatest specific heat capacity and latent heat of vaporization known among liquids, and has a higher thermal conductivity than any other liquid except mercury. It also has a high heat of fusion. One consequence of these properties is that water at the Earth's surface tends to moderate effects of hot or cold air temperatures.

Water Density and Viscosity

Liquid water contracts as it cools, reaching a maximum density (ρ) of 1000 kg m^{-3} at 3.98°C (Figure A2.1; Table A2.1). Between 3.98 and 0°C, water expands slightly, but once frozen, ice has a density of about 920 kg m^{-3}. Because of the temperature dependence of water density, warm water will float on top of cooler water at temperatures above 3.98°C, while below 3.98°C, cooler water will float on top of warmer water. As a result, ice will tend to form and remain at the surface of a water body. When it melts, the water becomes denser again and sinks below the surface. This process leads to spring mixing of surface and deeper water in many mid- to high-latitude lakes.

The viscosity of water (μ, often referred to as the *dynamic* viscosity) also varies with temperature (Figure A2.1; Table A2.1). In fact, the variation in water viscosity is much greater than that of density over the temperature range most commonly encountered in hydrology. For example, from 10 to 30°C, water density decreases from 999.73 kg m^{-3} to 997.07 kg m^{-3}, a decrease of 0.27%. Over the same temperature range, the viscosity of water decreases from 1.307×10^{-3} Pa · s (at 10°C) to 0.7975×10^{-3} Pa · s (at 30°C), a decrease of 39%. For this reason, hydrologists are often concerned with the temperature effects on water viscosity when making their calculations. In the SI system (see Appendix 1), the units of viscosity are Pa · s.

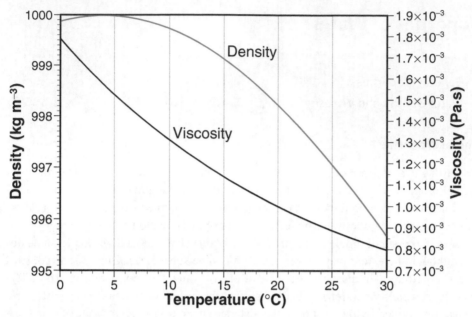

Figure A2.1 The density (ρ) and viscosity (μ) of water as a function of temperature, from 0 to 30°C.

Table A2.1 The density (ρ) and viscosity (μ) of water as a function of temperature, from 0 to 50°C

Temperature (°C)	Density (kg m^{-3})	Viscosity (Pa·s)
0	999.87	1.787×10^{-3}
3.98	1000	
5	999.99	1.519×10^{-3}
10	999.73	1.307×10^{-3}
15	999.13	1.139×10^{-3}
20	998.23	1.002×10^{-3}
25	997.07	8.904×10^{-4}
30	995.67	7.975×10^{-4}
40	992.24	6.529×10^{-4}
50	988.07	5.468×10^{-4}

APPENDIX 3

Basic Statistics in Hydrology

Often, hydrologists have only *point* measurements of quantities such as precipitation, which clearly vary temporally and/or spatially. Due to the complexity of processes such as precipitation, we must often take a probabilistic approach to studying them, based on observation rather than physical theory or mechanistic causes.

In a probabilistic approach, processes are viewed as being somewhat random, or as having some element of uncertainty or unpredictability. Common examples used to discuss probability are rolling dice or flipping a coin. In these cases, the outcome of the random process is a number (1 through 6 on each die) or "heads" or "tails." The numbers of dots on a die face and "heads" or "tails" are examples of *discrete random variables*, because only a discrete number of values are possible. Many hydrological measurements or quantities (i.e., average annual precipitation) are considered *continuous random variables*, because the quantity of interest may have any (reasonable) value.

Consider the data in Table A3.1. There seems to be a general tendency for annual precipitation to be around one meter, with two "extreme" values, 780 and 1192 mm. How would we use this information to make a prediction about future precipitation? With continuous random variables such as annual precipitation, the question we might ask is: What is the probability that precipitation will be greater than 1100 mm? To address this question, we need to consider how probability is "distributed," or what probabilities are associated with what measured or expected values.

Table A3.1 Annual precipitation for a ten-year period

Year	Annual precipitation (mm)
1975	1020
1976	987
1977	894
1978	1040
1979	995
1980	780
1981	1004
1982	930
1983	1192
1984	950

Continuous random variables are described using the *probability density function* (pdf), $f(x)$. The *cumulative distribution function* (cdf), $F(a)$, is the integral of the probability density function:

$$F(a) = P(x \leq a) = \int_{-\infty}^{a} f(x) dx. \tag{A3.1}$$

In other words, the cdf for a chosen value, a (1100 mm, in the example above), is the probability that the outcome of a random process, x (precipitation next year), will be *less than or equal to* the chosen value. Notice that the limits of integration are negative infinity and the value a; the integration is adding up all of the probabilities associated with all of the values less than and including a. Conversely, the *complementary cumulative distribution function*, often written as $G(a)$, is defined as:

$$G(a) = P(x > a) = 1 - F(a) = \int_{a}^{\infty} f(x) dx, \tag{A3.2}$$

which, in the example, provides the probability that precipitation will be *greater than* 1100 mm. We refer to this as the *exceedance probability*. Several relationships can be inferred from the definitions above:

$$P(a \leq x \leq b) = F(b) - F(a) = \int_{a}^{b} f(x) dx. \tag{A3.3}$$

$$P(x = a) = \int_{a}^{a} f(x) dx \equiv 0. \tag{A3.4}$$

$$P(-\infty \leq x \leq \infty) = \int_{-\infty}^{\infty} f(x) dx \equiv 1. \tag{A3.5}$$

Equation A3.3 describes the probability of the outcome of a random process being within a fixed range of values. Equation A3.4 simply points out that, with continuous random variables, the probability of a single definite outcome is zero; that is why we always refer to the probability of an outcome being greater than, or less than, or within a range of values. Equation A3.5 indicates that the total probability is equal to 1 (annual precipitation must have *some* value).

The distribution of a *sample* of values of a random variable can exhibit a variety of forms and can often be described by relatively simple probability density functions. Examples include the normal distribution, the log-normal distribution, and the exponential distribution.

The *normal distribution* can be described with two parameters: the *mean* (μ_x) and the *standard deviation* (σ_x). Given a sample of data (e.g., several decades of annual precipitation), the mean and standard deviation are approximately $\bar{x}$ (for μ_x) and s_x (for σ_x):

$$\bar{x} = \frac{\sum_{i=1}^{n} x_i}{n} \tag{A3.6}$$

$$s_x = \sqrt{\frac{\sum_{i=1}^{n}(x_i - \mu_x)^2}{n-1}}, \tag{A3.7}$$

where n is the number of observations. For the data in Table A3.1, the sample mean is 979 mm and the sample standard deviation is 101 mm. The pdf and cdf for the normal distribution are:

$$f(x;\mu_x,\sigma_x) = \frac{1}{\sqrt{2\pi}\sigma_x}e^{-\frac{(x-\mu_x)^2}{2\sigma_x^2}} \tag{A3.8}$$

$$F(x;\mu_x,\sigma_x) = P(X \le x) = \int_{-\infty}^{x} f(x;\mu_x,\sigma_x). \tag{A3.9}$$

The normal distribution is a "bell-shaped curve" (Figure A3.1). The mean provides a measure of the central tendency of the variable. For normally-distributed data, the mean equals the median (the middle value, if all the values are arranged from lowest to highest or vice versa). The probability that a value will lie within one standard deviation of the mean is about two-thirds (0.68) and the probability that a value will lie within two standard deviations of the mean is about 0.95 (refer to the gray and dark blue regions in Figure A3.1).

Often, data are *normalized* by calculating *z-values*. Data tables that provide the cumulative distribution function of the *standard normal distribution* are based on *z*-values. Normalization of data involves shifting the entire set of data by the amount required to bring

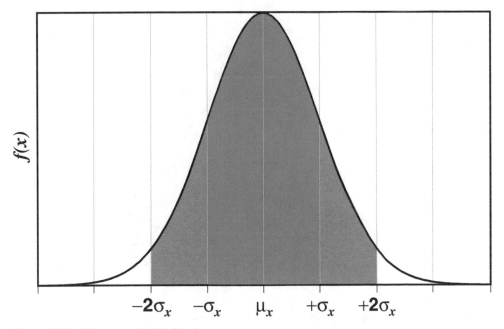

Figure A3.1 The normal distribution.

the mean to zero, and then shrinking or expanding the distribution so that the standard deviation is 1 (Figure A3.2). Values of the cumulative distribution function (Table A3.2) or complementary cumulative distribution function of the standard normal distribution can be used in calculations. The z-value represents the normalized outcome for which the probability is desired (1100 mm, in the earlier example). This value is calculated according to:

$$z = \frac{a - \bar{x}}{s_x}.$$
(A3.10)

For the example data in Table A3.1, with $a = 1100$ mm, $\bar{x} = 979$ mm, $s_x = 101$ mm, the calculated z-value using equation (A3.10) is 1.20. (Note that the z-value is dimensionless.) The cdf for this z-value (Table A3.2) is 0.8849 (i.e., $F(z) = 0.8849$), which indicates that the probability that precipitation in any year will be less than or equal to 1100 mm is about 0.9 (Figure A3.3).

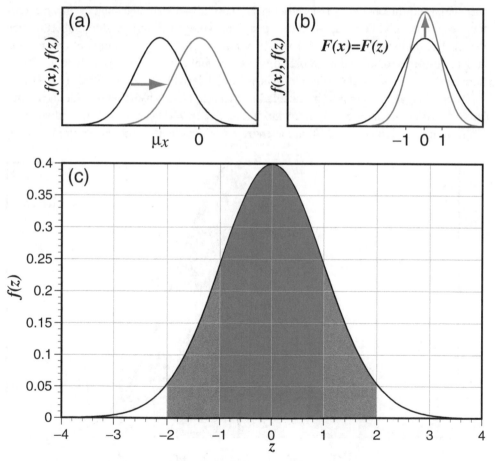

Figure A3.2 Transformation to the standard normal distribution (*c*) involves altering both the mean (*a*) and the standard deviation (*b*) of the original distribution.

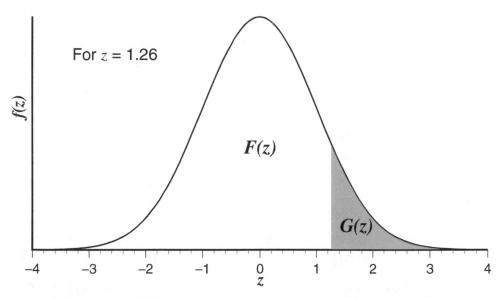

For $z = 1.26$

$f(z)$

$F(z)$

$G(z)$

−4   −3   −2   −1   0   1   2   3   4

z

Figure A3.3 The standard normal distribution showing the portions of the distribution given by the cumulative distribution function, $F(z)$, and complementary cumulative probability, $G(z)$.

To find the exceedance probability, we subtract this value from one [equation (A3.2)] to get approximately a 0.1 chance that precipitation will equal or exceed 1100 mm in any year. In other words, there is a 10% chance of annual precipitation exceeding 1100 mm. That is, on average, 1100 mm or more of precipitation occurs once every 10 years. The return period for annual precipitation equal to or greater than 1100 mm is 10 years:

$$T_{return} = \frac{1}{\text{exceedance probability}} = \frac{1}{0.1 \text{ yr}^{-1}} = 10 \text{ yr.} \tag{A3.11}$$

This entire procedure can also be used, in reverse, to determine the value or range of values associated with a given return period or exceedance probability. For example, using the example precipitation data, we can calculate the annual precipitation amount that we expect to be exceeded 1 out of 3 years. Problems such as this are common in water resource planning. From the given information, we know that $G(z)$ is equal to approximately 0.33. Table A3.2 can be used to find the z-value for $F(z) = 1 - G(z) = 0.67$, which is 0.44. Substituting this value into equation (A3.10) and rearranging to solve for a (we know the sample mean and standard deviation) gives a value of 1021 mm.

This brief review only scratches the surface of applications of statistics to hydrological problems, but does give some idea of how the methods might be used to predict possible behavior. Some of the concepts introduced, such as probability, probability distribution, exceedance probability, and return period, are used throughout the book.

Table A3.2 The cumulative distribution function of the standard normal distribution, $F(z)$, for $z > 0$

z	0.00	0.01	0.02	0.03	0.04	0.05	0.06	0.07	0.08	0.09
0.0	0.5000	0.5040	0.5080	0.5120	0.5160	0.5199	0.5239	0.5279	0.5319	0.5359
0.1	0.5398	0.5438	0.5478	0.5517	0.5557	0.5596	0.5636	0.5675	0.5714	0.5753
0.2	0.5793	0.5832	0.5871	0.5910	0.5948	0.5987	0.6026	0.6064	0.6103	0.6141
0.3	0.6179	0.6217	0.6255	0.6293	0.6331	0.6368	0.6406	0.6443	0.6480	0.6517
0.4	0.6554	0.6591	0.6628	0.6664	0.6700	0.6736	0.6772	0.6808	0.6844	0.6879
0.5	0.6915	0.6950	0.6985	0.7019	0.7054	0.7088	0.7123	0.7157	0.7190	0.7224
0.6	0.7257	0.7291	0.7324	0.7357	0.7389	0.7422	0.7454	0.7486	0.7517	0.7549
0.7	0.7580	0.7611	0.7642	0.7673	0.7704	0.7734	0.7764	0.7794	0.7823	0.7852
0.8	0.7881	0.7910	0.7939	0.7967	0.7995	0.8023	0.8051	0.8078	0.8106	0.8133
0.9	0.8159	0.8186	0.8212	0.8238	0.8264	0.8289	0.8315	0.8340	0.8365	0.8389
1.0	0.8413	0.8438	0.8461	0.8485	0.8508	0.8531	0.8554	0.8577	0.8599	0.8621
1.1	0.8643	0.8665	0.8686	0.8708	0.8729	0.8749	0.8770	0.8790	0.8810	0.8830
1.2	0.8849	0.8869	0.8888	0.8907	0.8925	0.8944	0.8962	0.8980	0.8997	0.9015
1.3	0.9032	0.9049	0.9066	0.9082	0.9099	0.9115	0.9131	0.9147	0.9162	0.9177
1.4	0.9192	0.9207	0.9222	0.9236	0.9251	0.9265	0.9279	0.9292	0.9306	0.9319
1.5	0.9332	0.9345	0.9357	0.9370	0.9382	0.9394	0.9406	0.9418	0.9429	0.9441
1.6	0.9452	0.9463	0.9474	0.9484	0.9495	0.9505	0.9515	0.9525	0.9535	0.9545
1.7	0.9554	0.9564	0.9573	0.9582	0.9591	0.9599	0.9608	0.9616	0.9625	0.9633
1.8	0.9641	0.9649	0.9656	0.9664	0.9671	0.9678	0.9686	0.9693	0.9699	0.9706
1.9	0.9713	0.9719	0.9726	0.9732	0.9738	0.9744	0.9750	0.9756	0.9761	0.9767
2.0	0.9772	0.9778	0.9783	0.9788	0.9793	0.9798	0.9803	0.9808	0.9812	0.9817
2.1	0.9821	0.9826	0.9830	0.9834	0.9838	0.9842	0.9846	0.9850	0.9854	0.9857
2.2	0.9861	0.9864	0.9868	0.9871	0.9875	0.9878	0.9881	0.9884	0.9887	0.9890
2.3	0.9893	0.9896	0.9898	0.9901	0.9904	0.9906	0.9909	0.9911	0.9913	0.9916
2.4	0.9918	0.9920	0.9922	0.9925	0.9927	0.9929	0.9931	0.9932	0.9934	0.9936
2.5	0.9938	0.9940	0.9941	0.9943	0.9945	0.9946	0.9948	0.9949	0.9951	0.9952
2.6	0.9953	0.9955	0.9956	0.9957	0.9959	0.9960	0.9961	0.9962	0.9963	0.9964
2.7	0.9965	0.9966	0.9967	0.9968	0.9969	0.9970	0.9971	0.9972	0.9973	0.9974
2.8	0.9974	0.9975	0.9976	0.9977	0.9977	0.9978	0.9979	0.9979	0.9980	0.9981
2.9	0.9981	0.9982	0.9982	0.9983	0.9984	0.9984	0.9985	0.9985	0.9986	0.9986

Note: To use the table, scan the left column to locate the z-value to the first decimal place, and then scan across the row to find $F(z)$ for the second decimal place. Note that for $z < 0$, use $F(z) = 1 - F(|z|)$.

Answers to Example Problems

Chapter 1

Problem 1.1. Precipitation is typically measured as a volume [L^3] per unit area [L^2], which has dimensions of length [L]. In the United States, the average annual precipitation varies from a minimum at Death Valley, California (1.6 inches), to a maximum on Mt. Waialeale on the island of Kauai in Hawaii (460 inches). What is the average annual precipitation (in millimeters, mm) at each of these locations?

From Table A1.4, 1.0 inch is equivalent to 25.4 mm. The conversion is as follows:

Death Valley: $1.6 \text{ in} \times \dfrac{25.4 \text{ mm}}{1 \text{ in}} = 41 \text{ mm}$

Mt. Waialeale: $460 \text{ in} \times \dfrac{25.4 \text{ mm}}{1 \text{ in}} = 1.17 \times 10^4 \text{ mm}$

Problem 1.2. In the United States, stream discharge is often measured in units of cubic feet per second (ft^3 s^{-1}, or "cfs"). In most other countries, discharge is measured in cubic meters per second (m^3 s^{-1}). What is the equivalent flow (in m^3 s^{-1}) of 18.2 ft^3 s^{-1}? (You might want to review Appendix 1 on units, dimensions, and conversions.)

From Table A1.7, 1.0 ft^3 s^{-1} is equivalent to 0.02832 m^3 s^{-1}. The conversion is as follow:

$$18.2 \text{ ft}^3 \text{ s}^{-1} \times \frac{0.02832 \text{ m}^3 \text{ s}^{-1}}{1 \text{ ft}^3 \text{ s}^{-1}} = 5.15 \times 10^{-1} \text{ m}^3 \text{ s}^{-1}$$

(Note that three significant figures appear in the solution, following the rules described in Appendix 1.)

Problem 1.3. In an average year, 1.0 meter of precipitation falls on a catchment with an area of 1000 (or 10^3) km².

A. What is the volume of water received during an average year in cubic meters?

$$\text{Volume} = \text{Depth} \times \text{Area} = 1.0\,\text{m} \times 10^3\,\text{km}^2 \times \frac{10^6\,\text{m}^2}{1\,\text{km}^2} = 10^9\,\text{m}^3.$$

B. In gallons?

$$\text{Volume} = 10^9\,\text{m}^3 \times \frac{264.2\,\text{gal}}{1\,\text{m}^3} = 3 \times 10^{11}\,\text{gal}.$$

Problem 1.4. The polar ice caps (area $= 1.6 \times 10^7$ km²) are estimated to contain a total equivalent volume of 2.4×10^7 km³ of liquid water. The average annual precipitation over the ice caps is estimated to be 5 inches per year. Estimate the residence time of water in the polar ice caps, assuming their volume remains constant in time.

The residence time for a reservoir at steady state is defined as the reservoir volume divided by the volumetric inflow or outflow rate:

$$V = 2.4 \times 10^7\,\text{km}^3 \times \frac{10^9\,\text{m}^3}{1\,\text{km}^3} = 2.4 \times 10^{16}\,\text{m}^3.$$

$$I = 5\,\text{in yr}^{-1} \times \frac{1\,\text{m}}{39.37\,\text{in}} \times 1.6 \times 10^7\,\text{km}^2 \times \frac{10^6\,\text{m}^2}{1\,\text{km}^2} = 2.032 \times 10^{12}\,\text{m}^3\,\text{yr}^{-1}.$$

$$\text{Residence time} = T_r = \frac{V}{I} = \frac{2.4 \times 10^{16}\,\text{m}^3}{2.032 \times 10^{12}\,\text{m}^3\,\text{yr}^{-1}} = 1.0 \times 10^4\,\text{yrs}.$$

Problem 1.5. In an average year, a small (area $= 3.0$ km²) agricultural catchment receives 950 mm of precipitation. The catchment is drained by a stream, and a continuous record of stream discharge is available. The total amount of surface-water runoff for the year, determined from the stream discharge record, is 1.1×10^6 m³.

A. What is the volume of water (in m³) evapotranspired for the year (assume no change in water stored in the catchment)?

$$\bar{r}_s = 1.1 \times 10^6\,\text{m}^3.$$

$$\bar{p} = \left(950\,\text{mm} \times \frac{1\,\text{m}}{1000\,\text{mm}} \times 3.0\,\text{km}^2 \times \frac{10^6\,\text{m}^2}{1\,\text{km}^2} \right) = 2.85 \times 10^6\,\text{m}^3.$$

$$\overline{et} = \bar{p} - \bar{r}_s = 2.85 \times 10^6\,\text{m}^3 - 1.1 \times 10^6\,\text{m}^3 = 1.75 \times 10^6\,\text{m}^3 = 1.8 \times 10^6\,\text{m}^3.$$

B. What is the depth of water (in mm) evapotranspired for the year (again, assuming no change in water stored in the catchment)?

To calculate the depth, divide the volume by the area:

$$\frac{1.75\times10^6 \text{ m}^3}{3.0 \text{ km}^2 \times \dfrac{10^6 \text{ m}^2}{1 \text{ km}^2}} = 0.58 \text{ m}\times\frac{1000 \text{ mm}}{1 \text{ m}} = 580 \text{ mm.}$$

(Note that the volume calculated in A was reported with two significant figures, hence 1.75×10^6 m³ became 1.8×10^6 m³. For B, the original result was used in the calculation, and the final answer was rounded to two significant figures.)

C. What is the runoff ratio ($\bar{r}_s/\bar{p}$) for the catchment?

$$\frac{\bar{r}_s}{\bar{p}} = \frac{1.1\times10^6 \text{ m}^3}{950 \text{ mm}\times\dfrac{1 \text{ m}}{1000 \text{ mm}}\times3.0 \text{ km}^2\times\dfrac{10^6 \text{ m}^2}{1 \text{ km}^2}} = \frac{1.1\times10^6 \text{ m}^3}{2.85\times10^6 \text{ m}^3} = 0.39.$$

Chapter 2

Problem 2.1. Two tipping bucket rain gages are used to collect the following rainfall data:

Time	Cumulative precipitation (mm) Station #1	Cumulative precipitation (mm) Station #2
4:00 a.m.	0.0	0.0
6:00 a.m.	0.0	0.0
8:00 a.m.	1.0	1.0
10:00 a.m.	4.0	3.0
12:00 noon	13	11
2:00 p.m.	17	15
4:00 p.m.	19	16
6:00 p.m.	19	17
8:00 p.m.	19	17
10:00 p.m.	19	17
12:00 midnight	19	17
2:00 a.m.	19	17
4:00 a.m.	19	17

A. Calculate the mean daily rainfall intensity for each station (mm hr^{-1}).

Station #1: intensity = 19 mm/24 hr = 0.79 mm hr^{-1}.
Station #2: intensity = 17 mm/24 hr = 0.71 mm hr^{-1}.

B. Calculate the maximum 2-hour rainfall intensity for each station (mm hr^{-1}).

By inspection, the greatest 2-hr rainfall occurred between 10:00 a.m. and noon:

Station #1: intensity = (13 − 4) mm/2.0 hr = 4.5 mm hr^{-1}.
Station #2: intensity = (11 − 3) mm/2.0 hr = 4.0 mm hr^{-1}.

C. Calculate the maximum 6-hour rainfall intensity for each station (mm hr^{-1}).

The greatest 6-hr rainfall occurred between 8:00 a.m. and 2:00 p.m.:

Station #1: intensity = (17 −1.0) mm/6.0 hr = 2.7 mm hr^{-1}.
Station #2: intensity = (15 − 1.0) mm/6.0 hr = 2.3 mm hr^{-1}.

D. Using the arithmetic average method and knowing that the drainage basin area is 176 mi^2, calculate the total volume of rainfall (m^3) delivered to the basin during the event.

$p = (19 + 17)/2.0 = 18$ mm.

$$V = A \times p = \left(176 \text{ mi}^2 \times \frac{2.590 \times 10^6 \text{ m}^2}{1 \text{ mi}^2} \right) \left(18 \text{ mm} \times \frac{1 \text{ m}}{1000 \text{ mm}} \right) = 8.2 \times 10^6 \text{ m}^3.$$

Problem 2.2. Measurement of changes in volume of water in an evaporation pan is a standard technique for estimating potential evapotranspiration. United States Class A evaporation pans are cylindrical with the following dimensions: depth = 10.0 inches and diameter = 47.5 inches. An evaporation pan can be considered a hydrological system with an inflow, outflow, and storage volume. Evaporation from pans is not the same as evaporation from natural surfaces for a variety of reasons. For example, water temperatures in shallow pans will be much more variable than temperatures in a nearby lake. Evaporation measured in pans is adjusted by a factor called a pan coefficient to convert to an estimate of potential evapotranspiration (see Brutsaert, 1982).

A. Calculate the cross-sectional area (m^2) of a United States Class A evaporation pan through which inflows and outflows of water can pass. Also, calculate the total storage volume of the pan (m^3).

$$\text{Cylinder area} = \pi r^2 = \pi (d/2)^2 = \pi \left(\frac{47.5 \text{ in}}{2} \times \frac{2.54 \text{ cm}}{1 \text{ in}} \times \frac{1 \text{ m}}{100 \text{ cm}} \right)^2 = 1.14 \text{ m}^2.$$

$$\text{Cylinder volume} = \text{area} \times \text{depth} = 1.14 \text{ m}^2 \times \left(10.0 \text{ in} \times \frac{2.54 \text{ cm}}{1 \text{ in}} \times \frac{1 \text{ m}}{100 \text{ cm}} \right) = 0.290 \text{ m}^3.$$

B. Initially, the pan contains 10.0 U.S. gallons of water. Calculate the depth of water in the pan (mm).

$$\text{Volume} = \left(10.0 \text{ gal} \times \frac{0.003785 \text{ m}^3}{1 \text{ gal}} \right) = 3.79 \times 10^{-2} \text{ m}^3.$$

Depth = volume/area = 3.79×10^{-2} m³/1.14 m² = 0.0332 m = 33.2 mm.

C. Assuming a water density of 997.07 kg m⁻³ (25°C), calculate the mass (kg) of water in the pan

Mass = $\rho \times$ volume = (997.07 kg m⁻³)(3.79 × 10⁻² m³) = 37.8 kg.

D. After 24 hours in an open field (no precipitation), the pan is checked and the volume of water left in the pan is determined to be 9.25 gallons. Calculate the average evaporation rate (mm hr⁻¹) from the pan.

$dV/dt = I - O = $ (pan area) $\times (p - et)$ or $et = - (dV/dt)/$(pan area),

where p and et are the average rate of precipitation and evaporation expressed as a depth per time,

$$\frac{dV}{dt} = -\left(\frac{0.75 \text{ gal}}{24 \text{ hr}} \times \frac{0.003785 \text{ m}^3}{1 \text{ gal}} \right) = -1.18 \times 10^{-4} \text{ m}^3 \text{ hr}^{-1}.$$

$et = -(-1.18 \times 10^{-4} \text{ m}^3 \text{ hr}^{-1})/(1.14 \text{ m}^2) = 1.04 \times 10^{-4} \text{ m hr}^{-1}.$

$et = 0.10 \text{ mm hr}^{-1}$

E. The pan is emptied and refilled with 10.0 gallons of water and left in an open field for another 24 hours. During this period, rain fell for a 3-hour period at a constant intensity of 2.5 mm hr⁻¹; after 24 hours, the volume of water in the pan was 11.50 gallons. Calculate the average evaporation rate (mm hr⁻¹) from the pan during this period.

$et = p - (dV/dt)/$(pan area).

The total precipitation for the 24-hour period is $(3\,hr)(2.5\,mm\ hr^{-1}) = 7.5\,mm$. Over the 24-hour period, this is equal to an average precipitation rate of:

$$p = (7.5\,mm)/(24\,hr) = 0.3125\,mm\ hr^{-1} = 3.125 \times 10^{-4}\ m\ hr^{-1}.$$

$$\frac{dV}{dt} = \left(\frac{1.50\ gal}{24\ hr} \times \frac{0.003785\ m^3}{1\ gal} \right) = 2.37 \times 10^{-4}\ m^3\ hr^{-1}.$$

$$et = (3.125 \times 10^{-4}\ m\ hr^{-1}) - (2.37 \times 10^{-4}\ m^3\ hr^{-1})/(1.14\ m^2).$$

$$et = 1.05 \times 10^{-4}\ m\ hr^{-1} = 0.11\ mm\ hr^{-1}.$$

F. If the evaporation rate calculated in E remains constant and no additional precipitation occurs, estimate the time (days) for the pan to empty as a result of evaporation.

$$et = p - (dV/dt)/(\text{pan area}).$$

$$dt = -\frac{dV}{(et)(\text{pan area})}.$$

$$dV = -11.50\ gal \times \frac{0.003785\ m^3}{1\ gal} = -4.35 \times 10^{-2}\ m^3.$$

$$dt = \frac{-4.35 \times 10^{-2}\ m^3}{(1.05 \times 10^{-4}\ m\ hr^{-1})(1.14\ m^2)} = 363\ hr = 15\ days.$$

Problem 2.3. For Problem 2.2 part D, calculate the flux of latent heat from the water in the pan to the atmosphere ($W\ m^{-2}$). Use a water density, $\rho_w = 1000.0\,kg\ m^{-3}$.

$$et = \frac{E_l}{\rho_w \lambda_v}, \text{ or } E_l = et\ \rho_w \lambda_v.$$

$$et = 1.04 \times 10^{-4}\ m\ hr^{-1} \times \frac{1\ hr}{3600\ s} = 2.89 \times 10^{-8}\ m^3\ s^{-1}.$$

$$E_l = (2.89 \times 10^{-8}\ m\ s^{-1})(1000\ kg\ m^{-3})(2.45 \times 10^6\ J\ kg^{-1}) = 70.8\ W\ m^{-2}.$$

Problem 2.4. A small (area = 300 ha) catchment in Iowa absorbs a mean $R_n = 330\ W\ m^{-2}$ during the month of June. In this problem, you will apply the energy balance approach to estimate evapotranspiration from the catchment during the month of June.

A. Write a complete energy balance equation (i.e., including all terms) for the catchment for the month of June.

$$\frac{dQ}{dt} = R_n - G - H - E_l.$$

B. Neglecting conduction to the ground (G) and the change in energy stored ($dQ/dt = 0$), simplify your energy balance equation for the catchment so that it can be solved for the latent heat flux, E_l. Also, replace the term H (the sensible heat flux) with $B \times E_l$ (where B is the Bowen ratio).

$$R_n - H - E_l = 0.$$

$$BE_l + E_l = R_n.$$

$$E_l = \frac{R_n}{B+1}.$$

C. Using a mean Bowen ratio of 0.20 for the catchment, calculate the mean daily flux of latent heat to the atmosphere (W m^{-2}) and the mean evapotranspiration rate (mm day^{-1}) from the catchment. Use a water density, $\rho_w = 1000.0$ kg m^{-3}.

$$E_l = \frac{330 \text{ W m}^{-2}}{1.20} = 275 \text{ W m}^{-2} = 280 \text{ W m}^{-2}.$$

$$et = \frac{E_l}{\rho_w \lambda_v} = \frac{275 \text{ W m}^{-2}}{(1000 \text{ kg m}^{-3})(2.45 \times 10^6 \text{ J kg}^{-1})} = 1.12 \times 10^{-7} \text{ m s}^{-1}.$$

$$et = 1.12 \times 10^{-7} \text{ m s}^{-1} \times \frac{1000 \text{ mm}}{1 \text{ m}} \times \frac{86400 \text{ s}}{1 \text{ day}} = 9.68 \text{ mm day}^{-1} = 9.7 \text{ mm day}^{-1}.$$

D. Calculate the total evapotranspiration from the catchment during the month of June (mm).

Total evapotranspiration $= 9.68$ mm day$^{-1} \times 30$ days $= 290$ mm.

Chapter 3

Problem 3.1. The following questions make use of the hydrostatic equation.

A. What is the gage pressure (Pa) at a depth of 10.0 m in a lake with a water temperature of 15°C?

At 15°C $\rho = 999.1$ kg m^{-3}, so

$$p = \rho g d = (999.1 \text{ kg m}^{-3})(9.81 \text{ m s}^{-2})(10.0 \text{ m}) = 9.80 \times 10^4 \text{ kg m}^{-1} \text{ s}^{-2} = 98.0 \text{ kPa}.$$

B. Would the pressure change significantly if the water temperature was 22°C instead?

At 25°C $\rho = 997.1$ kg m^{-3}, so

$$p = \rho g d = (997.1 \text{ kg m}^{-3})(9.81 \text{ m s}^{-2})(10.0 \text{ m}) = 9.78 \times 10^4 \text{ kg m}^{-1} \text{ s}^{-2} = 97.8 \text{ kPa}.$$

There is a small difference in the third significant figure.

C. At what depth (m) is the gage pressure 300 kPa?

$$d = p/\rho g = 300 \text{ kPa}/9.80 \text{ kN m}^{-3} = (3.00 \times 10^5 \text{ kg m}^{-1} \text{ s}^{-2})/(9.80 \times 10^3 \text{ kg m}^{-2} \text{ s}^{-2}).$$

$$d = 30.6 \text{ m}.$$

D. What depth (m) of mercury, with a unit weight of 133 kN m^{-3}, would be required to produce a pressure of 300 kPa?

$$d = p/(\rho g)_{\text{Hg}} = 300 \text{ kPa}/133 \text{ kN m}^{-3} = 2.26 \text{ m}.$$

Problem 3.2. A plate is pulled over a horizontal layer of water that is 10.0 mm deep (Figure 3.1). The temperature of the water is 20°C. If the plate exerts a shear stress of 0.01 N m^{-2} on the upper surface of the water, what is the speed (m s^{-1}) of the plate?

From Equation 3.1, $u_{\text{plate}} = (F/A)(d/\mu)$. F/A is the shear stress, equal in this case to 0.01 N m^{-2} or 0.01 kg m^{-1} s^{-2}. The viscosity at 20°C is 1.00×10^{-3} Pa · s (Table 3.1) or 1.00×10^{-3} kg m^{-1} s^{-1}. Therefore,

$$u_{plate} = (0.01 \text{ kg m}^{-1} \text{ s}^{-2})(0.010 \text{ m}/1.00 \times 10^{-3} \text{ kg m}^{-1} \text{ s}^{-1}) = 0.10 \text{ m s}^{-1}.$$

Problem 3.3. Observations show that flow in a circular pipe of diameter D remains laminar up to a Reynolds number $\mathbf{R}_{pipe} = \rho U D/\mu = 2000$ (Figure 3.10). What about flows in other flow in channels or pipes with different geometries? Consider the example of flow between two flat plates shown in Figure 3.1. In this case, the appropriate length scale for the Reynolds number is $L = d$. It also makes sense to use plate speed rather than mean flow velocity speed as the characteristic velocity in $\mathbf{R}$, giving $\mathbf{R}_{plate} = \rho u_{plate} d/\mu$. Changes in length and velocity scales can alter the upper limit for laminar flow, but for the case of flow between two parallel plates, flow is again laminar up to $\mathbf{R}_{plate} = 2000$. [A parameter called the hydraulic radius (Chapter 4.5) can be used to find values of $\mathbf{R}$ corresponding to the laminar –turbulent transition for different flow cross-sections. This is explored further in an example problem in Chapter 4.]

A. For 20°C water between 2 plates separated by a distance of 4.0 mm (i.e., $d = 4.0$ mm; Figure 3.1), what is the maximum speed that the upper plate can move and still maintain laminar flow?

Rearranging $R_{plate} = \rho u_{plate} d/\mu$ to solve for u_{plate} gives $u_{plate} = R_{plate} \mu/(\rho d)$. The maximum value of u_{plate} to maintain laminar flow is found when $R_{plate} = 2000$:

$$u_{plate} = \frac{(2000)(1.00 \times 10^{-3} \text{ kg m}^{-1} \text{ s}^{-1})}{(1000 \text{ kg m}^{-3})(4.0 \times 10^{-3} \text{ m})} = 0.5 \text{ m s}^{-1}.$$

B. Is the flow in Problem 2 laminar? [Note: if not, equation 3.1 is no longer correct.]

$$R_{plate} = \frac{\rho u_{plate} d}{\mu} = \frac{(1000 \text{ kg m}^{-3})(0.10 \text{ m s}^{-1})(10.0 \times 10^{-3} \text{ m})}{(1.00 \times 10^{-3} \text{ kg m}^{-1} \text{ s}^{-1})} = 1000.$$

Since $R_{plate} < 2000$, the flow is laminar.

Problem 3.4. Surface temperature in a river is measured by a thermometer drifting with the water at a rate of 1 km hr $^{-1}$. The water in the river as a whole is warming at a rate of 0.2°C hr $^{-1}$, and the temperature along the stream increases by 0.1°C every kilometer in the downstream direction. What change in temperature (°C) does the thermometer record in 6 hours?

$$\frac{dT}{dt} = \frac{\partial T}{\partial t} + u \frac{\partial T}{\partial x} = (0.2°\text{C hr}^{-1}) + (1000 \text{ m hr}^{-1})(1 \times 10^{-4}°\text{C m}^{-1}) = 0.3°\text{C hr}^{-1}.$$

The change over 6 hours is $(6 \text{ hr})(0.3°\text{C hr}^{-1}) = 1.8°\text{C}$.

Problem 3.5. A tank like the one pictured in Figure 3.6 is filled to a constant level of 0.70 m. The center of the outflow opening near the bottom is 0.10 m above the bottom of the tank. What is the velocity (m s^{-1}) of flow exiting from the outflow opening?

The depth between the water surface in the tank and the outflow opening is: $d = z_1 - z_2 = 0.70 \text{ m} - 0.10 \text{ m} = 0.60 \text{ m}$. From Equation 3.22:

$$u_2 = (2gd)^{1/2} = [(2)(9.81 \text{ m s}^{-2})(0.60 \text{ m})]^{1/2} = 3.4 \text{ m s}^{-1}.$$

This is the velocity of flow exiting the tank.

Problem 3.6. The pressure drop through a well-designed constriction can be used to measure the velocity of flow through a pipe. If the pressure drop from a 0.1-m diameter cross section to a 0.05-m diameter cross section is 7.5 kPa, what is the velocity (m s^{-1}) in the 0.1-m diameter section of the pipe? Hint: Use the conservation of mass equation to relate the velocity at the smaller cross section to that at the larger cross section.

From the Bernoulli equation

$$\frac{p_1 - p_2}{\rho g} = -\frac{U_1^2 - U_2^2}{2g}.$$

From conservation of mass,

$$U_2 = U_1 A_1 / A_2 = U_1 (D_1 / D_2)^2 = U_1 (0.1/0.05)^2 = 4U_1.$$

Substituting $U_2 = 4U_1$ into the Bernoulli equation gives:

$$\frac{p_1 - p_2}{\rho g} = -\frac{U_1^2 - (4U_1)^2}{2g} = \frac{15U_1^2}{2g}.$$

Solving for U_1:

$$U_1 = \left[\frac{2(p_1 - p_2)}{15\rho} \right]^{1/2} = \left[\frac{2(7.5 \times 10^3 \text{ Pa})}{1.5 \times 10^4 \text{ kg m}^{-3}} \right]^{1/2} = 1 \text{ m s}^{-1}.$$

Problem 3.7. A steady discharge of 2.0×10^{-4} m³ s⁻¹ is flowing through a 20-mm diameter hose. The viscosity of the water is 1.0×10^{-3} Pa · s, and the density of the water is 1000.0 kg m⁻³.

A. Calculate the Reynolds number. Is the flow laminar or turbulent?

$$U = Q/A = (2.0 \times 10^{-4} \text{ m}^3 \text{ s}^{-1})/[\pi(1.0 \times 10^{-2} \text{ m}^2)] = 0.64 \text{ m s}^{-1}$$

$$R = \rho U D / \mu = (1000.0 \text{ kg m}^{-3})(0.64 \text{ m s}^{-1})(2.0 \times 10^{-2} \text{ m})/(1.0 \times 10^{-3} \text{ Pa} \cdot \text{s}) = 1.3 \times 10^4$$

Because R > 4000, the flow is turbulent.

B. What is the friction factor and the head loss per unit length for this flow (both are dimensionless)?

From Figure 3.10, $f = 0.03$. We could also calculate f using Equation 3.39, because the flow is turbulent and $4000 < R < 100,000$. The head loss per unit length may be calculated as:

$$\frac{h_L}{L} = f \frac{1}{D} \frac{U^2}{2g} = 0.03 \frac{1}{(2 \times 10^{-2} \text{ m})} \frac{(0.64 \text{ m s}^{-1})^2}{2(9.81 \text{ m s}^{-2})} = 0.03.$$

C. What is the change in pressure (Pa) over a 10-m length of the hose?

$$p_1 - p_2 = \rho g(h_L/L)L = (9.81 \text{ kN m}^{-3})(0.03)(10 \text{ m}) = 3 \text{ kPa}.$$

Problem 3.8. Lava, with a density of 2700 kg m^{-3} and viscosity of 1.0×10^3 Pa · s flows through a conduit that is circular in cross-section. The diameter of the conduit is 1.0 m. Flow of lava through the conduit is driven by a pressure gradient of -2.0 kPa m^{-1}. Assuming the flow is laminar, what is the discharge of lava ($\text{m}^3 \text{ s}^{-1}$)? Is the assumption of laminar flow valid?

$$U = -\frac{dp}{dx}\frac{D^2}{32\mu} = -(-2.0 \times 10^3 \text{ Pa m}^{-1})\frac{(1.0 \text{ m})^2}{32(1.0 \times 10^3 \text{ Pa} \cdot \text{s})} = 0.063 \text{ m s}^{-1}.$$

$$Q = UA = U(\pi r^2) = (0.063 \text{ m s}^{-1})[\pi(0.50 \text{ m})^2] = 0.049 \text{ m}^3 \text{ s}^{-1}.$$

$$R = \frac{\rho U D}{\mu} = \frac{(2700 \text{ kg m}^{-3})(0.063 \text{ m s}^{-1})(1.0 \text{ m})}{(1.0 \times 10^3 \text{ Pa} \cdot \text{s})} = 0.17.$$

The Reynolds number is much less than 2000, so the assumption that the flow is laminar is valid.

Chapter 4

Problem 4.1. Water flows in a 3.00-m wide rectangular channel. The water depth is 1.50 m and the discharge is 1.50 $\text{m}^3 \text{ s}^{-1}$. The channel bottom drops smoothly by 0.100 m over a short distance (a step down in the bottom) with no head loss or change in the width of the channel.

A. Calculate the specific discharge ($\text{m}^2 \text{ s}^{-1}$) and specific energy (m) at the upstream station.

$$q_w = \frac{Q}{w} = \frac{1.50 \text{ m}^3 \text{ s}^{-1}}{3.00 \text{ m}} = 0.500 \text{ m}^2 \text{ s}^{-1}.$$

$$E = \frac{U^2}{2g} + h = \frac{q_w^2}{2gh^2} + h = \frac{(0.500 \text{ m}^2 \text{ s}^{-1})^2}{2(9.81 \text{ m s}^{-2})(1.50 \text{ m})^2} + 1.50 \text{ m} = 1.51 \text{ m}.$$

B. Calculate the specific discharge ($\text{m}^2 \text{ s}^{-1}$) and specific energy (m) at the downstream station.

The specific discharge, q_w, is constant because there is no change in channel width.

$$E_2 = E_1 + 0.100 \text{ m} = 1.61 \text{ m}.$$

Note that the channel *drops*, so the specific energy *increases*.

C. Calculate the water depth (m) at the downstream station

Determine the upstream flow criticality by calculating the Froude number:

$U = q_w/h = 0.500 \text{ m}^2 \text{ s}^{-1}/1.50 \text{ m} = 0.333 \text{ m s}^{-1}.$

$$F = \frac{U}{\sqrt{gh}} = \frac{0.333 \text{ m s}^{-1}}{\sqrt{(9.81 \text{ m s}^{-2})(1.50 \text{ m})}} = 0.086.$$

The downstream flow will be subcritical also, because we are moving along the subcritical limb of the specific energy diagram in the direction of increasing *E*. To find the water depth at the downstream location, we use an iterative procedure:

$$h^{(n)} = E - \frac{q_w^2}{2g(h^{(n-1)})^2},$$

where "n" indicates the iteration number. Choose an initial "guess" for the downstream depth, $h_2^{(0)}$, to be 2.0 m and substitute this into the right side of the iteration equation to get $h_2^{(1)}$:

$$h_2^{(1)} = 1.606 \text{ m} - \frac{(0.5 \text{ m}^2 \text{ s}^{-1})^2}{2(9.81 \text{ m s}^{-2})(2.0 \text{ m})^2} = 1.6028 \text{ m}.$$

Continue the iteration by substituting the newly calculated value of $h_2^{(1)}$ into the right side of the iteration equation:

$$h_2^{(2)} = 1.606 \text{ m} - \frac{(0.5 \text{ m}^2 \text{ s}^{-1})^2}{2(9.81 \text{ m s}^{-2})(1.6028 \text{ m})^2} = 1.601 \text{ m}.$$

Continuing the iteration, we find that a good estimate for h_2 is 1.6 m.

Problem 4.2. A discharge of 2.0 m^3 s^{-1} is carried in a canal with the cross section shown in Figure 4.15. The canal is 1400 m long and drops 0.50 m in elevation over that distance. Manning's n for the channel is estimated to be 0.020. What is the value of w (m) for this canal?

Discharge is mean velocity times cross-sectional area. From Manning's equation:

$$Q = Uwh = \frac{k}{n} R_H^{2/3} S^{1/2} wh = \frac{k}{n} R_H^{2/3} S^{1/2} \frac{w^2}{2.5}.$$

where the hydraulic radius is given by:

$$R_H = \frac{wh}{2h+w} = \frac{w^2}{2.5[w+2(w/2.5)]} = \frac{w^2}{4.5w} = \frac{w}{4.5}.$$

Therefore,

$$2\frac{\text{m}^2}{\text{s}} = \left(\frac{1\ \text{m}^{1/3}\ \text{s}^{-1}}{0.020}\right)\left(\frac{w}{4.5}\right)^{2/3}\left(\frac{0.5\ \text{m}}{1400\ \text{m}}\right)^{1/2}\left(\frac{w^2}{2.5}\right),$$

from which $w^2 w^{2/3} = 14.42$, so $w = 2.7$ m.

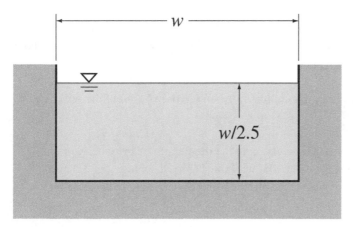

Figure 4.15 Canal cross section for Problem 4.2.

Problem 4.3. As noted in Problem 3.3, channels and pipes of varying geometry will be characterized by different formulations of the Reynolds number. For example, $\mathbf{R}_{pipe} = \rho U D / \mu$, where D is pipe diameter, whereas a more appropriate form for channels might be $\mathbf{R} = \rho U R_H / \mu$, where R_H is hydraulic radius. Use of a different length scale in $\mathbf{R}$ will generally alter the upper limit of $\mathbf{R}$ for laminar flow relative to that found for pipes. In this problem, we reformulate $\mathbf{R}$ using hydraulic radius as the length scale to obtain a more general form of $\mathbf{R}$ that we can use to relate the laminar-turbulent transition in pipes to other flow geometries.

A. Find an expression for the hydraulic radius R_H in terms of pipe diameter D for a pipe with a circular cross-section.

$$R_H = \frac{A}{P} = \frac{\pi(D/2)^2}{\pi D} = \frac{D}{4}.$$

B. Rearrange the expression found in 3A to get a relationship for D in terms of R_H. Substitute this into $\mathbf{R}_{pipe} = \rho U D / \mu$ to get an expression for $\mathbf{R}$ in terms of R_H.

Rearranging the answer from 4.3A gives $D = 4R_H$. Substituting this into $\mathbf{R}_{pipe}$ gives:

$$\mathbf{R}_{pipe} = \frac{\rho U (4R_H)}{\mu} = 4\frac{\rho U R_H}{\mu}.$$

C. Set the expression found in 3B equal to 2000, the critical Reynolds number for pipes. Rearrange this to find the equivalent critical value for $\mathbf{R}_{RH} = \rho U R_H / \mu$.

$$2000 = 4\frac{\rho U R_H}{\mu} \text{ or } \mathbf{R}_{RH-crit} = \frac{\rho U R_H}{\mu} = \frac{2000}{4} = 500.$$

D. Use the equation you developed in 3C to find the critical Reynolds number for the transition from laminar to turbulent flow in a relatively wide, rectangular channel (such that $R_H \approx h$).

If $R_H \approx h$, then the critical Reynolds number for the channel flow will be 500 as in 4.3C.

Problem 4.4. Use Equation 4.32 to estimate the depth and mean velocity of a flow in a channel with a slope $S = 0.003$, width $w = 15$ m, discharge $Q = 1.0$ m³ s⁻¹ and Manning's $n = 0.075$. Does the assumption that $R_H \approx h$ seem reasonable for this flow?

Specific discharge $q_w = Q/w = (1.0$ m³ s⁻¹)/(15 m) = 0.067 m² s⁻¹. Rearranging Equation (4.32) to solve for h gives:

$$h = \left(\frac{q_w n}{kS^{1/2}}\right)^{3/5} = \left(\frac{(0.067 \text{ m}^2 \text{ s}^{-1})(0.075)}{(1 \text{ m}^{1/3} \text{ s}^{-1})(0.003)^{1/2}}\right)^{3/5} = 0.24 \text{ m}.$$

$$U = q_w/h = (0.067 \text{ m}^2 \text{ s}^{-1})/(0.24 \text{ m}) = 0.28 \text{ m s}^{-1}.$$

For $w = 15$ m and $h = 0.24$ m, $R_H = (15$ m)(0.24 m)/[(15 m) + 2(0.24 m)] = 0.23 m. This is within 5% of the value of h, so the assumption that $R_H \approx h$ is reasonable for this flow.

Chapter 5

Problem 5.1. If the flood used in the reservoir example delivered the same volume of water in a shorter amount of time (shorter duration with higher peak discharge), as given by the inflow hydrograph in the table below, how would the outflow hydrograph change? Complete the table below, assuming the initial conditions and other reservoir parameters remain the same.

Step n	Time t_n (days)	I_n	$I_n + I_{n+1}$	$\frac{2V_n}{\Delta t} - O_n$	$\frac{2V_{n+1}}{\Delta t} + O_{n+1}$	O_{n+1}	t_{n+1}
		A	B	C	D	E	
1	0.00	0.7	4.7	342.6	347.3	0.66	0.25
2	0.25	4.0	22.0	346.0	368.0	5.31	0.50
3	0.50	18.0	28.0	357.3	385.3	10.0	0.75
4	0.75	10.0	15.6	365.3	380.9	8.75	1.00
5	1.00	5.6	8.6	363.4	372.0	6.35	1.25
6	1.25	3.0	4.5	359.3	363.8	4.30	1.50
7	1.50	1.5	2.2	355.2	357.4	2.81	1.75
8	1.75	0.7	1.4	351.8	353.2	1.88	2.0

The completed table for this larger inflow shows that the peak outflow is a smaller fraction of the peak inflow ($O_{peak}/I_{peak} = 0.56$) compared to the example in Table 5.3 and Figure 5.10 ($O_{peak}/I_{peak} = 0.63$).

Problem 5.2. The Muskingum routing coefficients for a stream reach are determined to be: $C_0 = 0.26$, $C_1 = 0.55$, $C_2 = 0.19$. For the inflow hydrograph given in the table below, complete the calculation of the predicted outflow hydrograph.

Time (hr)	Inflow (m³ s⁻¹)	Outflow (m³ s⁻¹)
0000	10	10
0600	50	**20**
1200	130	**65**
1800	110	**112**
2400	70	**100**

Problem 5.3. In Figure 5.14, a 40-year annual series (1950–1989) of floods on the Eel River, California, is plotted against the fraction $[r/(n+1)]$ of floods with discharges greater than or equal to each value

A. Fit a line through the data and determine the return period of an 8000 m³ s⁻¹ flood.

The value of the exceedance probability indicated by a best-fit line for $Q = 8000$ m³ s⁻¹ is approximately 0.20 (Figure PS.1). The return period $T_{return} = 1/0.20 = 5$ yr.

B. Estimate the magnitude of the 100-year flood.

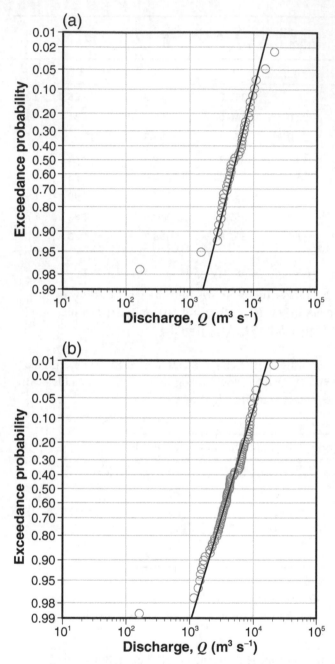

Figure PS.1 Probability plot for Eel River, California, peak annual discharge data, 1950–1989 (*a*), and 1917–1996 (*b*).

A 100-year flood has an exceedance probability of 1/100=0.01. From the best-fit line (Figure PS.1), the corresponding discharge is approximately 1.7×10^4 m^3 s^{-1}. Note that this is smaller than the largest discharge on record (over 2.1×10^4 m^3 s^{-1}, in 1965). If the annual series data for the Eel River were truly log-normally distributed, this would suggest that the 1965 flood has a return period longer than the 40-year record length. In fact, the full record for the Eel River begins in 1917. The 1965 flood is the largest of the 82-year record from 1917 to 1989. The best-fit line through the 82-year record still puts the 100-year flood at approximately 1.7×10^4 m^3 s^{-1}.

Chapter 6

Problem 6.1. You are charged with designing a very simple filtration system for a community water supply, using cylindrical sand columns ($K = 5.0$ m day^{-1}). The filter needs to be 3.0 m long to adequately trap particulates in the water, and since the system will be driven by gravity, the pressure heads at the top and bottom of the (vertically oriented) filter will be zero.

A. What diameter filter is required to treat 4.0×10^3 gallons of water per day? Is this value feasible (anything larger than about 1 m is not feasible)?

Because the pressure heads at the top and bottom of the filter are zero, the hydraulic gradient must be equal to 1. Therefore:

$$Q = 4.0 \times 10^3 \text{ gal day}^{-1} \times \frac{1 \text{ m}^3}{264.2 \text{ gal}} = 15.1 \text{ m}^3 \text{ day}^{-1}.$$

$$Q = -KA\frac{dh}{dz} = 5.0 \text{ m day}^{-1} \times \pi \left(\frac{D}{2}\right)^2,$$

or

$$D = \left(\frac{4}{\pi} \frac{15.1 \text{ m}^3 \text{ day}^{-1}}{5.0 \text{ m day}^{-1}}\right)^{1/2} = 2.0 \text{ m}.$$

This value is too large to be feasible.

B. Consider each of the alternatives and how you might modify your design:
  i. Lengthen the sand filter (how long?)
    Lengthening the sand filter will not increase the discharge, since the hydraulic gradient would still be equal to 1.

ii. Raise the hydraulic head at the inflow (how high?)

Increasing the hydraulic gradient from 1 to 4 would meet the requirement of 4.0×10^3 gallons per day of treated water. This would require a hydraulic head at the top of the 3.0-m long column of 12.0 m (almost 40 feet), which may be impractical.

iii. Use several filters (how many? what size?)

You could use four 1-m diameter filters in parallel to treat 4.0×10^3 gallons of water each day.

Problem 6.2. A permeameter is used to measure the hydraulic conductivity of a porous medium (see Figure 6.3). The permeameter is perfectly round in cross section, with a diameter of 50.0 mm. The following parameters are measured: $z_1 = 220.0$ mm, $z_2 = 150.0$ mm, $p_1/\rho g = 230.0$ mm, $p_2/\rho g = 280.0$ mm, with $L = 200.0$ mm, and a discharge at the lower end of the column $Q = 500.0$ mm^3 min^{-1}.

A. Which way is water flowing in the column? Is water flowing from high to low hydraulic head? From high to low pressure?

$$h_1 = p_1/\rho g + z_1 = 450.0 \text{ mm.}$$

$$h_2 = p_2/\rho g + z_2 = 430.0 \text{ mm.}$$

Therefore, the water is flowing from high to low hydraulic head, as we would expect. In this case, the water is flowing from low to high pressure; hydraulic head is decreasing along the flow because the elevation head is decreasing faster than the pressure head is increasing.

B. Calculate the specific discharge, q (mm min^{-1}).

$$q = Q/A = (500.0 \text{ mm}^3 \text{ min}^{-1})/[\pi(25.0 \text{ mm})^2] = 0.25 \text{ mm min}^{-1}$$

C. Calculate the hydraulic conductivity, K, of the material (m s^{-1})

$$K = -q\frac{dl}{dh} = -0.25 \text{ mm min}^{-1} \times [200.0 \text{ mm}/(430.0 \text{ mm} - 450.0 \text{ mm})].$$

$$K = 2.50 \text{ mm min}^{-1}.$$

$$K = 2.50 \text{ mm min}^{-1} \times 1 \text{ min}/60 \text{ s} \times 1 \text{ m}/1000 \text{ mm} = 4.2 \times 10^{-5} \text{ m s}^{-1}.$$

D. Calculate the intrinsic permeability, k (m^2).

Assume that the water bas a density $\rho = 1000.0$ kg m^{-3} and a viscosity $\mu = 1.0 \times 10^{-3}$ Pa·s. Then,

$$k = K\frac{\mu}{\rho g} = 4.2 \times 10^{-5} \text{ m s}^{-1} \times [1.0 \times 10^{-3}/(1000.0 \text{ kg m}^{-3} \times 9.81 \text{ m s}^{-2})] = 4.3 \times 10^{-12} \text{ m}^2.$$

Problem 6.3. Consider two piezometers placed side-by-side but open in different aquifers at depth (i.e., the two piezometers on the right side of Figure 6.6). The following measurements are made:

	Piezometer #1	Piezometer #2
Elevation of piezometer (m above mean sea level)	200	200
Depth of piezometer (m)	60	20
Depth to water in piezometer (m)	20	18

A. Calculate the elevation and hydraulic heads (relative to mean sea level), pressure head, and fluid pressure in the two piezometers, and fill in the table below.

	Piezometer #1	Piezometer #2
Elevation head, z (m)	140	180
Pressure head, $p/\rho g$ (m)	40	2
Hydraulic head, $h = p/\rho g + z$ (m)	180	182
Pressure, $p = \rho g d$ (Pa)	3.92×10^5	1.96×10^4

B. Calculate the *vertical* hydraulic gradient between the two piezometers. Is the flow of water upward or downward?

$$dh/dz = (h_1 - h_2)/(z_1 - z_2) = (180 - 182)/(140 - 180) = -2/-40 = 0.05.$$

Therefore, since the gradient is positive upward, the flow is downward (from piezometer #2 to piezometer #1)

Problem 6.4. Consider the flow net for a drainage problem shown in Figure 6.15. Drains such as pipes and culverts placed in a wet field may be used to remove groundwater by creating a "sink" or area of low hydraulic head. In the figure, a cross section through such a field is shown. The hydraulic conductivity, K, of the surficial material is 1.0×10^{-5} m s^{-1}. The thick black lines represent impermeable boundaries; a constant head is assigned to the top and lower left side. The cross section is 20 m long by 10 m deep. The gray lines are equipotentials and the blue lines are streamlines.

$h = 10$ m

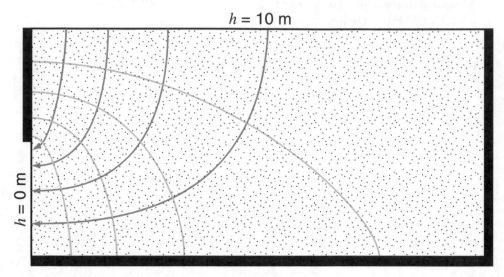

$h = 0$ m

Figure 6.15 Flow net for drainage problem (Problem 6.4).

A. Place labels on the equipotentials, indicating the value of the hydraulic head along the line.

   Beginning with the uppermost equipotential, they should be labeled as 8 m, 6 m, 4 m, and 2 m.

B. Calculate the discharge through each streamtube, and the total discharge or rate at which the field is drained (m^3 day^{-1} per m width of material).

   $Q_s = Kb\ dh = 10^{-5}$ m s$^{-1} \times 1$ m $\times 2$ m $= 2 \times 10^{-5}$ m^3 s$^{-1} \times 86400$ s day$^{-1} = 1.7$ m^3 day^{-1}.

   There are 5 streamtubes, so $Q = 5Q_s = 8.6$ m^3 day^{-1}.

Chapter 7

Problem 7.1. Determine the natural basin yield (m^3 yr^{-1} per meter basin width) for the following cases.

A. For the basin in Figure 7.4a with $L = 5000$ m and $K = 30$ m yr^{-1}.

   The basin is 5000 m long, so the equipotential spacing, *dh*, is 15 m. Therefore, the discharge through each streamtube, Q_s, is (30 m yr^{-1}) × (15 m) or 450 m^3 yr^{-1} per meter width. There are slightly more than 2 streamtubes, so the natural basin yield is slightly more than 900 m^3 yr^{-1} per meter basin width.

B. For the basin in Figure 7.4b with $L = 5000$ m and $K = 30$ m yr^{-1}.

Because L and K are the same as in part A, the discharge through each streamtube is the same, $Q_s = 450$ m^3 yr^{-1} per meter width. There are slightly less than 8 complete streamtubes, so the natural basin yield is slightly less than 3600 m^3 yr^{-1} per meter basin width.

C. For the basin in Figure 7.5a with $L = 200$ m and $K = 100$ m yr^{-1}. You might want to begin by calculating the discharge through each local flow system (dashed box).

The basin is 200 m long, so the equipotential spacing, dh, is 0.6 m. Therefore, the discharge through each streamtube, Q_s, is $(100$ m yr$^{-1}) \times (0.6$ m$)$ or 60 m^3 yr^{-1} per meter width. There are slightly more than 4 streamtubes within each local flow system, so the natural basin yield from each is slightly more than 240 m^3 yr^{-1} per meter basin width. There are 5 local flow systems in the basin, so the total natural basin yield is approximately 1200 m^3 yr^{-1} per meter basin width.

Problem 7.2. Answer the following questions for the basin in Figure 7.25.

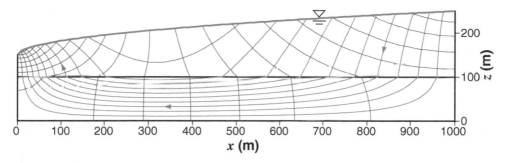

Figure 7.25 Flow net for Problem 7.2.

A. Is the lower unit an aquifer or an aquitard?

Groundwater flow is being focused into the lower unit, which indicates that it is an aquifer. There are other clues that indicate this as well. Downward flow across the upper unit indicates that it is less permeable than the lower unit; the flow is mostly horizontal in the aquifer (lower unit). The streamlines and equipotential lines form curvilinear squares in the upper unit, but not in the lower unit. This is a good indication that the hydraulic conductivity of the two units differs. In the lower unit, the streamlines and equipotentials form rectangles with their long axes along the streamlines. Therefore, the lower unit must be more permeable.

B. What fraction of the total basin yield passes through the lower unit?

The total number of streamtubes in the flow net is 11. Approximately 9.5 streamtubes (approximately 85% of the flow) pass through the lower unit.

Problem 7.3. Consider the flow net shown in Figure 7.26. The sides of the region are groundwater divides, the top boundary is the water table, and the bottom is an impermeable boundary. Streamlines are blue and equipotentials are gray.

The answers to parts A through D are shown in Figure PS.2.

A. Label the equipotentials with the appropriate value of hydraulic head (m).

B. Draw arrows on the streamlines indicating the direction of groundwater flow.

C. Label all recharge and discharge areas.

D. Indicate at least one area within the flow net where flow is relatively fast, and one area where flow is relatively slow.

E. Determine the water level (hydraulic head, m) in wells A and B.

The open portions of the wells intersect the 12.5 m (well A) and 30 m (well B) equipotentials.

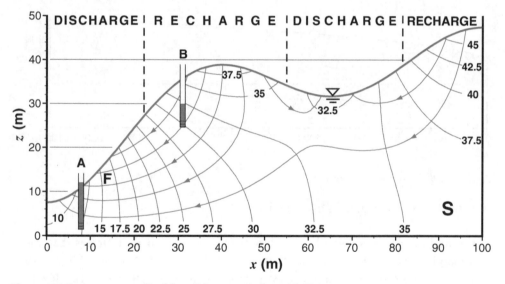

Figure PS.2 Answers to Problem 7.3, parts A through D.

Problem 7.4. Well hydrographs for unconfined aquifers can be used to determine the change in water stored within the aquifer. Consider the change in water-table level observed in Coffee County, Tennessee, between mid-May and August 2010 (Figure 7.9).

A. What is the change in water-table level during this time period (m)? Does this indicate an increase or a decrease in water stored within the aquifer?

The water-table level decreases by approximately 15 m over this time interval. This indicates that the amount of water in storage has decreased over this time interval, assuming that the observed change in water-table level is representative of that in the aquifer.

B. If the specific yield of the aquifer is 0.25, and the aquifer has an area of 600 km², what is the change in water stored over the same time interval (m³)?

Again, assuming that the change observed in the one well is representative of the aquifer as a whole, the change in water stored is found by multiplying the specific yield by the area and the change in water level:

$$\Delta V = S_y \times dh \times A = (0.25)(-15\,\text{m})(6.0 \times 10^8\,\text{m}^2) = 2.25 \times 10^9\,\text{m}^3.$$

Problem 7.5 Determine the recession constant, c (day⁻¹), for the two groundwater "reservoirs" contributing to Pescadero Creek during the spring and summer of 2010 (Figure 7.23).

The equation for the recession curve is (7.17). In logarithmic form this would be:

$$\ln Q - \ln Q_0 = -ct$$

or, rearranging to solve for c:

$$c = \frac{\ln Q_0 - \ln Q}{t}.$$

Choosing points on the two straight lines in Figure PS.3, we can solve this expression for c. For the early "reservoir," choosing June and July:

$$c = \frac{\ln Q_0 - \ln Q}{t} = \frac{\ln(0.40) - \ln(0.19)}{30\,\text{days}} = \frac{-0.92 - (-1.66)}{30\,\text{days}} = 2.5 \times 10^{-2}\,\text{days}^{-1}.$$

For the later "reservoir":

$$c = \frac{\ln Q_0 - \ln Q}{t} = \frac{\ln(0.14) - \ln(0.115)}{30\,\text{days}} = \frac{-1.97 - (-2.16)}{30\,\text{days}} = 6.3 \times 10^{-3}\,\text{days}^{-1}.$$

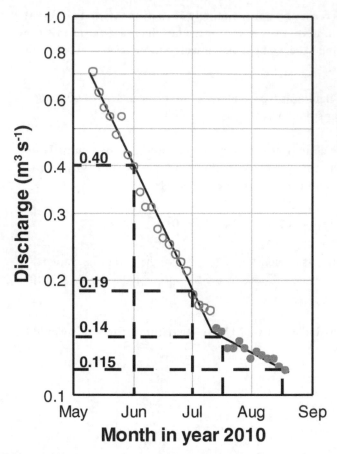

Figure PS.3 Groundwater recession analysis for Pescadero Creek, California, showing points used in analysis of recession constants (Problem 7.5).

Chapter 8

Problem 8.1. Tensiometers are installed at 0.4 m and 0.5 m above the water table in a uniform sandy soil with the moisture characteristic and hydraulic conductivity curves given in Figures 8.4 and 8.5, respectively. One set of tensiometer readings indicates that the capillary-pressure head at the first of these tensiometers is −0.45 m and at the second is −0.6 m.

A. What is the direction of water movement between the two tensiometers?

If the gradient in hydraulic head is positive, water flow is downward (specific discharge is negative). Conversely, if the gradient in hydraulic head is negative, water flow is upward (specific discharge is positive).

$$\frac{\partial h}{\partial z} = \frac{h_{0.5} - h_{0.4}}{0.1 \text{ m}} = \frac{(0.5 \text{ m} + \psi_{0.5}) - (0.4 \text{ m} + \psi_{0.4})}{0.1 \text{ m}} = \frac{[(0.5 - 0.6) - (0.4 - 0.45)]}{0.1 \text{ m}} = -0.5.$$

Therefore, the flow is upward.

B. Estimate the magnitude of the specific discharge between the two tensiometers.

From Darcy's law, $q = -K \times (-0.5) = 0.5K$. The estimate requires a value for K. From Figure 8.4, we determine that the moisture content is about 0.14 at the 0.4-m tensiometer and about 0.05 at the 0.5-m tensiometer. From Figure 8.5 we determine K values for the two tensiometer locations to be about 5×10^{-6} m s^{-1} and 6×10^{-8} m s^{-1}. We conclude that the upward specific discharge is between 2.5×10^{-6} and 3×10^{-8} m s^{-1}.

Problem 8.2. Heidmann et al. (1990) report that the moisture characteristic for a sandy loam soil in Arizona can be represented accurately by the equation $\psi = -963.7\theta^{-4.659}$, where ψ is in meters and θ is expressed as a percentage (e.g., 26% rather than 0.26). The saturation value of moisture content for this soil is estimated to be 27%.

A. Plot the moisture characteristic for this soil.

Choose values of moisture content from 2% to 27% and calculate capillary pressure head from the equation given. The moisture characteristic is the plot of capillary-pressure head versus moisture content (Figure PS.4).

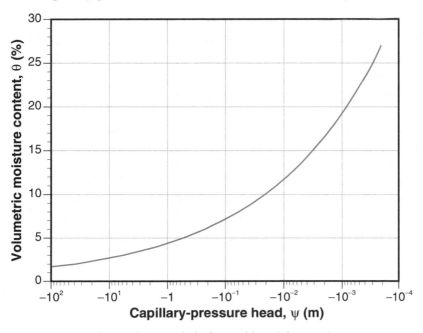

Figure PS.4 Moisture characteristic for Problem 8.2, part A.

B. What is the equilibrium profile of (a) capillary-pressure head and (b) moisture content above a static water table in this sand?

(a) The term equilibrium in this case is interpreted as zero flow. For zero flow, the hydraulic head is constant (no gradient) so $\psi = -z$. (b) The equilibrium profile of moisture content has the same shape as the moisture characteristic that is given by the equation presented by Heidmann et al. (1990) but with ψ replaced by $-z$ (Figure PS.5).

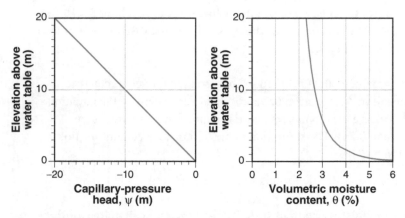

Figure PS.5 Distribution of capillary-pressure head (*left*) and volumetric moisture content (*right*) for Problem 8.2, part B.

C. Suppose that moisture content measurements from 10 m below the soil surface in this sandy loam to a depth of 50 m show a constant moisture content of 3%. The water table is at a depth of 80 m. Over the interval of the moisture measurements, how do capillary-pressure head and hydraulic head vary?

Moisture content is constant at 3% over the interval so capillary-pressure head is also constant: $\psi = -963.7 \times 3^{-4.659} = -5.77$ m. Hydraulic head is the sum of pressure and elevation heads. Elevation head varies from 30 m to 70 m to which is added the constant -5.77 m to get hydraulic head. Note that the gradient in hydraulic head is positive in this case (slope of h versus z curve is upward to the right; Figure PS.6), indicating downward water flow.

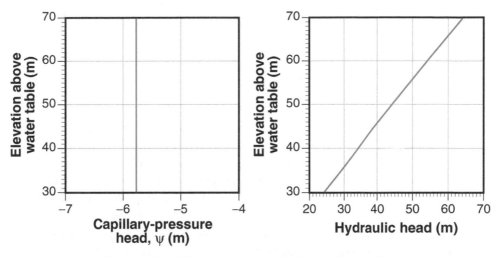

Figure PS.6 Distribution of capillary-pressure head (*left*) and hydraulic head (*right*) for Problem 8.2, part C.

Problem 8.3. Assume that a wetting front moves into the sandy loam soil described in Problem 8.2. The moisture content at the surface of the soil is held constant at 27%. The underlying moisture content is constant at 6%. If the saturated hydraulic conductivity for this soil is 3×10^{-6} m s^{-1}, estimate how long it will take the wetting front to move 1 m into the soil.

Assume that the Green-Ampt equation is valid for this case. Estimate the wetting front capillary-pressure head as $\psi = -963.7 \times 6^{-4.659} = -0.23$ m. Then, noting that $I = L_f \Delta \theta$,

$$t = \frac{L_f \Delta \theta}{K_S} + \frac{\psi_f \, \Delta \theta}{K_S} \ln\left[1 + \frac{L_f \Delta \theta}{(-\psi_f \, \Delta \theta)}\right]$$

and substituting appropriate numerical values, we get $t \approx 12$ hr.

Chapter 9

Problem 9.1. Irrigation is used in areas in which agricultural productivity can be enhanced by maintaining the soil water content well above the permanent wilting point (e.g., $\theta \geq 2\theta_w$) throughout the growing season. When irrigation water is applied, root zone soil moisture should not exceed field capacity θ_{fc}, because, if $\theta > \theta_{fc}$, water and fertilizer are lost from the root zone by drainage. With this management approach soil moisture remains below field capacity and the only losses of soil water from the root zone are by evapotranspiration. Assuming that no rainfall occurs, the temporal dynamics of depth-average soil moisture θ in the root zone can be expressed by the soil water balance equation.

$$Z\frac{d\theta}{dt} = -et,$$

where Z is the depth of the root zone. In other words, changes in the water stored in the control volume (i.e., the root zone) are due to the difference between water inputs and outputs (see Chapter 1). In this specific case there are no rainfall-induced inputs, while the only outflow is due to *et*. The irrigation period (i.e., the time between two consecutive applications of irrigation water) is calculated by integrating the soil water balance equation between time $t = 0$ when $\theta = \theta_{fc}$ (i.e., right after irrigation) and time t_i when $\theta = \theta_0$ (and it is time to irrigate again), assuming that the evapotranspiration rate linearly decreases from the potential rate PET at $\theta = \theta_{fc}$ to zero at $\theta = \theta_w$ (i.e., $et = \text{PET}(\theta - \theta_w)/(\theta_{fc} - \theta_w)$).

$$t_i = \int_{\theta_{fc}}^{\theta_0} -\frac{Z(\theta_{fc} - \theta_w)}{\text{PET}} \frac{d\theta}{\theta - \theta_w} = \frac{Z(\theta_{fc} - \theta_w)}{\text{PET}} \ln \frac{(\theta_{fc} - \theta_w)}{(\theta_0 - \theta_w)}.$$

Consider the case of a crop grown on a sandy soil with $\theta_{fc} = 0.19$ and a crop-specific wilting point, $\theta_w = 0.05$. The root zone is 40 cm deep. After each application of irrigation water, soil moisture is equal to θ_{fc}. Assuming that PET = 6 mm d^{-1}, calculate the irrigation period, t_i, i.e., how long it would take for soil moisture to decrease from field capacity to the value $\theta_0 = 2\theta_w$ in the absence of any rainfall input.

Using the above equation with PET = 6 mm d^{-1}, $\theta_{fc} = 0.19$, $\theta_w = 0.05$, $\theta_0 = 2\theta_w$, and $Z = 0.40$ m, we find: $t_i \approx 10$ days.

Problem 9.2. With the irrigation scheme described in the previous problem during each application of irrigation water soil moisture is increased from $\theta_0 = 2\theta_w$ to θ_{fc}. The volume of water (per unit area) required to make this change is $Z(\theta_{fc} - \theta_0)$. Calculate the amount of water per unit area that should be applied every time the area is irrigated. Assume that during the irrigation process about 20% of the irrigation water is lost in evaporation before it infiltrates.

During each application of irrigation water soil moisture is increased from $\theta_0 = 2\theta_w$ to θ_{fc}. The volume of water (per unit area) required to make this change is $Z(\theta_{fc} - \theta_0) = 36$ mm. Because of evaporation losses the irrigation method has an 80% efficiency (i.e., only 80% of the irrigation water infiltrates). Thus the amount of water that needs to be spent in irrigation is 36/0.80 = 45 mm every 10 days.

Chapter 10

Problem 10.1. A hydrologist studying runoff generation in a catchment measures the following chloride concentrations during the peak of a rainstorm event: $C_n = 4.5$ μmol L^{-1}, $C_o = 40.5$ μmol L^{-1}, and $C_t = 36.0$ μmol L^{-1}. What fractions of total streamflow are contributed by new water and old water?

From Equation 10.3,

$$Q_n = \left[\frac{(C_t - C_o)}{(C_n - C_o)}\right]Q_t = \left[\frac{(36.0 - 40.5)}{(4.5 - 40.5)}\right]Q_t = \frac{-4.5}{-36.0}Q_t = 0.13Q_t$$

or $Q_n/Q_t = 0.13 = 13\%$. Therefore, $Q_o/Q_t = 1 - Q_n/Q_t = 0.87 = 87\%$. Thus, 13% of the flow is contributed by new water and 87% by old water.

Problem 10.2. Data on hydraulic conductivity versus depth in a soil on a forested slope are presented by Harr (1977). Examine the data to determine whether the assumption that K decreases exponentially with depth is reasonable, and, if it is, estimate K_o and f.

Depth in soil (m)	Hydraulic conductivity, K (m s^{-1})
0.10	9.8×10^{-4}
0.30	1.1×10^{-3}
0.70	4.5×10^{-4}
1.10	4.9×10^{-4}
1.30	4.4×10^{-5}
1.50	6.1×10^{-5}

If K decreases exponentially with depth, the data should define a straight line on a plot of K on a logarithmic scale and z on a linear scale. The relationship can be inspected visually and the parameters of Equation 10.13 (K_o and f) can be found as the slope and intercept of a line drawn through the data (Figure PS.7).

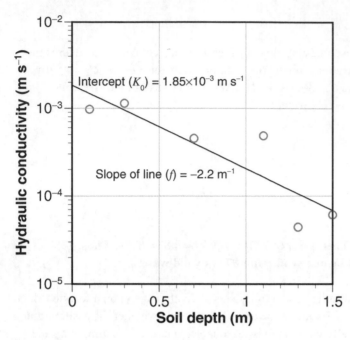

Figure PS.7 Data and fitted line for Problem 10.2.

Problem 10.3. Two "streamtube" segments of a catchment (i.e., portions of a flow net based on topography) have properties shown below:

	Segment 1	Segment 2
Upslope area, A (m²)	500	500
Length of contour at base of segment, c (m)	3.5	25
Slope at base of segment, tanβ	0.02	0.08

Calculate the topographic index for each segment and indicate which segment is more likely to produce saturation-excess overland flow.

Equation 10.5 is used:

$TI = \ln(a/\tan\beta)$.

For segment 1, TI = ln[(500 m²/3.5 m)/0.02] = 8.9. (Note that the units of the topographic index in this case are ln(m).) For segment 2, TI = ln[(500/25)/0.08] = 5.5. Segment 1 has a larger TI and is more likely to produce saturation-excess overland flow.

Problem 10.4. The soil of a given catchment has a porosity (ϕ) of 0.4 and hydraulic conductivity as described in Problem 10.2. The catchment has an average value of the topographic index of 3.5. Under conditions where the average saturation deficit of the catchment is 100 mm, calculate the subsurface flow to the stream, $q_{subsurface}$. If the throughfall rate at this time is 4×10^{-3} mm s^{-1} (14 mm hr^{-1}), what fraction of the catchment would have to be saturated if saturation-excess overland flow were to be equal to one-half of the subsurface flow?

Equation 10.24 can be used:

$$\bar{q}_{subsurface} = T_{max} e^{-\lambda} e^{-(\bar{s}/m)}.$$

$T_{max} = K_o/f = 1.85 \times 10^{-3}$ m s^{-1}/2.2 m^{-1} = 8.4×10^{-4} m^2 s^{-1}. The soil parameter, m, is $\phi/f = 0.4/2.2$ m^{-1} = 0.2 m. Thus, $\bar{q}_{subsurface} = 1.5 \times 10^{-5}$ m s^{-1} = 1.5×10^{-2} mm s^{-1}. Even if the entire catchment were saturated (in which case $\bar{s}$ would be zero, not 100 mm) the contribution of overland flow cannot be half of the subsurface flow.

Glossary

actual evapotranspiration the real rate of *evapotranspiration* from a land surface. {Section 2.4}

advection the transport of a contaminant in groundwater due to the physical flow of water {Section 7.6}

alternate depths the two physically allowable water depths for flow in a channel for a given *specific energy* and a given *specific discharge*. {Section 4.2}

anisohydric plants plants that tend to maintain the stomata open and sustain higher photosynthetic uptakes even when soil moisture is low. {Section 9.2.1}

anisotropic a material whose properties (such as *intrinsic permeability*) depend on the direction of measurement. {Section 6.5.2}

aquiclude a saturated geological formation that may contain water but does not transmit significant quantities. {Section 6.4}

aquifer a saturated geological formation that contains and transmits significant quantities of water under normal field conditions. {Section 6.4}

aquifer test an experiment designed to measure the in situ properties of an aquifer, based on the response of the water level in a well to pumping or injecting water. {Section 6.3.1}

aquifuge a saturated geological formation that neither contains nor transmits significant quantities of water. {Section 6.4}

aquitard a saturated geological formation that is of relatively low permeability; see *aquiclude, aquifuge*. {Section 6.4}

artesian aquifer see *confined aquifer*. {Section 6.4}

assimilative capacity the ability of a river, lake, aquifer, or other water body to process waste discharges such that they are harmless. {Section 1.2}

average linear velocity $\bar{v} = q/\phi$ [L T^{-1}], the average velocity of fluid within the pores of a porous medium, equal to the *specific discharge* divided by the porosity. {Section 6.3.1 and Section 7.6}

baseflow background low-flow conditions in a stream {Section 5.2}; see also {Sections 7.6 and 10.2}

basin aspect ratio the ratio of basin length (the direction parallel to flow) to basin depth or thickness (above a low-permeability unit). {Section 7.3.1}

Bernoulli equation $(u^2/2g) + z + (p/\rho g) = H$, stating that the sum of the pressure, elevation, and velocity heads in a frictionless fluid is constant along a *streamline*. {Section 3.5.2}

body forces forces that act uniformly on each fluid element; examples include gravitational forces and electromagnetic forces. {Section 3.3}

Bowen ratio B [dimensionless], the ratio of the *sensible heat* flux (H) to the *latent heat flux* (E_l). {Section 2.4.2}

capillary barrier coarse layers of sediments that impede the movement of water under unsaturated conditions. {Section 8.11}

capillary forces forces that are exerted on soil water due to the strong attraction of water by soil minerals; see *capillary-pressure head*. {Section 8.2.1}

capillary fringe the zone immediately above the water table in which the pores are filled with water but the water is under pressure less than atmospheric. {Section 8.1}

capillary-pressure head ψ [L], the (negative) pressure head in the unsaturated zone that develops across curved air-water menisci because water is attracted more strongly to soil minerals than it is to other water molecules. {Section 8.2.1}

catchment an area of land, bounded by a *divide*, in which water flowing across the surface will drain into a stream or river and flow out of the area through a specified point on that stream or river. {Section 1.4.2}

cavitation formation of vapor bubbles in (liquid) water under low (negative) pressure (i.e., strong suction). {Section 9.2.1}

Chézy equation $U = C\sqrt{SR_H}$, an equation relating channel mean velocity to channel slope and *hydraulic radius* via a roughness coefficient C called the *Chézy number*. {Section 4.5}

Chézy number C [$L^{1/2}\,T^{-1}$], an empirical roughness coefficient used in the *Chézy equation* describing open channel flow. {Section 4.5}

coalescence a physical process occurring in clouds by which falling drops of water repeatedly collide with other drops or droplets, thus forming larger drops. {Section 2.2}

combination method evapotranspiration methods that combine the energy balance and *mass-transfer methods*. {Section 2.4}

cone of depression the spatial pattern of *drawdown* in an *aquifer* as the result of pumping. {Section 7.4.1}

confined aquifer a permeable formation whose upper boundary is an *aquitard*; water in a well within a confined aquifer will rise above the top of the aquifer. {Section 6.4}

conservation of mass $dM/dt = I' - O'$, the law that states that for any particular compartment (usually referred to as a control volume), the time rate of change of mass stored within the compartment is equal to the difference between the inflow rate and the outflow rate. {Section 1.4}

continentality effect the decrease in mean annual precipitation with increasing distance from the oceans. {Section 2.2}

continuity equation an expression of *conservation of mass* in a flow, stating that the inflow rate minus the outflow rate equals the change in storage. {Section 3.5.3}

contributing area the area of *catchment* upslope from a given block that contributes inflow to that block. {Section 10.5.1}

convective acceleration the spatial component of acceleration at a fixed time, e.g., $u(\partial u/\partial x)$ for flow in the x-direction [L T^{-2}]. {Section 3.5.1}

critical flow flow that occurs at the minimum value of *specific energy* allowed for a given *specific discharge*. The *Froude number* equals 1 for critical flow. {Section 4.2.2}

Darcy's law $q = -K(dh/dl)$, states that the *specific discharge* through a porous medium is proportional to the *hydraulic gradient*. {Section 6.3}

density ρ [M L^{-3}], the mass per unit volume of a substance, defined at a point. {Section 3.2}

diffusion a transport process driven by concentration gradients. {Section 2.4.2}

discharge $Q = UA$ [L^3 T^{-1}], the volume flux of water. {Section 1.4.2}

discharge area a region in which *groundwater* is moving upward across the *water table*, thereby discharging into the *unsaturated zone* above, or to the land surface or a surface-water body such as a lake or stream. {Section 7.2}

dispersion the transport of a *groundwater* contaminant by processes that mix waters that flow along paths with various linear flow velocities. {Section 7.6}

displacement height z_d [L] a parameter expressing the effect of canopy height on the near surface air flow. {Section 2.4.2}

divide the boundary of a *catchment*. Typically topographic highs or ridges, a divide separates an area of land that should drain toward a particular point on a stream or river from surrounding land areas that do not. {Section 1.4.2}

drawdown the change in water level in a pumping well or nearby observation well or piezometer. {Section 7.4.1}

effective stress σ_e [M L^{-1} T^{-2}], an upward stress (force per unit area) exerted by aquifer solids in the subsurface. {Section 7.4.2}

elevation head z [L], a component of the total head that may be thought of as the potential energy per unit fluid weight. {Section 3.5.2}

equipotential a line of constant *hydraulic head*; equipotentials and *streamlines* together constitute a *flow net*. {Section 6.5}

evaporation the physical process involving a phase change from liquid to vapor by which water is returned to the atmosphere. {Sections 2.1 and 2.4}

evapotranspiration *et* [L T^{-1}], the sum of all processes by which water changes phase (from solid or liquid) to vapor and is returned to the atmosphere. {Section 2.4}

exceedance probability [dimensionless], the relative frequency associated with a random variable attaining a value greater than some specified value. {Section 2.2.3}

field capacity the relatively constant moisture content that a sandy soil tends to attain following drainage. {Section 8.1}

flood a peak in stream discharge due to inflow of water following a rain or snowmelt event. {Section 5.2}

flood routing determination of river discharge at a point based on knowledge of the discharge at some upstream location (inflow) and the characteristics of the intervening river channel or reservoir. {Section 5.1}

flow net a two-dimensional map of *equipotentials* (lines of constant *hydraulic head*) and *streamlines* in a region of *groundwater* flow. {Section 6.5}

fluid a substance that has no resistance to deformation when subjected to a shearing force. {Section 3.2}

free surface the upper boundary of an open channel flow, between the water and the atmosphere. {Section 4.2}

frequency analysis a statistical technique used by hydrologists for estimating the average rate at which floods, droughts, storms, stores, rainfall events, etc., of a specified magnitude recur. {Section 2.2.3}

friction factor f [dimensionless], an empirically determined, dimensionless quantity relating head loss to flow properties such as velocity and diameter or depth [$U^2L/(2gD)$ for pipe flow]. A friction factor diagram gives values of friction factor f as a function of *Reynolds number*. {Section 3.6.1}

friction slope the slope of the imaginary line that is a distance $U^2/2g$ above the water surface in an open channel. {Section 4.4}

Froude number $\mathbf{F} = U/\sqrt{gh}$, a dimensionless number that is used to define critical, subcritical, and supercritical flows in channels. {Section 4.2.2}

gage pressure the fluid pressure relative to atmospheric pressure. The absolute pressure is the sum of the gage pressure and atmospheric pressure. {Section 3.4}

gaging station a facility installed at a selected river site to collect long-term records of flow depth as a function of time. {Section 5.2}

groundwater water found in the *saturated zone* of the subsurface. {Section 6.1}

groundwater divide an impermeable vertical or near-vertical boundary or a ridge in the potentiometric surface separating groundwater flow systems. {Section 7.2}

groundwater mining prolonged pumping of groundwater at rates that exceed the rate of replenishment by recharge. {Section 7.1}

groundwater recession a decline in groundwater input to a stream through time. {Section 7.8}

groundwater runoff groundwater that discharges to streams or to the ocean. {Section 1.3}

head loss h_L [L], the loss in head due to friction within the fluid and between the fluid and the side walls of a pipe or channel. {Section 3.6}

heterogeneity the condition of spatial variation in physical properties; often applied to the *intrinsic permeability* of rocks and soil; adjectival form is heterogeneous. {Section 6.5.2}

hollow-stem auger auger for drilling wells that has a hollow core that allows placement of a casing and construction of the well inside the augers, which are then removed, letting the casing remain in place. {Section 6.4.1}

homogeneity the spatial constancy of physical properties, such as density or *intrinsic permeability;* adjectival form is homogeneous. {Sections 3.2 and 6.5.1}

Hortonian overland flow see *infiltration-excess overland flow.* {Section 10.4.2}

hydraulic conductivity $K = k(\rho g/\mu)$ [L T^{-1}], the ability of a porous medium to transmit fluid, dependent on both fluid and porous medium properties. {Section 6.3}

hydraulic descent nocturnal transfer of water from the shallow (and wetter) to the deep (and drier) soil through the root system. {Section 9.2.3}

hydraulic gradient *dh/dl* [dimensionless], the change in *hydraulic head* per unit distance; the driving force in flow through porous media. {Section 6.3}

hydraulic head $h = (p/\rho g) + z$ [L], the mechanical energy per unit fluid weight, used in the study of flow through porous media. {Section 6.2}

hydraulic lift nocturnal transfer of water from the deep (and wetter) to the shallow (and drier) soil through the root system. {Section 9.2.3}

hydraulic radius R_H [L], the ratio of the cross-sectional area of flow in a channel to the wetted perimeter. {Section 4.5}

hydrograph a continuous record of streamflow (stage or discharge) as a function of time. {Section 5.2}

hydrological cycle the global-scale, endless recirculatory process linking water in the atmosphere, on the continents, and in the oceans. {Section 1.3}

hydrology the study of the occurrence and movement of water on and beneath the surface of the Earth, the properties of water, and its relationship with the living and material components of the environment. {Section 1.1}

hydrophytes plants that grow partly or completely submerged by water. {Section 9.2.2}

hydrostatic equation an equation describing the relationship between pressure and depth in a fluid at rest. For a homogeneous fluid, the hydrostatic equation is $p = \rho g d$. {Section 3.4}

hyetograph a graph of precipitation versus time. {Section 2.2.1}

hysteresis the loop-like curve that relates pairs of hydraulic properties of an unsaturated porous medium because *volumetric moisture content, capillary-pressure head*, and *hydraulic conductivity* covary along different curves depending on whether the soil is imbibing water or draining. {Section 8.9}

incompressible fluid a fluid for which density is not a function of pressure. {Section 3.2}

infiltration the movement of rain or melting snow into the soil at the Earth's surface. {Section 8.8}

infiltration capacity the maximum rate at which water can infiltrate into a soil. {Section 8.8}

infiltration-excess overland flow also known as Hortonian overland flow (after its proponent, Robert Horton), a mechanism of runoff generation in which the *infiltration* capacity of a *catchment* or portion of a catchment is exceeded by the rainfall intensity, which results in ponding of precipitation at the soil surface and flow across the surface either in sheets or in small rivulets. {Section 10.4.2}

infiltrometer a device to measure *infiltration* rates with water ponded inside a ring driven into the ground. {Section 8.8.1}

interception storage the process by which precipitation (either liquid or solid or both) is temporarily stored either on vegetation surfaces (canopy interception) or on litter surfaces (litter interception); intercepted water either can return to the atmosphere as evaporation or can become *stemflow* or *throughfall*. {Section 2.3}

intermediate flow system a *groundwater* flow system that is smaller than a *regional flow system* and that is characterized by flow from a water-table high to a non-adjacent water-table low. {Section 7.3.2}

internal energy E_u [M L^2 T^{-2}], the component of the total energy of a substance that is due to the kinetic and potential energy of the individual molecules. {Section 2.4.2}

intrinsic permeability $k = K(\mu/\rho g)$ [L^2], the ability of a porous medium to transmit fluid, independent of the fluid properties. {Section 6.3.1}

isohydric plants plants that tend to close the stomata and reduce photosynthesis as the soil becomes drier. {Section 9.2.1}

isohyet line of equal precipitation, in the isohyetal method of estimating areal precipitation. {Section 2.2.1}

isohyetal method a technique for estimating areal precipitation to a catchment based on representing precipitation structure using lines of equal precipitation known as isohyets. {Section 2.2.1}

isotropic a material whose properties (such as *intrinsic permeability*) do not depend on the direction of measurement. {Section 6.5.2}

laminar flow a smooth, regular flow in which disturbances are damped out by viscous forces. Laminar flows in pipes and channels occur at Reynolds numbers less than 2000 and 500, respectively. {Sections 3.1 and 3.7}

land subsidence the decline of the land surface produced by pumping groundwater wells. {Section 7.7}

Laplace equation an expression of conservation of mass combined with *Darcy's law* that describes steady two-dimensional groundwater flow in a *homogeneous* region. Flow nets are graphical solutions to the Laplace equation. {Section 7.2}

latent heat [M L^2 T^{-2}], the portion of the *internal energy* of a substance that cannot be "sensed" (i.e., is not proportional to absolute temperature); latent heat is the internal energy that is released or absorbed during a phase change at constant temperature. {Section 2.4.2}

latent heat flux E_l [M T^{-3}], the rate per unit area at which latent heat is transferred to the atmosphere. {Section 2.4.2}

latent heat of vaporization λ_v [L^2 T^{-2}], the amount of energy per unit mass absorbed during a phase change from liquid to vapor at constant temperature. For evaporation of water at 0°C, $\lambda_v = 2.5 \times 10^6$ J kg^{-1}. {Section 2.4.2}

leaky aquifer an *aquifer* that is not perfectly confined but that has leakage across the surrounding confining layers. {Section 7.4.2}

local acceleration the change of velocity with time at a fixed location, $\partial u/\partial t$ [L T^{-2}]. {Section 3.5.1}

local flow system a groundwater flow system characterized by flow from a water-table high to an adjacent low. {Section 7.3.2}

macropore a relatively large pore, such as a soil pipe, animal burrow or shrinkage crack, in an otherwise fine-grained soil. {Section 8.9}

Manning coefficient n [dimensionless], an empirical channel roughness used in *Manning's equation* describing open channel flow. {Section 4.5}

Manning's equation $U = kR_H^{\frac{2}{3}}S^{\frac{1}{2}}/n$, an equation commonly used to calculate channel mean velocity, based on channel geometry and roughness; n is the *Manning coefficient*. {Section 4.5}

manometer a device used to measure fluid pressure, consisting of a tube filled with fluid and open at one end; the hydrostatic equation is used to relate the pressure at the open end to an unknown pressure at the measurement point. {Section 3.4}

mass-transfer methods methods providing a discrete representation of diffusion processes such as evaporation or transpiration. {Section 2.4}

mean velocity $U = Q/A$ [L T^{-1}], the cross-sectionally averaged fluid velocity. {Section 3.5.3}

moisture characteristic the relationship between moisture content and *capillary-pressure head* for a porous medium. {Section 8.3}

Muskingum method a numerical method for routing a flood through a river channel. Given an inflow hydrograph, the method predicts the outflow hydrograph for a given stream reach. {Section 5.4.2}

natural basin yield the average rate of discharge from a basin under natural or undisturbed conditions (i.e., in the absence of anthropogenic groundwater withdrawals or changes in climate or vegetation). {Section 7.2}

normal stress a force per unit area oriented perpendicular to a surface of a fluid or solid object. Pressure is a normal stress. {Section 3.3}

numerical method a method for solving equations by transforming them into one or more algebraic equations that can be solved more easily than the original equation, usually on a computer. {Section 5.4}

occult precipitation the condensation of dew and/or deposition of water droplets from fog and low clouds onto leaf surfaces and the subsequent dripping of this water down to the forest floor. {Section 2.2}

orographic effect the increase in mean annual precipitation with elevation typically observed in many mountain ranges around the world. {Section 2.2}

osmometer a device used to measure the osmotic pressure. {Box 9.1}

osmosis transport of water molecules through semipermeable membranes from lower to higher solute concentration areas. {Box 9.1}

osmotic compensation increase in xylem water potential attained by increasing solute concentration in xylem water, thereby decreasing the xylem osmotic potential. {Section, 9.2.1}

osmotic pressure Ω [L], pressure that develops in the presence of concentration gradients across a semipermeable membrane {Box 9.1}

paired watersheds a set of two watersheds of similar size, in the same region, and with similar topographic and hydrogeological features but different land cover conditions; typically used to investigate the impact of land cover on streamflow. {Section 9.3.2}

Penman's method see *combination method*. {Section 2.4.4}

permanent wilting point the driest soil moisture conditions that a plant can withstand without wilting. {Section 9.2.1}

permeameter a device used to measure the flow rate through and the *hydraulic conductivity* of a porous medium. {Section 6.3.1}

phreatophytes deep rooted plants that draw part of their water from beneath the water table. {Section 9.2.2}

piezometer a single tube manometer used to measure *pressure head* (and thereby *hydraulic head*) at a point in the subsurface. {Section 6.4}

piezometric surface see *potentiometric surface*. {Section 6.4}

Poiseuille's law $U = -(dp/dx)(D^2/32\mu)$, an equation for the velocity of a laminar pipe flow. {Section 3.7.1}

porosity $\phi = V_v/V_t$ [dimensionless], the fraction of the total volume of a porous medium occupied by void space. {Section 6.3.1}

porous medium a rock, sediment, or soil that contains pores or void spaces. {Section 6.2}

potential evapotranspiration PET, the maximum rate of evapotranspiration from a vegetated catchment under conditions of unlimited moisture supply. {Section 2.4.2}

potentiometric surface a surface that depicts the distribution of *hydraulic heads* in a *confined aquifer*; the water in a well or *piezometer* penetrating a confined aquifer defines the surface. {Section 6.4}

precipitation the dominant process by which water vapor in the atmosphere is returned to the Earth's surface either as liquid drops (e.g., rain) or solid particles (e.g., snow) under the influence of gravity. {Section 2.2}

precipitation intensity [L T^{-1}], a measure of the rate of precipitation, commonly computed for a specified duration. {Section 2.2.3}

precipitation recycling water that enters the atmosphere from evapotranspiration in a region and later contributes to precipitation in the same region. {Sections 1.3 and 2.2}

pressure p [M L^{-1} T^{-2}], the force per unit area acting perpendicular to a surface, or normal stress. {Section 3.3}

pressure head $p/\rho g$ [L], a component of the total head that may be thought of as the "flow work" or the work due to pressure per unit fluid weight. {Section 3.5.2}

pressure transducer an instrument that converts pressure into a proportional electrical signal that can be recorded. {Section 6.4.2}

pumping test a technique to estimate aquifer transmissivity and storativity by pumping water from one well and observing induced changes in water level over time in another well. {Section 7.5.2}

quickflow one of two components into which a flood hydrograph can be separated (the other is known as *baseflow*). {Section 10.2}

rain shadow low rainfall area on the leeward side of a mountain range. {Section 2.2}

rating curve a relationship between *stage* and *discharge* used to convert continuous measurements of stream depth (stage hydrograph) to a discharge hydrograph. {Section 5.2}

reach a segment of a stream or river channel. {Section 4.2}

recharge area a region in which water is crossing the *water table* downward, hence recharging the groundwater system. {Section 7.2}

recurrence interval the interval between two events associated with a random variable attaining a value greater than some specified value. {Section 2.2.4}

regional flow system a groundwater flow system characterized by flow from a regional water-table high to a regional water-table low. {Section 7.3.2}

residence time $T_r = V/I$ [T], a measure of the average time a molecule of water spends in a reservoir. The residence time defined for steady-state systems is equal to the reservoir volume divided by the inflow or outflow rate. {Section 1.4.1}

return flow the process by which groundwater reemerges from the soil at a saturated area and flows downslope as overland flow. {Section 10.4.4}

return period T_{return} [T], a measure of how often (on average) an event (precipitation, flood, etc.) will occur that is greater than some chosen value; the inverse of the exceedance probability. {Section 2.2.3}

Reynolds number $\mathbf{R} = \rho UD/\mu$ [dimensionless], a dimensionless number representing the ratio of inertial to viscous forces in a flow. Flows with low Reynolds numbers (<2000 for pipe flow) are laminar; flows with high Reynolds numbers (>4000 for pipe flow) are turbulent. Flows of different fluids with the same Reynolds number will be similar. {Section 3.7}

Richards' equation an expression of mass conservation in the unsaturated zone, incorporating *Darcy's law*. Solutions to the Richards' equation provide a history of the pressure distribution in a vertical soil column. {Section 8.5}

roughness irregularities or protrusions along the boundary of a flow (e.g., inner pipe wall or stream channel bed) that increase frictional flow resistance. {Sections 3.7.2, 4.4 and 4.5}

roughness height z_o [L], a parameter that accounts for the roughness of the land surface. {Section 2.4.2}

runoff ratio the ratio of average annual surface runoff to average annual precipitation for a given land area ($\bar{r}_s/\bar{p}$). {Section 1.4.1}

saturated zone a region of the subsurface where pores are completely filled with water; the saturated zone is bounded at the top by the *water table*. {Section 6.4}

saturation value of moisture content the volumetric moisture content when all pores of a soil or rock are filled with water. {Section 8.1}

saturation vapor pressure e_{sat} [M L^{-1} T^{-2}], in a system in which both liquid water and water vapor are present, the partial pressure exerted by the water vapor during an equilibrium condition in which the rates of vaporization and condensation are equal. {Section 2.2}

saturation-excess overland flow a mechanism of runoff generation that is particularly important in vegetated catchments in humid regions in which a shallow water table intersects the ground surface, causing ponding of water at the soil surface and flow across the surface either in sheets or in small rivulets. {Section 10.4.2}

sensible heat [M L^2 T^{-2}], that portion of the *internal energy* of a substance that can be sensed (i.e., is proportional to absolute temperature). {Section 2.4.2}

shallow subsurface stormflow a mechanism of runoff generation whereby water flows through a shallow, permeable soil horizon, such as when a perched water table forms above a layer of the soil with low permeability; some of the flow may occur along preferred pathways known as *macropores*. {Section 10.4.3}

shear stress τ [M L^{-1} T^{-2}], a tangential force per unit area acting on the surface of a solid or fluid. {Sections 3.2 and 3.3}

slug test a technique to estimate hydraulic conductivity of a formation by removing a "slug" of water from a well and observing the rate of recovery of the water level in the well. {Section 7.5.1}

snowpack accumulated snow that melts seasonally in many mountain regions but which, in some instances, is permanent and forms glaciers. {Section 1.3}

soil horizon a soil layer defined on the basis of physical and chemical properties and on the history of its formation. {Section 8.9}

soil moisture water that is held in soils and rocks under pressures less than atmospheric; water in the *unsaturated zone*. {Section 8.1}

soil water potential h_s [L], the hydraulic head of soil water. {Section 9.2.1}

solar energy energy deriving from radiation from the sun. It is the driver of the hydrological cycle. {Section 1.3}

specific contributing area upslope *contributing area* per unit contour length, $a = A/c$. {Section 10.5.2}

specific discharge (groundwater flow) the discharge per unit cross-sectional area of flow through porous media, $q = Q/A$ [L T^{-1}]. {Section 6.3}

specific discharge (open channel flow) the discharge per unit width of channel in a rectangular open channel, $q_w = Q/w = Uh$ [L^2 T^{-1}]. {Section 4.2}

specific energy [L], the energy per unit weight of water in a channel with respect to the channel bottom as datum. The total energy H [L] is the sum of the bottom elevation and the specific energy. {Section 4.2}

specific energy diagram a graph of *specific energy* versus water depth for a given *specific discharge*. The diagram shows the physically allowable water depths for a given specific discharge and a given specific energy. {Section 4.2}

specific heat capacity c_p [$L^2 \, \Theta^{-1} \, T^{-2}$], a proportionality constant that relates the change in *internal energy* of a substance to a change in absolute temperature. {Section 2.4.2}

specific yield S_y [dimensionless], the volume of water produced from an *unconfined aquifer* per unit aquifer area per unit decline in the *water table*. {Section 7.4.1}

stage the depth of flow in a stream. {Section 5.2}

steady flow a flow that is constant in time at each location in the flow. In steady flow, the *local acceleration* is zero. {Section 3.5.1}

stemflow a physical process by which water is transferred from *interception storage* to the soil surface by flowing along the stem or trunk of a tree. {Section 2.3}

stomata tiny pores in the leaves of vascular plants by which gases (including carbon dioxide, oxygen, and water) are exchanged with the atmosphere. {Section 2.4}

storativity S [dimensionless], the volume of water produced from a *confined aquifer* per unit aquifer area per unit decline in the *potentiometric surface*. {Section 7.4.2}

streamline a path defined by the motion of fluid elements in a flow; at any point along a streamline, the flow direction is tangent to the streamline. Conceptually, water may not cross streamlines. In describing groundwater flow, streamlines and *equipotentials* together constitute a *flow net*. {Sections 3.5.2 and 6.5}

streamtube a region within a *flow net* between two *streamlines*. {Section 6.5.1}

stress a force per unit area (SI units: N m^{-2} or Pa). {Section 3.3}

subcritical flow the (relatively) deep and slow flow corresponding to a given *specific energy* and a given *specific discharge*. The *Froude number* is less than unity for subcritical flow. {Section 4.2.2}

sublimation the physical process by which water in the solid phase changes to water vapor and is directly returned to the atmosphere. {Section 2.4.2}

supercritical flow the (relatively) shallow and rapid flow corresponding to a given *specific energy* and a given *specific discharge*. The *Froude number* exceeds unity for supercritical flow. {Section 4.2.2}

surface forces forces that act through direct contact on specific surfaces of a fluid or solid body. {Section 3.3}

surface runoff water from rainfall or snowmelt that runs over the surface of the Earth in sheets, rivulets, streams, and rivers. {Section 1.3}

tensiometer a device for measuring negative *capillary-pressure heads* in soils. {Section 8.7}

throughfall a physical process by which water is transferred from *interception storage* to the soil surface by dripping off the leaves of the canopy. {Section 2.3}

time domain reflectometry (TDR) a method for measuring soil moisture by timing the movement of a high-frequency electromagnetic wave reflected from the open end of two steel rods inserted into the soil. {Section 2.4.2}

TOPMODEL a hydrological catchment model based on land-surface topography. {Section 10.6}

topographic index $TI = \ln(a/\tan\beta)$, where a is the upslope contributing area per contour length and $\tan\beta$ is the local slope; used by the catchment model *TOPMODEL* to calculate the water balance for individual blocks within a catchment. {Section 10.5.2}

total head H [L], the sum of the pressure, elevation, and velocity heads in a frictionless fluid. {Section 3.5.2}

total stress σ_T [M L^{-1} T^{-2}], the weight (a force) of material overlying a plane of unit cross-sectional area in the subsurface. {Section 7.4.2}

transmissivity $T = Kb$ [L^2 T^{-1}], a measure of the ability of an *aquifer* of thickness b to transmit water. {Section 6.5.1}

transpiration the physical process by which water changes phase from liquid to vapor, is released through the stomata of a plant, and returns to the atmosphere. {Section 2.1}

turbulent flow flow with rapid, irregular fluctuations of velocity in space and time. Turbulent flows in pipes and channels occur at large *Reynolds numbers*. {Sections 3.1 and 3.7}

unconfined aquifer a permeable formation whose upper boundary is the *water table*. {Section 6.4}

uniform flow a flow that does not change from place to place along the flow path; the convective acceleration is zero. {Section 3.5.1}

unit weight γ[M L^{-2} T^{-2}], the gravitational force per unit volume, ρg, acting on a fluid or solid (SI units: N m^{-3}). {Section 3.2}

unsaturated zone the zone in a soil or rock between the Earth's surface and the *water table*; pores in the unsaturated zone are partly filled with water and partly filled with air. {Section 8.1}

vadose zone see *unsaturated zone*. {Section 8.1}

vapor pressure e [M L^{-1}T^{-2}], the actual partial pressure exerted by a vapor within an air mass; related to the concentration of water vapor in air. {Section 2.2}

variable contributing area concept the idea that catchment areas where *saturation-excess overland flow* develops expand and contract with time over a storm. {Section 10.4.2}

velocity head [L], a component of the total head that may be thought of as the kinetic energy per unit fluid weight. {Section 3.5.2}

velocity profile the variation of velocity with distance away from a boundary. {Section 3.2}

viscosity μ [M L^{-1}T^{-1}], a measure of a fluid's ability to resist deformation. Fluids of high viscosity flow more slowly than low viscosity fluids, everything else being equal (SI units: Pa $\cdot$ s). {Section 3.2}

volumetric moisture content θ [dimensionless], the volume of water held in a soil or rock per bulk volume of the sample. {Section 8.1}

water budget a calculation of the inflows, outflows, and change in storage for a particular control volume (such as a lake or a catchment) over a particular time period. {Section 1.4}

waterlogging saturation of the shallow soil resulting from the rising of the water table to the ground surface. {Section 9.2.2}

water table a surface separating the *saturated* and *unsaturated zones* of the subsurface, defined as a surface at which the fluid pressure is atmospheric (or zero gage pressure). {Section 6.4}

water-table aquifer see *unconfined aquifer*. {Section 6.4}

water vapor density [M L^{-3}], mass of water vapor per unit volume. {Section 2.4.2}

weir an artificial obstruction such as a step or dam over which all the water in a channel must flow and that can be used to measure stream discharge. {Section 4.3}

well casing a pipe, typically steel or PVC, that serves to line a well. {Section 6.4.1}

well hydrograph a record of the variation in water level in a well through time. {Section 7.4.1}

well screen a section at the end of a well casing that allows water to enter the well; construction may be holes drilled or slots cut in a pipe or may be a wire mesh material. {Section 6.4.1}

xylem capillary tubes within the plant. {Section 9.2}

xylem water potential h_x [L], the sum of elevation head, capillary pressure head, and osmotic potential. {Section 9.2.1}

References

Allan, J.A. 1998. Virtual water: A strategic resource global solutions to regional deficits. *Ground Water* 36:545–546.

Andreassian, V. 2004. Waters and forests: From historical controversy to scientific debate. *Journal of Hydrology* 291:1–27.

Averyt, K., J. Fisher, A. Huber-Lee, A. Lewis, J. Macknick, N. Madden, J. Rogers, and S. Tellinghuisen. 2011. *Freshwater use by U.S. power plants: Electricity's thirst for a precious resource.* A report of the Energy and Water in a Warming World initiative. Cambridge, MA: Union of Concerned Scientists.

Barber, N.L. 2009. Summary of estimated water use in the United States in 2005. U.S. Geological Survey Fact Sheet 2009–3098.

Barnes, H.H., Jr. 1967. *Roughness characteristics of natural channels.* U.S. Geological Survey Water-Supply Paper 1849.

Bazemore, D.E. 1993. The role of soil water in stormflow generation in a forested headwater catchment. M.S. Thesis, Department of Environmental Sciences, University of Virginia.

Beven, K., and P. Germann. 1982. Macropores and water flow in soils. *Water Resources Research* 18:1311–1325.

Beven, K.J. 1978. The hydrological response of headwater and sideslope areas. *Hydrological Sciences Bulletin* 23:419–437.

Beven, K.J., J. Gilman, and M. Newson. 1979. Flow and flow routing in upland channel networks. *Hydrological Science Bulletin* 24:303–325.

Beven, K.J., and M.K. Kirkby 1979. A physically based, variable contributing area model of basin hydrology. *Hydrological Sciences Bulletin* 24:43–69.

Biswas, A.K. 1972. *History of hydrology.* Amsterdam: North-Holland.

Bonan, G. 2008. *Ecological climatology.* Cambridge, UK: Cambridge University Press.

Boyer, E.W., G.M. Hornberger, K.E. Bencala, and D.M. McKnight. 1997. Response characteristics of DOC flushing in an alpine catchment. *Hydrological Processes* 11:1635–1647.

Brovkin, V., M. Claussen, V. Petoukhov, and A. Ganopolski. 1998. On the stability of the atmosphere-vegetation system in the Sahara-Sahel Region. *Journal of Geophysical Research.* 103:31613–31624.

Brown A.E., L. Zhang, T.A. McMahon, A.W. Western, and R.A. Vertessy. 2005. A review of paired catchment studies for determining changes in water yield resulting from alterations in vegetation. *Journal of Hydrology* 310:28–61.

Brutsaert, W. 1982. *Evaporation into the atmosphere.* Dordrecht, the Netherlands: D. Reidel.

Brutsaert, W. 2005. *Hydrology: An introduction.* New York: Cambridge University Press.

Burgess, S.S.O., M.A. Adams, N.C. Turner, and C.K. Ong. 1998. The redistribution of soil water by tree root systems. *Oecologia* 115:306–311.

Burns, D.A., J.A. Lynch, B.J. Cosby, M.E. Fenn, and J.S. Baron, and US EPA Clean Air Markets Div. 2011. *National Acid Precipitation Assessment Program report to Congress 2011: An integrated assessment.* Washington, DC. National Science and Technology Council.

Caldwell, M.M., and J.H. Richards. 1989. Hydraulic lift water efflux from upper roots improves effectiveness of water uptake by deep roots. *Oecologia* 79:1–5.

Campbell, G.S., and J.M. Norman. 1998. *An introduction to environmental biophysics.* New York: Springer-Verlag.

Carr, J.A., P. D'Odorico, F. Laio, and L. Ridolfi. 2012. On the temporal variability of the virtual water network. *Geophysical Research Letters.* doi: 39, L06404.

Chanson, H. 2004. *Environmental hydraulics of open channel flows.* Oxford: Elsevier-Butterworth-Heinemann.

Childs, E.C. 1969. *The physical basis of soil water phenomena.* New York: John Wiley & Sons.

Christenson, S.C., and I.M. Cozzarelli. 2003. The Norman Landfill Environmental Research Site: What happens to the waste in landfills? U.S. Geological Survey Fact Sheet 040-03. http://pubs.usgs.gov/fs/fs-040-03/.

Clapp, R.B. 1977. Infiltration in relation to rainfall intensity: An investigation of an empirical equation using simulated data. M.S. Thesis, Department of Environmental Sciences, University of Virginia.

Cooper, H.H., and C.E. Jacob. 1946. A generalized graphical method for evaluating formation constants and summarizing well field history. *Transactions of the American Geophysical Union* 27:526–534.

Cosby, B.J., G.M. Hornberger, D.M. Wolock, and P.F. Ryan. 1987. Calibration and coupling of conceptual rainfall-runoff/chemical flux models for long-term simulation of catchment response to acidic deposition. In *Systems analysis in water quality management,* edited by M.B. Back, pp. 151–160. Oxford: Pergamon Press.

Court, A. 1960. Reliability of hourly precipitation data. *Journal of Geophysical Research* 65:4017–4024.

Darcy, H. 1856. *Les fontaines publiques de la ville de Dijon.* Paris: Victor Dalmont.

DePaul, V.T., D.E. Rice, and O.S. Zapecza. 2008. Water-level changes in aquifers of the Atlantic Coastal Plain, predevelopment to 2000: U.S. Geological Survey Scientific Investigations Report 2007-5247.

DeWalle, D., and A. Rango. 2008. *Principles of snow hydrology.* New York: Cambridge University Press.

Dietrich, W.E., C.J. Wilson, D.R. Montgomery, J. McKean, and R. Bauer. 1992. Erosion thresholds and land surface morphology. *Geology* 20:675–679.

D'Odorico, P., F. Laio, and L. Ridolfi. 2010. Does globalization of water reduce societal resilience to drought? *Geophysical Research Letters* 37:L13403.

Domenico, P. A., and F.W. Schwartz. 1990. *Physical and chemical hydrogeology.* New York: John Wiley & Sons.

Dunne, T., and R.D. Black. 1970. Partial area contributions to storm runoff in a small New England watershed. *Water Resources Research* 6:1296–1311.

Dunne, T., and L.B. Leopold. 1978. *Water in environmental planning.* San Francisco: W. H. Freeman.

Dunne, T., T.R. Moore, and C.H. Taylor. 1975. Recognition and prediction of runoff-producing zones in humid regions. *Hydrological Sciences Bulletin* 20:305–327.

Eagleson, P.S. 1970. *Dynamic hydrology.* New York: McGraw-Hill.

Eltahir A.E.B., and R.L. Bras. 1996. Precipitation recycling. *Reviews of Geophysics* 34:367–378.

Falkenmark, M., and J. Rockström. 2006. The new blue and green water paradigm: Breaking new ground for water resources planning and management. *Journal of Water Resources Planning and Management* 132:129–132.

Falkenmark, M., A. Berntell, A. Jägerskog, J. Lundqvist, M. Matz, and H. Tropp. 2007. *On the verge of a new water scarcity: A call for good governance and human ingenuity.* SIWI Policy Brief.

Falkenmark, M., and J. Rockström. 2004. *Balancing water for humans and nature.* London: Earthscan.

Food and Agriculture Organization of the United Nations (FAO). 2010. Global forest resource assessment 2010: Key findings. http://www.fao.org/forestry/fra/fra2010/en/.

Fedoroff, N.V., D.S. Battisti, R.N. Beachy, P.J.M. Cooper, D.A. Fischhoff, C.N. Hodges, V.C. Knauf, D. Lobell, B.J. Mazur, D. Molden, M.P. Reynolds, P.C. Ronald, M.W. Rosegrant, P.A. Sanchez, A. Vonshak, J.-K. Zhu. 2010. Radically rethinking agriculture for the 21st century. *Science* 327:833–834.

Fetter C.W. 2000. *Applied hydrogeology,* 4th ed. Upper Saddle River, NJ: Prentice Hall.

Foley J.A., N. Ramankutty, K.A. Brauman, E.S. Cassidy, J.S. Gerber, M. Johnston, N.D. Mueller, C. O'Connell, D.K. Ray, P.C. West, C. Balzer, E.M. Bennett, S.R. Carpenter, J. Hill, C. Monfreda, S. Polasky, J. Rockström, J. Sheehan, S. Siebert, D. Tilman, and D.P.M. Zaks. 2011. Solutions for a cultivated planet. *Nature* 478:337–342.

Fraedrich, K, A. Kleidon, and F. Lunkeit. 1999. A green planet versus a desert world: Estimating the effect of vegetation extremes on the atmosphere. *Journal of Climate* 12: 3156–3163.

Freeze, R.A. 1971. Three-dimensional, transient, saturated-unsaturated flow in a groundwater basin. *Water Resources Research* 7:929–941.

Freeze, R.A. 1972a. Role of subsurface flow in generating surface runoff: 1. Baseflow contributions to channel flow. *Water Resources Research* 8:609–623.

Freeze, R.A. 1972b. Role of subsurface flow in generating surface runoff: 2. Upstream source areas. *Water Resources Research* 8:1272–1283.

Freeze, R.A., and J.A. Cherry. 1979. *Groundwater.* New York: Prentice-Hall.

Freeze, R. A., and P. A. Witherspoon. 1966. Theoretical analysis of regional groundwater flow: 1. Analytical and numerical solutions to the mathematical model. *Water Resources Research* 2:641–656.

Freeze, R. A., and P. A. Witherspoon. 1967. Theoretical analysis of regional groundwater flow: 2. Effect of water-table configuration and subsurface permeability variation. *Water Resources Research* 3:623–634.

Freeze, R. A., and P. A. Witherspoon. 1968. Theoretical analysis of regional groundwater flow: 3. Quantitative interpretations. *Water Resources Research* 4:581–590.

Gedney, N., P.M. Cox, R.A. Betts, O. Boucher, C. Huntingtford, and P.A. Stott. 2006. Detection of a direct carbon dioxide effect in continental river runoff records. *Nature* 439:835–838.

Giordano, M. 2009. Global groundwater? Issues and solutions. *Annual Review of Environment and Resources* 34:153–178.

Gleick, P.H. (Editor). 1993. *Water in crisis: A guide to the world's fresh water resources.* New York: Oxford University Press.

Green, W.H., and G.A. Ampt. 1911. Studies on soil physics: 1. The flow of air and water through soils. *Journal of Agricultural Science* 4:1–24.

Guymon, G.L. 1994. *Unsaturated zone hydrology.* Englewood Cliffs, NJ: PTR Prentice Hall.

Haan, C.T. 2002. *Statistical methods in hydrology,* New York: J. Wiley & Sons.

Hamilton, P.A., and R.J. Shedlock. 1992. *Are fertilizers and pesticides in the groundwater? A case study of the Delmarva Peninsula: Delaware, Maryland, and Virginia.* U.S. Geological Survey Circular 1080.

Harr, R.D., 1977. Water flux in a soil and subsoil on a steep forested slope. *Journal of Hydrology* 33:37–58.

Harris, D.M., J.J. McDonnell, and A. Rodhe. 1995. Hydrograph separation using continuous open system isotope mixing. *Water Resources Research* 31:157–171.

Heidmann, L.J., M.G. Harrington, and R.M. King. 1990. *Comparison of moisture retention curves in representative basaltic and sedimentary soils in Arizona prepared by two methods.* Research Note RM-500, USDA Forest Service, Rocky Mountain Forest and Range Experiment Station.

Helm, D.C. 1982. Conceptual aspects of subsidence due to fluid withdrawal. In *Recent trends in hydrogeology,* edited by T. N. Narasimhan. Geological Society of America Special Paper 189, pp. 103–139.

Helton, J.C., D. R. Anderson, M.G Marietta, and R. Rechard. 1997. Performance assessment for the waste isolation pilot plant: From regulation to calculation for 40 CFR 191.13. *Operations Research* 45:157–177.

Hoekstra, A.Y. and A.K. Chapagain. 2008. *Globalization of water: Sharing the planet's freshwater resources.* Malden, MA: Wiley-Blackwell.

Hornberger, G.M., K.E. Bencala, and D.M. McKnight. 1994. Hydrological controls on dissolved organic carbon during snowmelt in the Snake River near Montezuma, Colorado. *Biogeochemistry* 25:147–165.

Hornberger, G.M., K.J. Beven, and P.F. Germann. 1990. Inferences about solute transport in macroporous forest soils from time series models. *Geoderma* 46:249–262.

Hornberger, G.M., and E.W. Boyer. 1995. Recent advances in watershed modeling. In *U.S. National Report to IUGG, 1991–1995, Contributions in Hydrology,* pp 949–957. Washington: American Geophysical Union.

Horton, R.E. 1933. The role of infiltration in the hydrologic cycle. *Transactions of the American Geophysical Union* 14:446–460.

Hsieh, P.A. 2001. Topodrive and Particleflow: Two computer models for simulation and visualization of groundwater flow and transport of fluid particles in two dimensions. U.S. Geological Survey Open File Report 01-286.

Hubbert, M.K. 1940. The theory of ground-water motion. *Journal of Geology* 48:785–944.

Jonkman, S.N. 2005. Global perspectives on loss of human life caused by floods. *Natural Hazards* 34:151–175.

Jury, W.A., and H. Vaux. 2005. The role of science in solving the world's emerging water problems. *Proceedings of the National Academy of Sciences* (*PNAS*) 102:15715–15720.

Jury, W.A., and H.J. Vaux, Jr. 2007. The emerging global water crisis: Managing scarcity and conflict between water users. *Advances in Agronomy* 95:1–76.

Katul, G.G., R. Oren, S. Manzoni, C. Higgins, and M.B. Parlange. 2012. Evapotranspiration: A process driving mass transport and energy exchange in the soil-plant-atmosphere-climate system. *Reviews of Geophysics* 50:RG3002. doi:10.1029/2011RG000366.

Klein, G. 2005. California's Water-Energy Relationship. California Energy Commission, Final Staff Report, CEC-700-2005-011-SF.

Kohn, M.S. 2012. Bathymetry of Totten Reservoir, Montezuma County, Colorado, 2011: U.S. Geological Survey Scientific Investigations Map 3203. http://pubs.usgs.gov/sim /3203/.

Konikow, L.F. 2011. Contribution of global groundwater depletion since 1900 to sea-level rise. *Geophysical Research Letters* 38:L17401.

Konikow, L.F., T.E. Reilly, P.M. Barlow, and C.I. Voss. 2006. Groundwater modeling. In *The handbook of groundwater engineering*, 2nd ed., edited by J. W. Delleur, Chapter 23. Boca Raton: CRC Press.

L'vovich, M.I. 1979. *World water resources and their future*, translated by R.L. Nace. Washington: American Geophysical Union.

Labat, D., Y. Goddéris, J.L. Probst, and J.L. Guyot. 2004. Evidence for global runoff increase related to climate warming. *Advances in Water Resources* 27:631–642.

Leakey, A.D.B, E.A. Ainsworth, C.J. Barnacchi, A. Rogers, S.P. Long, and D.R. Ort. 2009. Elevated CO_2 effects on plant carbon, nitrogen, and water relations: Six important lessons from FACE. *Journal of Experimental Botany* 60:2859–2876.

Lee, J.-E., R.S. Oliveira, T.E. Dawson, and I. Fung. 2005. Root functioning modifies seasonal climate. *Proceedings of the National Academy of Sciences* 102:17576–17581.

Leopold, L.B. 1994. *A view of the river*. Cambridge, MA: Harvard University Press.

Leopold, L.B., and M.G. Wolman. 1957. *River channel patterns: Braided, meandering, and straight*. U.S. Geological Survey Professional Paper 282-B, pp. 39–85.

Leopold, L.B., M.G. Wolman, and J.P. Miller. 1964. *Fluvial processes in geomorphology*. San Francisco: W. H. Freeman.

Luckey, R.R., and Becker, M.F. 1999. Hydrogeology, water use, and simulation of flow in the High Plains aquifer in northwestern Oklahoma, southeastern Colorado, southwestern Kansas, northeastern New Mexico, and northwestern Texas: U.S. Geological Survey Water-Resources Investigations Report 99-4104.

Maidment, D.R. 1993. Hydrology. In *Handbook of Hydrology,* edited by D. R. Maidment, pp. 1.1–1.15. New York: McGraw-Hill.

Margat, J., S. Foster, and A. Droubi. 2006. Concept and importance of non-renewable resources. In *Non-renewable Groundwater Resources: A Guidebook on Socially-Sustainable Management for Water-Policy Makers,* edited by S. Foster, and D. P. Loucks, IHP-VI Ser. Groundwater, vol. 10, pp. 13–24. Paris: UNESCO.

McDowell, N., W.T. Pockman, C.D. Allen, D.D. Breshears, N. Cobb, T. Kolb, J. Klaut, J. Sperry, A. West, D.G. Williams, and E.A. Yepez. 2008. Mechanisms of plant survival and mortality during drought: Why do some plants survive while others succumb to drought? *New Phytologist* 178:719–739.

McGuire, V.L. 2011. Water-level changes in the High Plains aquifer, predevelopment to 2009, 2007–08, and 2008–09, and change in water in storage, predevelopment to 2009: U.S. Geological Survey Scientific Investigations Report 2011-5089. http://pubs.usgs.gov/sir/2011/5089/.

Meehl, G.A., T.F. Stocker, W.D. Collins, P. Friedlingstein, A.T. Gaye, J.M. Gregory, A. Kitoh, R. Knutti, J.M. Murphy, A. Noda, S.C.B. Raper, I.G. Watterson, A.J. Weaver, and Z.-C. Zhao. 2007. Global climate projections. In *Climate Change 2007: The Physical Science Basis. Contribution of Working Group I to the Fourth Assessment Report of the Intergovernmental Panel on Climate Change,* edited by S. Solomon, D. Qin, M. Manning, Z. Chen, M. Marquis, K.B. Averyt, M. Tignor and H.L. Miller. Cambridge, UK: Cambridge University Press.

Meinzer, O.E. 1923. *The occurrence of ground water in the United States.* U.S. Geological Survey Water-Supply Paper 489.

Mekonnen, M.M., and A.Y. Hoekstra. 2010. The green, blue and grey water footprint of crops and derived crop products. Value of Water Research Report Series No. 47, Delft, the Netherlands: UNESCO-IHE. http://www.waterfootprint.org/Reports /Report47-WaterFootprintCrops-Vol1.pdf.

Mekonnen, M.M., and A.Y. Hoekstra. 2012. The Blue Water footprint of electricity from Hydropower. *Hydrology and Earth System Sciences* 16:179–187.

Miah, M.M., and K. R. Rushton. 1997. Exploitation of alluvial aquifers having an overlying zone of low permeability: Examples from Bangladesh. *Hydrological Sciences Journal* 42:67–79.

Milly, P.C.D., K.A. Dunne, and A.V. Vecchia. 2005. Global pattern of trends in streamflow and water availability in a changing climate. *Nature* 438:347–350.

Molle, F, P. Jayakody, R. Ariyaratne, and H.S. Somatilake. 2008. Irrigation versus hydropower: Sectoral conflicts in southern Sri Lanka. *Water Policy* 10 (Supplement 1): 37–50. doi:10.2166/wp.2008.051.

Moran, M.S., and R.D. Jackson. 1991. Assessing the spatial distribution of evapotranspiration using remotely sensed inputs. *Journal of Environmental Quality* 20: 725–737.

National Research Council (NRC). 1995a. *Flood risk management and the American River Basin.* Washington: The National Academies Press.

National Research Council (NRC). 1995b. *Mexico City's water supply: Improving the outlook for sustainability.* Washington: The National Academies Press.

National Research Council (NRC). 1995c. *Ward Valley: An examination of seven issues in earth sciences and ecology.* Washington: The National Academies Press.

National Research Council (NRC). 2000. *Seeing into the Earth: Noninvasive characterization of the shallow subsurface for environmental and engineering applications.* Washington: The National Academies Press.

National Research Council (NRC). 2007. *Elevation data for floodplain mapping.* Washington: The National Academies Press.

National Research Council (NRC). 2012a. *Challenges and opportunities in the hydrologic sciences.* Washington: The National Academies Press.

National Research Council (NRC). 2012b. *Climate change: Evidence, impacts, and choices.* Washington: The National Academies Press.

National Research Council (NRC). 2012c. *Water reuse: Potential for expanding the nation's water supply through reuse of municipal wastewater.* Washington: The National Academies Press.

Ortega-Guerrero, A., J.A. Cherry, and D.L. Rudolph. 1993. Large-scale aquitard consolidation near Mexico City. *Ground Water* 31:708–718.

Perrone, D., and G.M. Hornberger. 2014. Water, food, and energy security: Scrambling for resources or solutions? *WIREs Water* 1:49–68. doi: 10.1002/wat2.1004.

Qu, Y., and C.J. Duffy. 2007. A semidiscrete finite volume formulation for multiprocess watershed simulation. *Water Resources Research* 43:W08419. doi:10.1029/2006WR005752.

Robson, A., K.J. Beven, and C. Neal. 1992. Towards identifying sources of subsurface flow: A comparison of components identified by a physically based runoff model and those determined by chemical mixing techniques. *Hydrological Processes* 6:199–214.

Rockström, J., M. Lannerstad, and M. Falkenmark. 2007. Assessing the water challenge of a new green revolution in developing countries. *Proceedings of the National Academy of Sciences (PNAS)* 104:6253–6260.

Rosegrant, M.W., C. Ringler, and T. Zhu. 2009. Water for agriculture: Maintaining food security under growing scarcity. *Annual Review of Environment and Resources* 34:205–222.

Runyan, C.W., P. D'Odorico, and D. Lawrence. 2012. Physical and biological feedbacks on deforestation. *Reviews of Geophysics* 50:RG4006.

Scanlon, B. R., I. Jolly, M. Sophocleous, and L. Zhang. 2007. Global impacts of conversions from natural to agricultural ecosystems on water resources: Quantity versus quality. *Water Resources Research* 43:W03437.

Scanlon, B.R., K.E. Keese, A.L. Flint, L.E. Flint, C.B. Gaye, W.M. Edmunds, and I. Simmers. 2006. Global synthesis of groundwater recharge in semiarid and arid regions. *Hydrological Processes* 20:3335–3370.

Scholl, M.A., and S.C. Christenson. 1998. Spatial variation in hydraulic conductivity determined by slug tests in the Canadian River alluvium near the Norman Landfill, Norman, Oklahoma. Water Resources Investigations Report 97-4292.

Shah, T., D. Molden, R. Sakthivadivel, and D. Seckler. 2000. *The global groundwater situation: Overview of opportunities and challenges.* Colombo, Sri Lanka: International Water Management Institute.

Shiklomanov, I.A. 1999. World water resources and their use: A joint SHI/UNESCO product. http://webworld.unesco.org/water/ihp/db/shiklomanov/index.shtml.

Siebert, S., J. Burke, J.M. Faures, K. Frenken, J. Hoogeveen, P. Döll, and F.T. Portmann. 2010. Groundwater use for irrigation: A global inventory. *Hydrology and Earth System Sciences* 14:1863–1880.

Sklash, M.G., and R.N. Farvolden. 1979. The role of groundwater in storm runoff. *Journal of Hydrology* 43:45–65.

Skopp, J., M.D. Jawson, and J.W. Doran. 1990. Steady-state aerobic microbial activity as a function of soil water content. *Soil Science Society of America Journal* 54:1619–1625.

Smith, J.A., M.L. Baeck, M. Steiner, and A.J. Miller. 1996. Catastrophic rainfall from an upslope thunderstorm in the central Appalachians: The Rapidan storm of June 27, 1995. *Water Resources Research* 32:3099–3113.

Smith, J.A., D.J. Seo, M.L. Baeck, and M.D. Hudlow. 1996. An intercomparison study of NEXRAD precipitation estimates. *Water Resources Research* 32:2035–2045.

Smith, T.R. 1974. A derivation of the hydraulic geometry of steady-state channels from conservation principles and sediment transport laws. *Journal of Geology* 82:98–104.

Soeder, D.J., J.P. Raffensperger, and M.R. Nardi. 2007. Effects of withdrawals on ground-water levels in southern Maryland and the adjacent Eastern Shore, 1980–2005. U.S. Geological Survey Scientific Investigations Report 2007-5249.

Stednick, J.D. 1996. Monitoring the effects of timber harvest on annual water yield. *Journal of Hydrology* 176:79–95.

Tilman D., C. Balzer, J. Hill, and B.L. Befort. 2011. Global food demand and the sustainable intensification of agriculture. *Proceedings of the National Academy of Sciences* (*PNAS*) 108:20260–20264.

Tóth, J. 1962. A theory of groundwater motion in small drainage basins in central Alberta, Canada. *Journal of Geophysical Research* 67:4375–4387.

Tóth, J. 1963. A theoretical analysis of groundwater flow in small drainage basins. *Journal of Geophysical Research* 68:4795–4812.

Wada,Y., L.P.H. van Beek, C.M. van Kempen, J.W.T.M. Reckman, S. Vasak, and M.F.P. Bierkens. 2010. Global depletion of groundwater resources. *Geophysical Research Letters* 37:L20402.

Winograd, I.J. 1981. Radioactive waste disposal in thick unsaturated zones. *Science* 212:1457–1464.

Winter, T.C. 1983. The interaction of lakes with variably saturated porous media. *Water Resources Research* 19:1203–1218.

Wolman, M.G. 1955. *The natural channel of Brandywine Creek Pennsylvania*. U.S. Geological Survey Professional Paper 271.

Wolock, D.M. 1993. *Simulating the variable-source-area concept of streamflow generation with the watershed model TOPMODEL*. U.S. Geological Survey Water-Resources Investigations Report 93-4124.

Index

Page numbers in italics refer to figures and tables.